ASIAN AND PACIFIC COASTS 2003

Proceedings of the 2nd International Conference

ASIAN AND PACIFIC COASTS 2003

Proceedings of the 2nd International Conference

APAC2003

Makuhari, Japan 29 February – 4 March 2004

Editors

Yoshimi Goda
Yokohama National University, Japan

Wataru Kioka
Nagoya Institute of Technology, Japan

Kazuo Nadaoka
Tokyo Institute of Technology, Japan

World Scientific

NEW JERSEY • LONDON • SINGAPORE • SHANGHAI • HONG KONG • TAIPEI • BANGALORE

Published by

World Scientific Publishing Co. Pte. Ltd.
5 Toh Tuck Link, Singapore 596224
USA office: Suite 202, 1060 Main Street, River Edge, NJ 07661
UK office: 57 Shelton Street, Covent Garden, London WC2H 9HE

British Library Cataloguing-in-Publication Data
A catalogue record for this book is available from the British Library.

ASIAN AND PACIFIC COASTS 2003
Proceedings of the 2nd International Conference

ISBN 981-238-558-4 (pbk)

Printed by FuIsland Offset Printing (S) Pte Ltd, Singapore

PREFACE

This book collects the abstracts presented at the second International Conference on *Asian and Pacific Coasts* (*APAC*), held in Chiba, Japan on February 29 – March 4, 2004. The volume starts with the paper of the keynote lecture. The included Proceedings on CD-ROM contains all full length papers directly reproduced from material submitted by the authors, allowing for high quality color graphics and photographs.

The series of conferences was conceived and organized with the aim to promote technological progress and activities, international technical transfer and cooperation, and opportunities for engineers and researchers to maintain and improve scientific and technical competence in the field of coastal engineering and other related fields. It is further intended to provide an international forum for the academic and technical activities, cooperation, opportunity and fellowship among researchers and engineers, and to exchange experience of coastal and port engineering development, and coastal environmental problems among Asian and Pacific countries. The first conference on *Asian and Pacific Coastal Engineering* (*APACE*) was held in Dalian, China in 2001. The conference name has been slightly changed to *Asian and Pacific Coasts* (*APAC*). It aims at emphasizing that the subject area of the conference may not be limited to the classical topics of coastal engineering but be extended to related field including coastal environments, marine ecology, coastal oceanography, and fishery science and engineering.

Key conference topics include:

-Coastal oceanography and meteorology (waves, currents, tides, tsunami, etc.)
-Coastal sedimentary processes and their control measures
-Design and investigation of coastal and harbor structures
-Coastal environmental problems and marine ecology
-Coastal fishery problems and resource management
-Global environmental problems
-Coastal zone management and planning
-Remote sensing / laboratory and field measurement techniques
-Miscellaneous coastal problems.

This conference was jointly organized by the Coastal Engineering Committee, Japan Society of Civil Engineers, the Chinese Ocean Engineering Society and the Korean Society of Coastal and Ocean Engineers. It was organized in cooperation with the Coastal, Oceans, Ports and Rivers Institute of the American Society of Civil Engineers (COPRI), the International Association of Hydraulic Engineering and Research (IAHR), the Oceanographic Society of Japan, the Japanese Society of Fisheries Science and the Japanese Society of Fisheries Engineering. Finally the organizing committee should like to express their appreciation for the support afforded by the sponsors, the Service Center of Port Engineering (SCOPE), Coastal Development Institute of Technology (CDIT), the Overseas Coastal Area Development Institute of Japan (OCDI), Waterfront Vitalization and Environment Research Center (WAVE), Japan Dredging and Reclamation Engineering Association, Japan Ocean Development Construction Association, Japan Port and Airport Construction Association, Japan Institute of Construction Engineering, Foundation for Riverfront Improvement and Restoration, Cold Region Port and Harbor Engineering Research Center, JSCE Scientific Promotion Fund, Chiba Convention Bureau and International Center, and the Federation of Electric Power Companies of Japan.

Editors
Y. Goda
W. Kioka
K. Nadaoka

CONTENTS

CLIMATE CHANGE AND SEA-LEVEL RISE: CHALLENGES TO COASTAL SCIENCE AND ENGINEERING

Nobuo Mimura[1]

This paper presents an overview for the current status of the impacts studies on climate change and sea-level rise focusing on the Asia and Pacific region. International research efforts such as IPCC have revealed the past trend of global warming and its future. Based on these estimates, climate change and sea-level rise could impose a wide range of impacts on the coastal zones; some of the typical impacts shown here are coastal erosion, increased flood risk, and impacts on infrastructure. A region-wide vulnerability assessment for the Asia and Pacific region is also introduced. The assessment consists of three steps; setting scenarios for sea-level rise and other external forces, assessing primary impacts in the coastal zones such as inundated and flooded areas, and assessing secondary impacts on natural and socio-economic systems. The spatial distribution of affected areas and population are estimated using global databases. Low-lying areas in the region such as deltas of Mekong River, Ganges and Brahmaputra Rivers, and Yangtze River, and islands in the Pacific are most seriously inundated. The estimated amounts of inundated/flooded areas and population are quite large, indicating that the Asia and Pacific region is very vulnerable to sea-level rise and climate change. Lastly, countermeasures against these adverse effects will be examined as adaptation. Through such introduction, this paper tries to indicate new challenges to coastal science and engineering which are raised by climate change and sea-level rise.

Keywords: global warming; climate change; sea-level rise; impacts on the coastal zones; vulnerability; adaptation; Asia and the Pacific

1. Introduction

Global warming has been drawn strong attention of the people worldwide. It brings about climate change and sea-level rise, which in turn would impose serious impacts on the natural environment and human society. Coastal zone is among the focused areas in terms of their serious impacts. Primary impacts of sea-level rise are inundation of low-lying areas, exacerbation of flooding, accelerated beach erosion, and salt water intrusion to rivers and ground water aquifers. Since there are many large deltas, coastal wetlands, coral reefs, and small islands in the Asia and Pacific region, it is crucial to predict the degree and extent of the possible impacts of

1 Professor, Center for Water Environment Studies, Ibaraki University, Nakanarusawa 4-12-1, Hitachi, Ibaraki 316-8511, Japan. mimura@mx.ibaraki.ac.jp

sea-level rise and climate change as a basis for preparing responses to global warming. This region also face high grow of population and economic development, therefore it is further important how to ensure the safety and environmental well-being of the region's coastal zones in this century. This is a new challenge to coastal science and engineering to achieve sustainable development in the region's coastal zones.

2. Present Status of Climate Change Research

The science, impact assessment, and response strategies to global warming have been reviewed and summarized by the Inter-Governmental Panel on Climate Change(IPCC) organized by WMO and UNEP in 1988. The Panel provides scientific information to the world efforts to combat against global warming, such as international negotiations on UNFCCC (international treaty for the prevention of global warming) and the Kyoto Protocol by publishing comprehensive reports three times so far. The distinguish feature of the IPCC's Third Assessment Repots(IPCC WGI; WGII; WGIII, 2001) is that the increased temperature during the 20th century was attributed to the enhanced human-induced emission of green-house gases(GHGs) like CO_2 with more well-marked evidences than the past.

The atmospheric concentration of CO_2 has increased 31% in the past 250 years to reach 375ppm today, while it was 280ppm before the industrial revolution started in the middle of the 18th century. As a result, the global average surface temperature had increased 0.6±0.2 degree in the 20th century. The 1990s is the warmest decade in the past 1000 years, and 1998 marked the year of the highest mean temperature(Fig.1). IPCC indicated that the recent increasing trend of the mean temperature could be reproduced only by considering both natural factors such as changes solar radiation, volcanic eruption, and anthropogenic emission of GHGs. Thus it concluded that the warming actualized during the past 50 years was caused by the human activities.

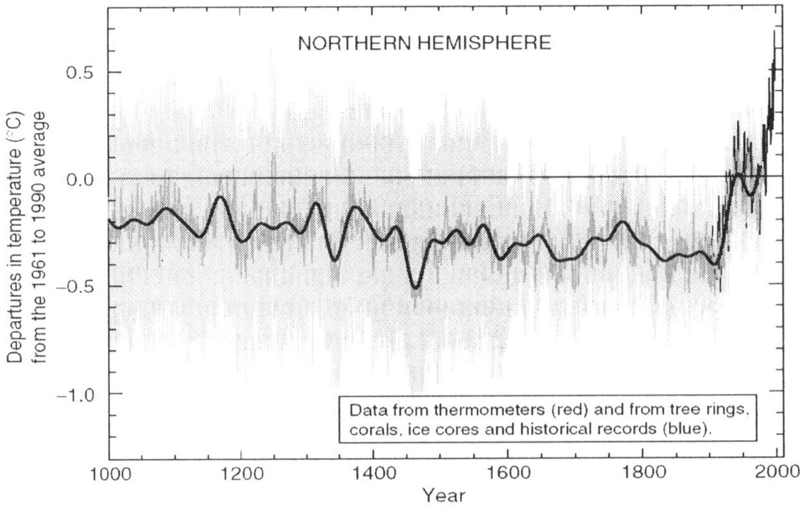

Figure 1 Trends of surface mean temperature in the past 1000 years(IPCC WGI, 2001)

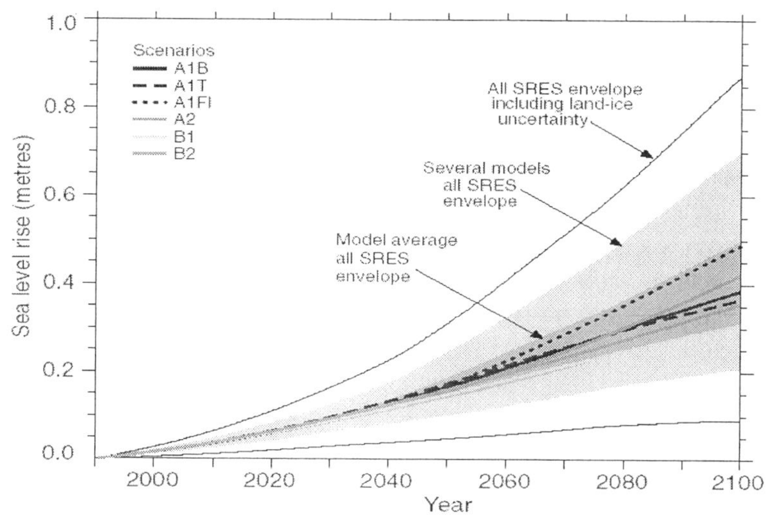

Figure 2 Prediction of sea-level rise in the 21st century(IPCC WGI, 2001)

The IPCC Third Reports also presented future predictions on the global warming and climate change. Future trend of the global temperature depends on the emission of GHGs including CO_2 as a major contributor, which in turn is determined by the shape of the future society and its socioeconomic activities. If human society develops economic activities without consideration of environmental protection, CO_2 emission will increase enormously, while if the world pays high attention to the prevention of global warming, the future activities of industries and life style of the people will be different and discharge less CO_2. In order to take such different directions into the prediction, IPCC(2000) established several typical scenarios for the future society, known as the IPCC SRES scenarios. Based on these scenarios, future trends of global warming have been estimated using general circulation models(GCMs). The results scatter because of the difference of the emission scenarios and models used, and the global mean temperature will increase from 1.4 to 5.8 degree by 2100.

IPCC predicted that sea-level rise would amount to 0.09 to 0.88m according to the degree of warming mainly due to the thermal expansion of ocean surface water and melting of land-based ice sheets and ice caps(Fig.2). Regarding the rainfall, the amount will increase during the 21st century and the annual variability will also increase. It is uncertain whether the tropical cyclones or typhoons, which we are most concerned for coastal disasters, may change its strength and frequency of occurrence. Some models predict decrease of the number of typhoons by 20 to 40 % in the warmer world, while others give an increase of its generation. We have to wait until more definite prediction is given to obtain a reliable future picture on this matter.

3. Impacts of Climate Change and Sea-Level Rise-Case for the Japanese Coasts

3.1 *Exposure units in the coastal zones*

Impacts of climate change and sea-level rise will spread in a wide range of areas sectors existing in the coastal zones. These areas and sectors exposed to the hazardous forces are called exposure systems, which are basically natural and man-made. The natural systems in the coastal zones consist of variety of land forms and ecosystems, such as sandy beaches, rocky coasts, cliffs, tidal flats, coral reefs, and wetlands including mangroves. These natural systems will respond to the climate change and sea-level rise in their own ways, in which some systems are seriously damaged and even disappear in the worst case.

On the man-made side, mega/large cities are located on the coasts and usually the population density in the coastal lands is higher than inland areas. In Japan, 46% of the national population is concentrated in the coastal cities and villages. This is a general trend in the Asia and Pacific region, and the population increasing with a high growth rate tends to live on the coasts in the future. Many ports and fishing harbors, industries, power stations, tourism facilities, and other infrastructures are also concentrated to the coasts. These are all exposure systems to the future threats of climate change and sea-level rise. In the following, we will see how the future impacts on typical systems have been studied.

3.2 *Coastal erosion*

A common impact of sea-level rise(SLR) is acceleration of coastal erosion. If the mean sea level rises, sandy beaches adjust themselves to the increased sea level resulting in a new equilibrium profile. When we consider only the cross-shore profile change, sediments on the foreshore have to be transported to offshore to form the new equilibrium profile; as a result the shoreline retreats. This is called the Bruun Rule, and based on this concept we can calculate the amount of shoreline retreat. Mimura et al.(1995, 1996) applied this method to the Japanese coastlines to give the national assessment for the beach erosion. As seen in Fig.3, 56% of the currently existing Japanese sandy beaches(190 km^2 in area) would be eroded by 30cm SLR, furthermore the number will increase to over 90% for 1m SLR.

Japan has already been suffering from serious coastal erosion because of decrease of sediment supply from rivers and blockage of longshore sediment transports by large port and coastal structures such as breakwaters and groins. If the effects of sea-level rise and climate change superpose on the present trend, the coastal erosion will be amplified. Japan will need much more efforts to the protect coasts and national lands, however, in the worst case the national lands are surrounded by man-made walls with no sandy beaches in front of them.

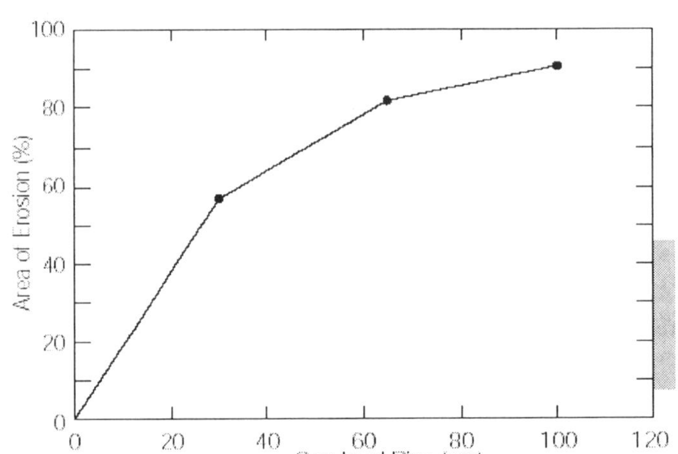

Figure 3 Beach erosion due to sea-level rise(Mimura et al, 1995)

Climate change and sea-level rise may cause other adverse effects; changes in deposition of riverine sediments at the river mouth, and inundation and erosion of coastal wetlands and mangroves. These precious ecosystems may survive by migrating towards inland in parallel with rising sea level. In many places, however, there are no rooms for them to retreat, as seawalls already exist behind the wetlands. In the face of a real threat of SLR, we need to consider the saving measures like evacuation corridors for their survival.

3.3 *Coastal disaster potential*

As a macroscopic index for the societal impacts, the areas and population at risk imposed by sea-level rise and coastal flooding has been estimated. In Japan, currently 861km^2 of areas are already under the mean sea level, which is called "zero-meter area", where two million people live. The areas and population at risk increase 2.7 times(2,339km^2) and 2 times(4.1million) due to 1m SLR. If storm surges superpose on the increased sea level, the numbers will further increase(Matsui et al., 1992). As mentioned before, the changes in typhoon is still unclear. However, as a scenario study, Tsutsui and Isobe(1992) showed that an intensified typhoon with a larger central pressure decrease of 15hPa will cause a 1.2 times higher storm surge in Tokyo Bay resulting in an expansion of risk zones for storm surges. A similar estimate for the whole Asia and Pacific region will be discussed later.

3.4 *Infrastructures in the coastal zones*

The adverse effects of global warming will appear on the coastal infrastructures through a range of causes such as sea-level rise and changes in climatic and oceanic conditions. These effects induce lessening function of various facilities such as ports, fishing harbors, man-made islands, landfills, disaster prevention structures, drainage and sewage systems, and other infrastructures. These effects are widely examined in the Coastal Engineering Committee, JSCE(1994).

An example is the increased instability of breakwaters. The safety factor against sliding may decrease due to sea-level rise. This change depends on their site conditions like water depth and wave breaking. The decrease will sometimes become more than 10% in the worst case.

Another attention should be paid to the increased ground water table. If the water table increases following sea-revel rise, ground looses its supporting capacity and resistance against the liquefaction. These effects will cause increased earthquake damages on river dikes, port facilities, bridges, large buildings, etc. For countries which face the earthquake threats like Japan, such combined effects may become important.

Against such adverse effects, many countries will try to respond to them by upgrading the existing infrastructures. The costs needed to maintain the present safety standards and functions against future sea-level rise were estimated for the coastal infrastructures in Japan. For the facilities of ports and port-connected coats, we need to reinforce the breakwaters, revetments, quays, storage houses, water gates etc, and the cost for them were estimated to be 11.5 trillion Japanese Yen or about 96 billion US$(Kitajima et al., 1993). Moreover, as there are about 3000 fishing harbors and disaster prevention facilities on the remaining parts of the Japanese coastlines, the total cost will be more than 20 trillion Japanese Yen or 167 Billion US$. These figures are an apparent indicator for the extent of the possible impacts of sea-level rise and climate change.

4. Impacts on Asia and Pacific region

4.1 *Region-wide assessment(Sato et al, 2000; Mimura et al., 2001)*

There are a number of vulnerability assessment studies in global, regional, and national scales(e.g. Hoozemans et al, 1993; Nicholls et al., 1999). However, a constraint is that most studies look at only static changes in external forces, such as increase of mean sea level. Since the coastal environment is strongly affected by tropical cyclones and storm surges, it is important to take into account them in the vulnerability assessment. At the same time, the Asia and Pacific region is expected to face high growth of population and economic activities. For the coming decades, population and human activities in the coastal zone must increase remarkably. Therefore, incorporation of these changes into vulnerability assessments is essential to make them more realistic. Furthermore, many previous studies are confronted with lack of data, even if they try to respond to the above problems.

In order to overcome such constraints, global datasets on climatic, environmental, and societal information were employed. There are global and regional datasets for the basic geographic conditions, such as distribution of the land elevation, land use/cover, and population, and environmental parameters, such as coral reefs and wetlands. Collecting and using these datasets, an region-wide assessment of vulnerability for Asia and the Pacific was performed through integration by a geographic information system(GIS).

The target area of this study is the whole Asia and Pacific region, which covers 30E to 165S, and 90N to 60S. Its land area and population are about 6.5 million km^2 and 3.8 billion people. Necessary information for the assessment was collected from global and regional databases as shown in Table 1. The spatial resolution of the data ranges from 1 degree×1 degree to 0.5 minute×0.5 minute in the longitudinal and latitudinal directions. All data were arranged uniformly on a common grid of 1 minute×1 minute in the GIS.

Table 1 Global databases used

Parameter	Name	Organization	Resolution/Remarks
Elevation	GTOPO 30	EROS Data Center	1:1,000,000 1/120 deg.
Population	Gridded Population of the World	CIESIN	1/12 deg.
Tide	Tide Table No.1 & 2	Hydrographic Dept. Maritime Safety Agency	1852 points
Cyclone	World-wide Consolidated Tropical Cyclones	World Weather Disk (US NOAA)	1842-1989
Wetlands	Major World Ecosystem Complexes	Global GRASS (US NOAA)	1 deg.
Coral Reefs	Reef Base 3	World Conservation Monitoring Center, ICLARM	Over 7,000 coral reefs
Infrastructure	Digital Chart of the World	ESRI	1:1,000,000- 1:25,000,000

4.2 *External forces*

The external forces considered were sea-level rise, astronomical tide, and storm surge induced by tropical cyclones. By combining these forces, four scenarios of mean water level at two time points, i.e. the present and 2100, were set; 1) high tide level at present (no sea-level rise), 2) storm surge level at present (high tide+storm surge, no sea-level rise), 3) high tide level in 2100 (high tide+1m sea-level rise), and 4) storm surge level in 2100 (high tide+storm surge +1m sea-level rise). The levels for high tide and storm surge represent the water levels for the permanent inundation and episodic flooding.

Estimate of the future sea-level rise are given by IPCC. In this study, 1m sea-level rise was taken for simplicity. The high tide levels in 1852 tidal stations(Hydrographic Department, Marine Safety Agency, 1999a, b) was used to interpolate the distribution of the high tide level along the whole coastlines in the Asia and Pacific region.

The height of storm surge was calculated by Eq.(1), which represents suck-up of sea water and wind set-up by a tropical cyclone, but does not include the effect of wave set-up.

$$SS_{RP} = 0.991 \cdot \Delta P_{RP} + \frac{k \cdot U_{RP}^2}{10^3 \cdot S} \ln \frac{h_0}{h'} \tag{1}$$

where, SS_{RP} is storm surge height, ΔP_{RP} pressure decrease at the center of a cyclone, U_{RP} wind speed, k=4.8×10^{-2}, S mean sea bottom slope on the coast, h' water depth where the storm surge is calculated(=5 m), and h_0 water depth where the wind set-up begins.

The parameters necessary to calculate the storm surge, such as pressure distribution and cyclone track, wind speed, and the velocity of the cyclone movement were estimated using the World-wide Consolidated Tropical Cyclones edited and issued by U.S. NOAA. Storm surges were calculated for all the cyclones recorded for 40 years from 1949 to 1989 in the region. Then the highest surge level in each coastal segment was taken as the storm surge scenario along the coastlines. This means that this study assumed that, global warming will not change future tropical cyclones, thus storm surges, from those of the past 40 years.

4.3 *Areas and population affected by inundation and flooding*

The areas affected by high tide and storm surge were calculated assuming that the areas below the water levels calculated above would be inundated or flooded. In reality, where coastal dikes and seawalls exist, they should prevent the inundation and flooding. Moreover, the flooded areas by storm surge are usually limited to narrow coastal areas due to the short retention time of a cyclone. Therefore, the area calculated in the study is the possible maximum area of inundation or flooding, which represents the potential risk zone along the coastlines.

Then the population in the affected areas was calculated. As the reference, distribution of the population in 1994 was taken. For 2100, the population distribution is estimated by multiplying the growth rate estimate of World Bank by the present distribution country by country.

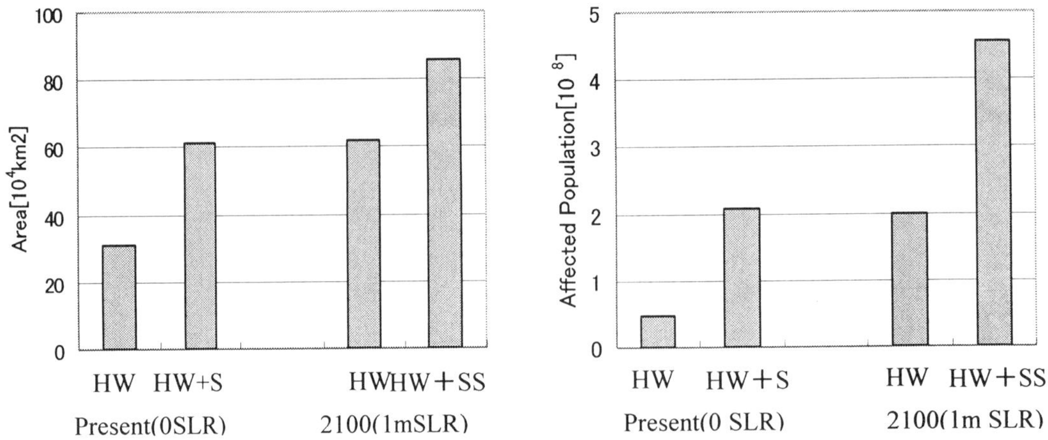

Figure 4 Areas and population affected by sea-level rise and storm surges

4.4 *Risks for the Asia and Pacific region*

As mentioned above, the total land area of the target region is about 6.7 million km^2. The population in 1994 in the region is about 3.8 billion which is estimated to increase to 7.6 billion in 2100. Even today, the areas below high tide level and storm surge level, i.e. inundated and flooded areas, are 311 thousand km^2 or 0.48% of the total area and 611 thousand km^2 or 0.94% , respectively, as shown in Fig.4. They increase to 618 and 858 thousand km^2 (0.98% and 1.32%) by 1m sea-level rise. The increased flooded area amounts to 247 thousand km^2.

Regarding the affected people today, about 47 million people or 1.21% of the total population lives in the area below high tide level, while 270 million people or 5.33% live below storm surge level. These show that the Asia and Pacific region is already vulnerable to flooding by storm surges. If the mean sea level rises by 1 m and the present population is assumed to be unchanged, the above populations increase to 160 and 258 million(2.73% and 6.61%) respectively. When the population growth by 2100 is taken into account, they become about 200 and 450 million people. The increase in the population in the flooded area reaches 249 million.

Figure 5 shows the distribution of the areas below storm surge level with 1m sea-level rise. The areas which may be affected seriously are deltas of Mekong River, Ganges and Brahmaputra Rivers, and Yangtze River, and the southern part of Papua New Guinea. The countries and areas where more than 10% of the national population is affected are Vietnam, Taiwan, Cambodia, Brunei, Bangladesh, Guam, etc.

The assessment has several weaknesses. Accuracy of the land elevation varies with areas, which reflects the accuracy of the source maps. The area below storm surge level is taken uniformly as the flooded area, while the flooding of storm surge occurs in the coastal belts as mentioned above. Though the resolution of the present analysis is rather fine for the regional analysis as 1 minute×1 minute, small islands in the Pacific, such as Kiribati and Tuvalu, cannot be represented precisely. They need to be studied on a country basis and to be incorporated to the analysis individually. In spite of such constraints, this study can give us the degree and extent of the possible impacts of future sea-level rise and storm surges.

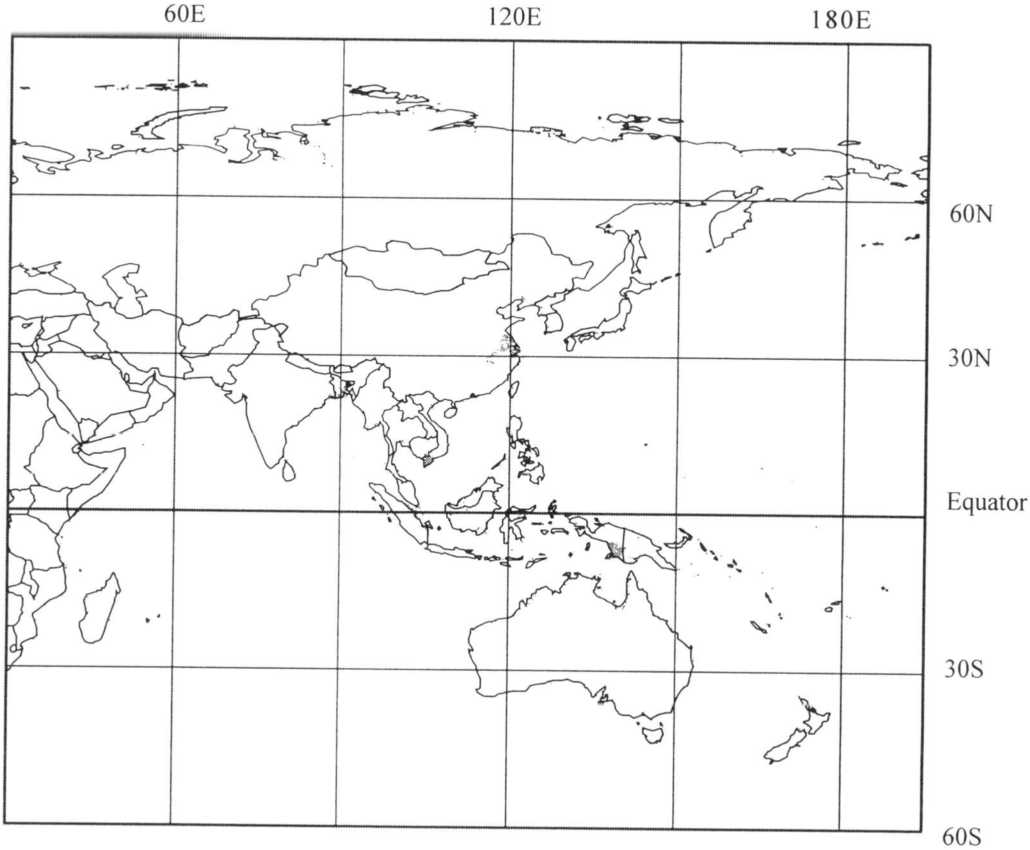

Fig.5 Distribution of flooded areas by sea-level rise and storm surges

5. Response Strategies -Adaptation to Climate Change and Sea-Level Rise

When considering the above-mentioned wide spread and serious impacts, we have to develop a long-term strategies to respond to climate change and sea-level rise. There are two basic categories for the responses; preventing global warming by regulation of the GHG emission (mitigation) and adaptation to the warmer world. As far as we cannot stop the global warming completely, we have to consider adaptive measures against climate change and sea-level rise.

The results introduced above can lead us to a consideration of adaptation strategies. Vulnerability is an overall concept for a society to cope with the adverse effects of sea-level rise and climate change. It consists of several components, such as susceptibility and resilience of exposure systems to external forces, and capacity of the society to adapt to them. The degree and extent of the possible impacts identified in the studies are quite large. On the other hand, the adaptive capacity is not so high in the developing countries in the Asia and Pacific region. Therefore, the impacts would overwhelm the social counteraction and adaptation capacity. The results indicate the necessity of the proactive consideration on how to strengthen the adaptive capacity in each country and the region on the whole. Furthermore, the international support is essential to this end(e.g., UNEP 1996).

IPCC proposed three options for coastal adaptation; planned retreat, accommodation, and protection. Planned retreat is an option to evacuate population and precious assets from areas at risk in the face of sea-level rise. Accommodation is that which allows people to continue to use the coastal areas with alteration of usages; e.g. raising houses on high poles, introduction of storm surge shelters and flood insurance, changes of crops to salt-resistant ones, and land use change from agriculture to fish farming. Protection is to defend the people's life and assets and socioeconomic activities in the vulnerable areas, which has hard and soft measures. As the studies on the adaptation and its measures just started, there are many unknown subjects for adaptation(e.g., Klein et al., 1999, IPCC WGII, 2001). Therefore, this is another area for the challenges for coastal science and engineering.

6. Conclusions

In this paper, we have explored the present status of the studies on the impacts of and vulnerability to climate change and sea-level rise targeting at the Asia and Pacific region. The impacts identified by the past studies are quite large and wide spread, which indicates that the Asia and Pacific region is very vulnerable to them. If we want to grasp the future threats more concretely and precisely, even at the individual coat level, we have to expand our views more widely, because the coastlines of Asia and the Pacific have much variety of physical, ecological and social properties. Natural systems of the coasts are ranging from coral reefs and mangroves in the tropical environment to ice-covered coastlines in the Arctic, and at the same time there also exist mega cities in the growth center of the world. A major task of coastal science and engineering is to understand the differential impacts of environmental changes on these coasts and their mechanisms. Therefore, further promotion of impact/vulnerability studies, both local and regional scales, are an important challenge to us.

The existing studies also give us a basis for the consideration on how to respond to climate change and sea-level rise by showing the distribution of the possible impacts and vulnerable areas. At the sane time, in the future of several decades, the society will also undergo considerable changes including those of population, industries, trades, and life styles. Although the future picture is uncertain, this means that the human society in the region will face not only climatic changes but also other environmental and societal changes. This is a complicated multiple stress situation. In the coastal zones, the pressure of changes will be stronger than other areas, as a large portion of the increased population will live in the coastal areas. Considering such situation, adaptation to these changes will be an inevitable policy direction to ensure the safety and well-being of human society against the future threats. For example, the Japanese Ministry of Land, Infrastructure and Transport has started examination on the coastal adaptation policies, even though it is still a preliminary stage(Research Committee, Ministry of Land, Infrastructure and Transport, 2002). Development of adaptation strategies in the national and regional polices obviously needs scientific understandings and engineering supports. Therefore, coastal science and engineering are expected to play an important role to support the development of policies and management framework. This is another challenge in front of us.

References

Coastal Engineering Committee, JSCE(1994). *Coastal Impacts of Global Warming –Present Status, Impacts and Response Strategies for Sea-Level Rise and Climate Change*, JSCE, 221p.(in Japanese).

Hoozemans, F.M.J., M. Marchand and H.A. Pennekamp(1993). *A Global Vulnerability Assessment for Population, Coastal Wetland and Rice Production on a Global Scale, 2nd Edition*, Delft Hydraulics.

Hydrographic Department, Marine Safety Agency(1999a). *1999 Tide Tables, Vol. 1, Japan and Its Vicinities*, 454p.

Hydrographic Department, Marine Safety Agency (1999b). *1999 Tide Tables, Vol. 2, Pacific and Indian Oceans*, 339p.

IPCC(2000). Special Report on Emission Scenarios, Cambridge University Press, 500p.

IPCC WGI(2001). *Climate Change 2001, Scientific Basis*, Cambridge University Press, 881p.

IPCC WGII(2001). *Climate Change 2001, Impact, Adaptation, and Vulnerability*, Cambridge University Press, 1032p.

IPCC WGIII(2001). *Climate Change 2001, Mitigation*, Cambridge University Press, 752p.

Kitajima, S., T. Ito, N. Mimura et al.(1993): Impacts of sea-level rise and cost estimate of counter- measures in Japan, *Proceedings of IPCC Eastern Hemisphere Workshop*, 115-124.

Klein, R.J.T., R.J. Nicholls, and N. Mimura(1999). Coastal adaptation to climate change: Cam the IPCC Technical Guidelines be applied?, *Mitigation and Adaptation Strategies for Global Change*, 4, 239-252.

Matsui, T., H. Tateishi, M. Isobe, A. Watanabe, N. Mimura and R. Shibazaki(1992). Prediction for inundation of Japanese coastal areas due to sea-level rise, *Proceedings of Coastal Engineering*, JSCE, Vol.39, 1031-1035(in Japanese).

Mimura, N(2001). Distribution of vulnerability and adaptation in the Asia and Pacific region, *Global Change and Asia Pacific Coasts-Proceedings of APN/SURVAS/LOICZ Conference on Coastal Impacts of Climate Change and Adaptation*, APN and Ibaraki University, 21-26.

Mimura, N., M. Kiyohashi and E. Kawaguchi(1995). Morphologic response of the coasts to sea-level rise, *Proceedings of the Third Symposium on Global Environment*, JSCE, 97-102(in Japanese).

Mimura, N. and H. Nobuoka(1996). Verification of the Bruun Rule for the estimation of shoreline retreat caused by sea-level rise, *Coastal Dynamics 95*, ASCE, 607-616.

Nicholls J.N., F.M.J. Foozemans and M. Marchand(1999). Increasing flood risk and wetland losses due to global sea-level rise: regional and global analyses, *Global Environmental Change*, 9, S69-S87.

Research Committee, Ministry of Land, Infrastructure and Transport(2002). *Report of Research Committee on Conservation of National Land against Sea-Level Rise due to Global Warming*, Ministry of Land, Infrastructure and Transport(in Japanese).

Sato, K., N. Mimura and S. Machida(2000). Assessment of vulnerability of climate change to the coastal zones in the Asia and Pacific region, *Proceedings of Coastal Engineering*, JSCE, Vol.47 (in Japanese).

Tsutsui, J. and M. Isobe(1992). Prediction of storm surges in Tokyo Bay after global warming, *Proceedings of Japan Association for Coastal Zone Studies*, 4, 9-19(in Japanese).

UNEP(1996). *Handbook on Methods for Climate Change Impact Assessment and Adaptation Strategies*, UNEP.

A NEW TSUNAMI NUMERICAL SIMULATION WITH BOUSSINESQ-TYPE EQUATIONS APPLIED FOR THE 1983 NIHONKAI-CHUBU EARTHQUAKE TSUNAMI

Hiroyuki Iwase
Ecoh Co., Ltd., 10-35, Sakae-Cho, Kanagawa-ku, Yokohama, Japan 221-0522
Ph: +81-45-440-5931; Fax: +81-45-453-0263; Email: hiroyuki_iwase@ecoh.co.jp

Fumihiko Imamura
Disaster Control Research Center, Tohoku Univ., 06, Aoba, Aoba-ku, Sendai, Japan 980-8579
Ph: +81-22-217-7513; Fax: +81-22-217-7514; Email: imamura@tsunami2.civil.tohoku.ac.jp

1. Introduction

In the 1983 Nihonkai-Chubu Earthquake tsunami causing serious damage including 100 casualties around Japan Sea, soliton-disintegrated wave was remarkably observed in shallow water, which amplifies by the waveform curvature, dispersion, effect more than a normal tsunami. Since a former numerical model widely used for a tsunami model excludes such an effect, soliton-disintegrated waves typed tsunami can not be simulated well. In this study, a new tsunami numerical simulation with Boussinesq-type equations is introduced for generation, propagation in deep and shallow water, and is applied to the 1983 Nihonkai-Chubu earthquake in order to discuss its applicability.

2. Computation model

Governing equations of the present model are obtained by depth integrated Boussinesq-type equations (Madsen et. al[1]). A tsunami amplified in a shallow water breaks, so that the wave-breaking model as the dissipation effect of momentum should be taken into consideration. Otherwise, the model overestimates the runup on the shore. We refer the previous research work on it and reach the following numerical model as a new tsunami simulation with dispersion effect;

$$\frac{\partial M}{\partial t}+\frac{\partial}{\partial x}\left(\frac{M^2}{D}\right)+\frac{\partial}{\partial y}\left(\frac{MN}{D}\right)+gD\frac{\partial \eta}{\partial x}=\frac{2}{5}h^2\left(\frac{\partial^3 M}{\partial t\partial x^2}+\frac{\partial^2 N}{\partial t\partial x\partial y}\right)+\frac{1}{15}gh^3\left(\frac{\partial^3\eta}{\partial x^3}+\frac{\partial^3\eta}{\partial x\partial y^2}\right)$$
$$+\nu_\beta\left(\frac{\partial^2 M}{\partial x^2}+\frac{\partial^2 N}{\partial y^2}\right)-\frac{gn^2}{D^{7/3}}M\sqrt{M^2+N^2}$$

Where η is the tsunami level, M, N are discharge flux x and y directions, h is the still water depth, D is the total water depth, g is the gravitational acceleration, ν_β is eddy diffusion coefficient, n is the manning roughness coefficient. Governing equations were computed by the 2-step mixed finite difference scheme[2] for saving the computational CPU time and ensuring the stability condition . In the run-up regions less then 2m-depth we switch to the shallow water equations used by explicit scheme, which is normal model.

3. Reliability of computation model

Before applying the model to the 1983 tsunami, we carry out the simulation to be compared with the hydraulic experiments in order to calibrate the coefficient of the breaking. Fig.1 shows the comparison of wave profile of the result between the spatial wave height of hydraulic experiment (○) and computed (Solid line). Good agreement is obtained for the comparison of the maximum water level between the experiment and the computation results by introducing water breaking model. Fig.2 shows computational regions and distribution of tsunami initial profile for Nihonkai-Chubu earthquake tsunami. Fig.3 shows the time history of tsunami level at the 8m-depth point offshore at Minehama village where the maximum runup height was measured. The present model is able to compute the amplification at tsunami front by the waveform curvature effect, and the decay of tsunami wave height by water breaking effect. Fig.4 shows the flood region of tsunami on land computed by present model and shallow water model. It demonstrates that the result computed with Boussinesq-type equation became wider than the result computed by shallow water equation because of tsunami amplification by soliton-disintegrated waves. We can say that the present model produce accurate and stable result more than the previous one with the shallow water equation.

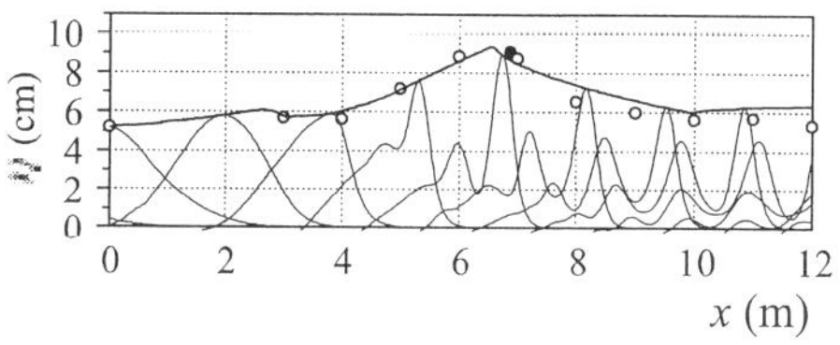

Fig.1 Spatial variation of water height

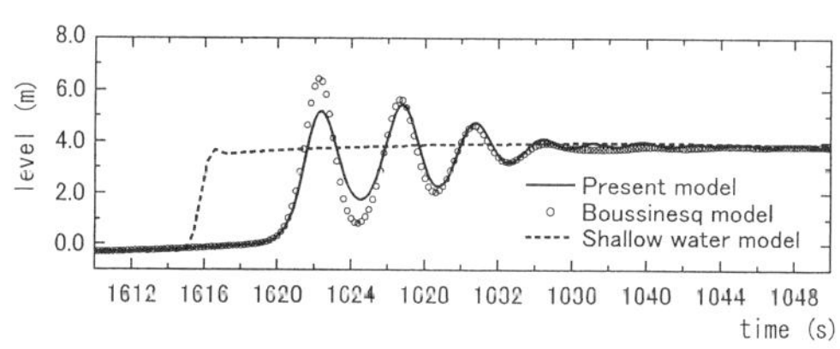

Fig.3 Time history of tsunami level

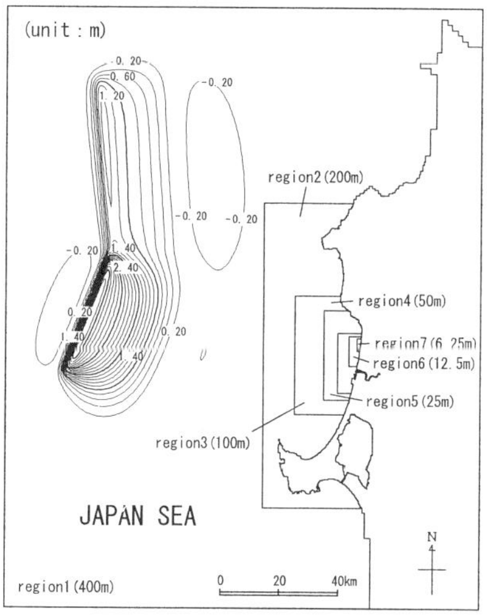

Fig.2 Computational regions and initial tsunami distribution

4. Conclusion

A tsunami numerical simulation model used Boussinesq-type equations was developed and the run-up of Ninonkai-Chubu earthquake tsunami was computed. The present model can estimate the soliton-disintegrated wave of tsunami in shallow water. The flood region of tsunami on land computed by present model became more widespread than the result computed by shallow water model.

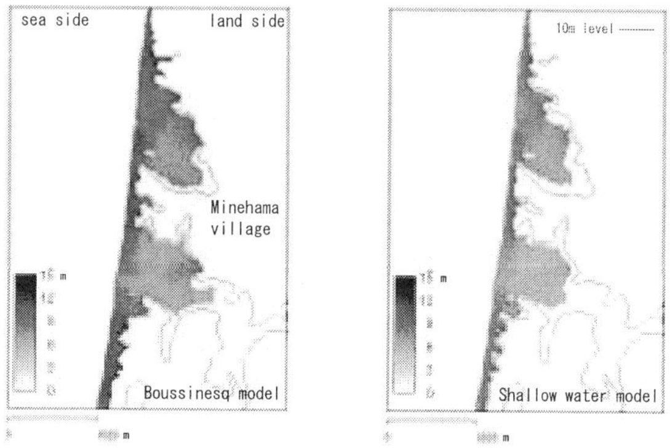

Fig.4 Flood regions of tsunami on land

References

[1] Madsen, P. A. and Sorensen O. R. (1992) : A New form of Boussinesq equations with improved linear dispersion characteristics, Part 2, A slowly-varying bathymetry, Coastal Eng., Vol. 18, pp. 183-204.

[2] H. Iwase., T. Mikami and C. Goto (1992) : Practical tsunami numerical simulation model by use of non-linear disperive long wave theory, Journal of Hydraulic, Coastal and Environmental Eng. ,JSCE ,No.600/ II -44, pp. 119-124.

HYDRAULIC EXPERIMENTS AND NUMERICAL MODEL OF TWO-LAYER FOR A LANDSLIDE-INDUCED TSUNAMI

Yoshinori SHIGIHARA[1], Daichi GOTO[2] and Fumihiko IMAMURA[3]

[1]Graduate student, Graduate School of Engineering, Tohoku University, Aoba06, Sendai, 980-8579, Japan

Email: shigi@tsunami2.civil.tohoku.ac.jp

[2]Graduate student, Graduate School of Engineering, Tohoku University, Aoba06, Sendai, 980-8579, Japan

Email: goto@tsunami2.civil.tohoku.ac.jp

[3]Professor, Graduate School of Engineering, Tohoku University, Aoba06, Sendai, 980-8579, Japan

Email: imamura@tsunami2.civil.tohoku.ac.jp

Tel: +81-22-217-7515 / Fax: +81-22-217-7514

Abstract

A tsunami generated by landslide is less studied than the more frequent tsunami generated by fault motion. From historical reports, it is clear that landslide-induced tsunamis killed many people and caused serious damage; about 1,500 casualties in the 1741 Oshima-oshima tsunami and 15,000 in the 1792 Ariake tsunami. This type of a tsunami can result in significant local wave amplification, energy concentration and/or radiation. A strong ground motion, failure of slope, volcanic activity and great societal impact, can trigger landslide-induced tsunami. The mechanisms of tsunami through hydraulic experiments and numerical simulation have been studied. For example Imamura and Lee (1998) and Matsumoto et al. (1998) proposed the two-layer model for interaction between landslide and tsunami, and computed the tsunami generation by using the numerical model to reproduce the historical tsunami in 1741([1] and [2]). Hermann et al. (2001) carried out the hydraulic experiment supposed rockslide impact of the Lituya bay tsunami in 1958, [3]. However, the mechanism of tsunami generation has not been clarified and modeled, since the physical mechanism when the sediment rushes into the water is still unknown and there are few examples of experimental research/analysis on the interaction between debris flow and tsunami. In this study, we carried out the hydraulic experiments in sloping one-dimensional open channels with the high-speed video camera and water gauges in order to measure the detail process and sequence of landslide motion and the generation of the tsunami by rushing into water and have clarified some processes of tsunami generation by landslides. Moreover, using the numerical model of two-layers with three kinds of shear stress, the tsunami by landslide try to be simulated and compared with the result from the hydraulic experiment.

Fig.1 shows the series of photographs debris flow rushing into water and the tsunami. The debris flow reaching the water surface and pushed the water generating the

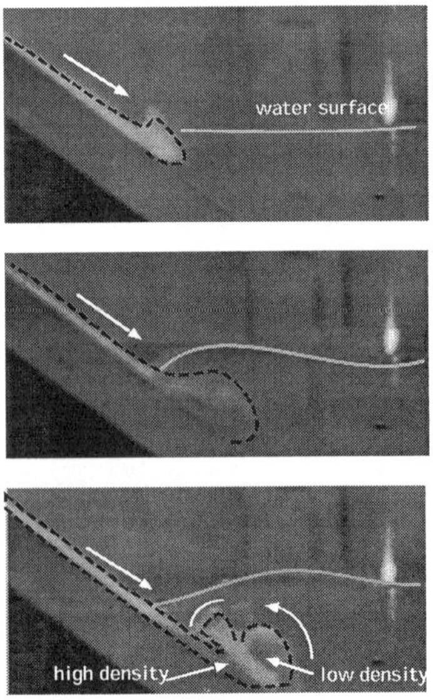

Fig. 1 Front of the debris flow rushing into water

positive wave. After that, the mixing with the water thickened the front of the debris flow. The debris flow going down toward the bottom pulled down the water surface, generating the negative wave. In this result, the effect of the interactive force occurring between the debris flow layer and the tsunami is significant. Moreover, the continuous flowing of the debris in the water lets the wave period of the tsunami short due to localize disturbance at the water surface, so that we confirmed a phenomena of wave frequency dispersion.

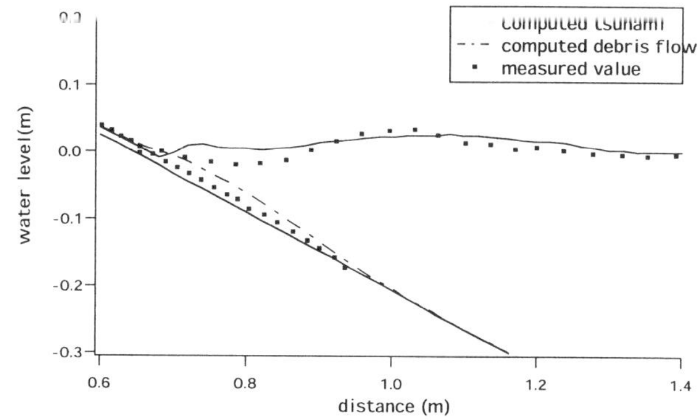

Fig. 2 Spatial profiles of wave at 0.5 sec. after the debris flow rushing into water

In order to simulate the experiment results, the basic equation of fluid with two layers is required. We applied nonlinear long-wave theory integrated over a layer with non-horizontal bottom/interface as the governing equation. Thus the model of two-layer introduced shear stress terms was performed and compared with the results of the experiment. As the shear stress terms, it is supposed that the bottom friction and the horizontal viscosity should be included in the lower layer for the debris flow, the dispersion should be in the upper layer for the tsunami, and the interactive force should be on each layers. Fig.2 shows spatial profiles of wave at 0.5 sec. after the debris flow rushing into water. Both computed the debris flow with the proposed coefficients and the tsunami height show good agreement with the measured, thus we concluded the Manning coefficient of 0.01 $s/m^{1/3}$ and the horizontal viscosity of 0.01 m^2/sec for landslide, and the interactive force of 0.2 for each layers. Moreover, the Manning coefficient of the debris flow on the dried should be different from those in the water in order to introduce by the mixing with water; the timing of the debris flow under the water and the tsunami is revised by introducing the large value of the roughness 0.12 $s/m^{1/3}$ for the debris under the water. However the following negative wave for the upper layer has not generated yet in this model. We could not include the dispersion effect into the model so far. Therefore, the effect of its term is needed to introduce in the upper layer.

Reference

[1] Imamura, F. and Lee, H. (1998) "State-of arts of numerical simulations for generation and propagation of tsunamis" Bulletin on Costal Oceanography, Vol. 36, 91-101, in Japanese.

[2] Matsumoto, T., Hashi, K., Imamura, F., and Shuto, N. (1998) "Development of tsunami model by debris flow" Bulletin on Costal Engineering, Vol. 45, 346-350, in Japanese.

[3] Hermann, M. F. Hager and H. E. Minor. (2001) "Lituya Bay case: Rockslide impact and wave run-up" J.Sci. Tsunami hazards, Vol. 19, No.1, 3-22.

STUDY ON THE EVALUATION OF TSUNAMI REDUCING OF COASTAL CONTROL FOREST FOR ACTUAL CONDITIONS

*Kenji HARADA, Graduate student, e-mail: harada@tsunami2.civil.tohoku.ac.jp

**Fumihiko IMAMURA, Professor, e-mail: imamura@tsunami2.civil.tohoku.ac.jp

Tsunami Engineering Laboratory, Disaster Control Research Center, Graduate School of Engineering, Tohoku University

Aoba 06, Sendai, JAPAN, 980-8579

Tel: +81-22-217-7515 / Fax: +81-22-217-7453

Abstract

The artificial coastal barriers such as seawalls and break waves have been constructed in Japan and have played an important role in protecting the city and coastal area from natural hazards such as tsunamis, tidal waves, high waves. However, the artificial coastal barriers need high cost of the construction and maintenance, change the present environment and have forced inconvenient to use the coastal area. Therefore, the countermeasures against tsunamis by using the artificial coastal barriers are not recommended for all coastal areas. It is required that a new countermeasures corresponding to every area is considering with the combination with artificial and natural functions for more appropriate management for natural disaster reduction and keeping good environment. One of a new way is to utilize a control forest along coast. However, the quantitative and concrete functions of coastal forest to reduce tsunamis are not established and formulated, so that no guidance to use control forest is available. Additionally, in order to take the countermeasures for the mitigation of tsunami damage, it is necessary to predict the situation at the time of a tsunami attack, which considering with tsunami control forest and other coastal structures. In order to use a coastal control forest positively and effectively as countermeasures against tsunamis, it is important to evaluate the hydrodynamic effect of tsunami control forest, and to clarify a disaster prevention function further. The present paper aims to provide the concrete functions of tsunami disaster prevention by control forest from old tsunami reports and other studies, and the quantitative guidance of tsunami reduction effect from numerical simulation including the resistance of tsunami control forest.

From old tsunami reports and other studies (Shuto, 1987, Ishikawa, 1988) the function of tsunami disaster prevention by control forest can be summarized as follows (see Fig.1) ; 1) to stop driftwood and other drifts, and prevents the secondary disaster of house damage by the attack of drifts. 2) to reduce tsunami energy, inundation depth, inundation area, tsunami current and hydraulic force at behind of forest by reflection and resistance of trees. 3) to provide a life-saving means by catching persons carried off by tsunamis. 4) to collect wind-blown sands and raises dunes, which act as a natural barrier against a tsunami. In order to evaluate these tsunami reduction effects quantitatively, tsunami numerical simulation considering with the resistance of forest by using with C_D and C_M for the some forest conditions as forest density and forest width of actual scale were carried out. The resistance coefficients of the coastal forest are selected as C_D=8.4Vo/V+0.66, C_M=1.7. These values modeled from the hydraulic experiment on coastal forest. From the calculated results, the spatial distribution of the maximum inundation depth, current and hydraulic forces can be shown quantitatively. As an example of result, fig.2 show the spatial distribution of maximum inundation depth for the different forest width (50, 100, 200, 400m). This figure shows that the inundation depth can be reduced with a function of the forest width. From the same simulation, these are obtained that the hydraulic force behind forest is reduced and the strong current in the coastal forest can be possible to scour the bed of trees. By use these results, the tsunami reduction effects can be evaluated quantitatively related to the tsunami and forest conditions.

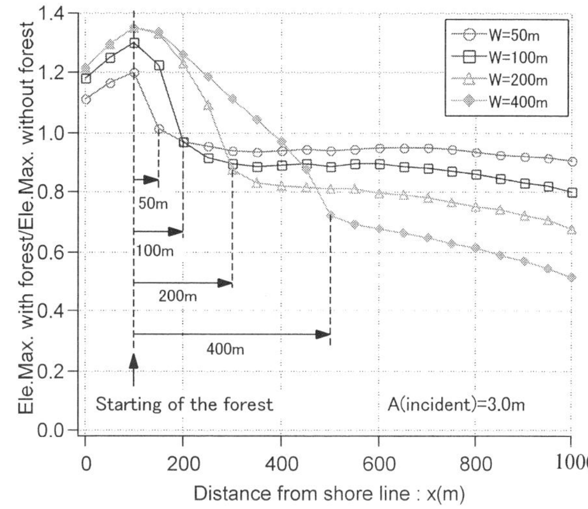

Fig.2 Spatial distribution of maximum inundation depth with forest width=50-400m.

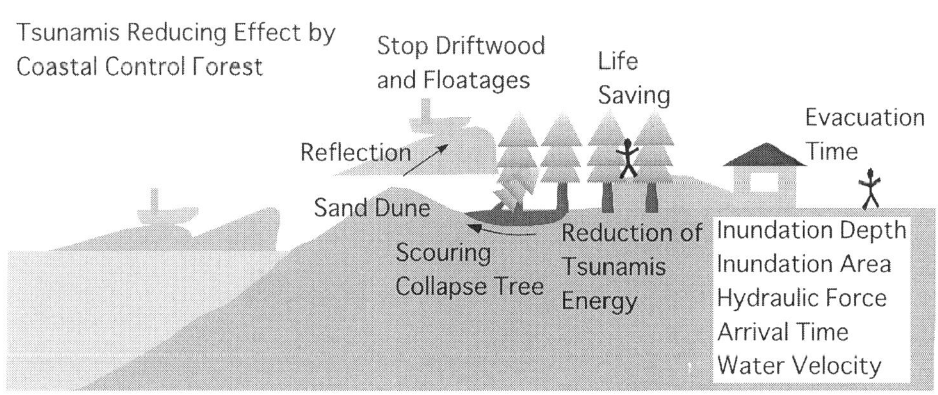

Fig.1 Tsunamis Reducing Effect by Coastal Control Forest

<u>Reference</u>

Shuto, N. (1987): Effectiveness and limit of tsunami control forests, coastal engineering in Japan, Vol. 30, No.1, pp.143-153.

Ishikawa, M. (1988): The forest functions on prevention of fogs, tsunamis and wind-blown sands, the sleeve notes to the function on public interests of forest, Vol.9, p.83.

Study on the accuracy of the tsunami numerical model around obstacles

Sung Jin HONG* and Fumihiko IMAMURA**

* *Graduate Student, Disaster Control Research Center, Graduate School of Engineering, Tohoku University, Aoba 06, Sendai 980-8579, Japan. (Email: hcsoul@tsunami2.civil.tohoku.ac.jp, Phone: +81-22-217-7515, Fax: +81-22-217-7514)*

** *Professor, Disaster Control Research Center, Graduate School of Engineering, Tohoku University, Aoba 06, Sendai 980-8579, Japan. (Email: imamura@tsunami2.civil.tohoku.ac.jp, Phone: +81-22-217-7515, Fax: +81-22-217-7514)*

ABSTRACT

The re-examination of the 1983 tsunami in Korea especially at Imwon Port was carried out by Hong and Imamura (2002). Although they have good agreement with the measured, the low accuracy of the simulation for run-up is found at 3 locations with the complex land-use and geometry. One of the reasons is considered to be related with the run-up simulation on the land with houses, building and so on. From those results, we need the advanced model for run-up on a complex land-use to represent appropriate friction coefficient of the bottom. In recent years, the Equivalent Roughness Model (ER-model) based on shallow water theory, which is utilized to obtain appropriate friction coefficient at the bottom following the percentage of the structures, is proposed by Aburaya and Imamura (2002). However, the quantitative verification about the ER-model has not thoroughly been clarified yet. Therefore, in this study, the accuracy of ER-model is studied through comparing with measured data, and its applicability is also discussed.

The grid system to calculate wave height and velocity in the numerical simulation is selected. The length is open-channel is 1200cm and the width is 30cm. The two obstacles are located at x=130cm and 150cm from the origin. The wave height is measured at four locations; x=50cm, 115cm, 140cm and 200cm in the channel, and the wave velocity at two locations; x=50cm and 200cm to compare with the result of calculation with. ER-model. Figure 1 shows the open-channel and the measuring/calculating points for wave height and wave velocity.

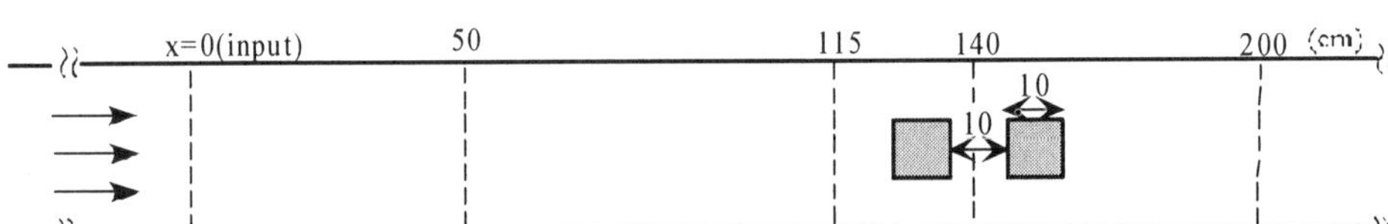

Figure 1. The open-channel and the measuring/calculating points for wave velocity and height

We discuss about accuracy and reliability of the ER-model with the different spatial grid size, Dx=7.5 ~ 50 cm, by comparing the calculated with measured. To analysis the accuracy or reliability, we should calculate the relative errors between: $Error\,(\%) = \left(\sum|C - M| / \sum M\right) \times 100$, where C is value of calculated wave profile and M is value of measured. Also, for discussion of the characteristic and accuracy, the open-channel is divided into three parts; (1) front of the obstacles, (2) middle around them and (3) back of them, because each propagation process should have different characteristic and numerical error. Generally we have fours kinds of errors in the simulation; (a) Physical modeling error, (b) Truncation error (numerical dispersion or dissipation), (c) Sampling error (descretization of wave profile), (d) Rounding error. In the case, we found that the rounding error and sampling errors can be negligible. Figure 2 shows the comparison of wave profiles at the front of obstacles (at x=50cm) as an example. As shown it, we find that the propagation of the tsunami wave doesn't have an influence by obstacles until t=1.6 sec, because the reflected wave doesn't reach the point. In general, the tsunami wave propagates with nonlinear behavior. For the characteristic, many researches verify that 2-D numerical model using shallow water theory can be applied to simulate such nonlinear behavior of propagation of a tsunami wave without obstacle, meaning that the physical modeling error is also small. Then, it is suggested that the truncation error is dominant and especially dissipation error might be large. Figure 3 shows the comparison of wave profiles at the middle of obstacles (at x=140cm). The reason for large errors in height and time at the middle more than the front is that a tsunami is affected and reflected by structures; reflection, diffraction and refraction, which phenomena can not be included fully in the 2-D model. Thus, the physical modeling error might be predominant in this region. Comparison of wave profile at point of the back of obstacles (at x=200cm) is shown Figure 4. The effect of obstacle would not appear in propagation of tsunami wave after passing through obstacle except for shortwaves behind the first wave crest, meaning physical error become small. As stated above, the accuracy of ER-model can be explained through characteristics of each part in the results of numerical simulation.

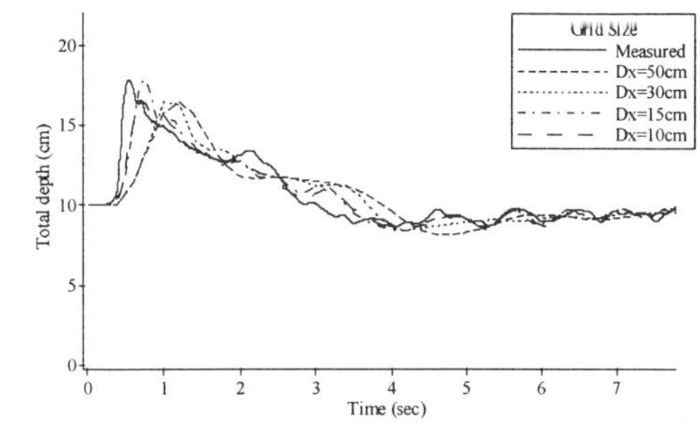

Figure 2. Measured and computed wave profiles at the front of obstacles

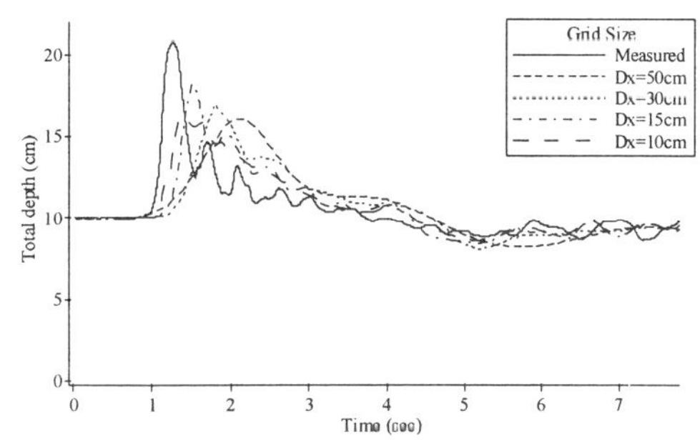

Figure 3. Measured and computed wave profiles at the middle of obstacles

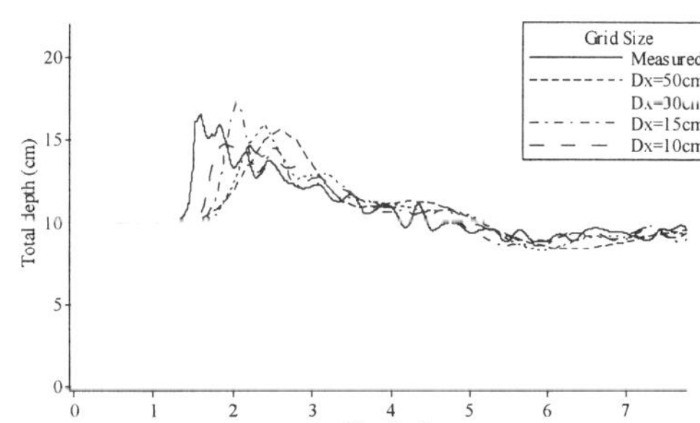

Figure 4. Measured and computed wave profiles at the back of obstacles

REFERENCES

Aburaya, T. and Imamura, F. (2002), *Proposal of Tsunami Inundation Simulation using Equivalent Roughness Model*, Proceedings of coastal engineering, JSCE, Vol. 49, pp. 276-280.

Hong, S. J. and Imamura, F. (2002), *Re-examination of the 1983 Tsunami with the Measured Data in Korea*, Technical meeting of Tohoku branch, JSCE, pp. 206-207.

FLOOD IN JAKARTA - LESSONS LEARNT FROM THE 2002 FLOOD

SUBANDONO DIPOSAPTONO[1], WIDI A. PRATIKTO[2], AKIRA MANO[3]

[1] Dr. Eng., Directorate General of Coastal and Small Islands, Ministry of Marine Affairs and Fisheries, Republic of Indonesia
Jl MT Haryono, Kav 52-53, Jakarta 12770, Indonesia. Phone: +61-21-217-79180163, Fax: +62-21-79180456.
E-mail: sbdn@cbn.net.id (*address for correspondence*)

[2] Professor, -ditto-

[3] Professor, Disaster Control Research Center, Graduate School of Engineering, Tohoku University, Aoba 06, Sendai 980-8579, Japan. Phone/ Fax: +81-22-217-7512. E-mail: mano@civil.tohoku.ac.jp

Extended Abstract

1. Introduction

In February 2002, Jakarta experienced one of the most severe floods in history, which caused widespread damage and affected more than 75% of the total population of the city. Duration of the flood was from the end of January to the beginning of February 2002. Jakarta City is frequently affected by floods; a third of northern Jakarta is flooded on average twice a year.

This paper is intended to present the flood problems in Jakarta, the recently flood in 2002, how such a flood could happen and lessons learnt from it to overcome future floods that most probably attack the Jakarta City with bigger (could be) magnitude.

2. Jakarta - A brief description of geography and climate

The history of Jakarta is start from 16th century, at that time the city name was Jayakarta. During Dutch colonialism, it changed to Batavia and then changed again by Indonesian government to Jakarta. Since that Jakarta City become the Capital City of Republic of Indonesia. Today's Jakarta stretches 661.51 square kilometres across alluvial lowland on the north coast of West Java, and is criss-crossed by 13 rivers, both natural and man-made.

Before the 2nd World War, the population of Jakarta City is estimated around 800.000 persons. Now, Jakarta become a metropolitan city and the population growth very rapidly and now the population is estimated around 8.4 million persons.

Jakarta is located in tropical area. We have only two-seasons; i.e., wet and dry seasons. The dry season is between May to September and the wet season, between October to April. The heaviest rains usually fall around January to February. The temperature trough the year averages 25 – 31 degrees Celsius and humidity averages a high 70%.

3. Jakarta Flood Control

Since the beginning, floods have always been a problem to the people of Jakarta City. Since 17th century and also in 19th century, floods became a serious problem of Batavia people that had population around 600.000 persons.

In 20th century, Prof.Ir. H. van Breen developed a plan to overcome floods of Batavia. The van Breen plan consists of developing flood canals and also flood control construction along the rivers. The flood canal consists of the Western Banjir canal and Eastern Banjir canal. The purpose of both canals is to divert the floodwater coming from outside Batavia City; i.e., Bogor area.

In 1965, Government of Indonesia developed a “master plan for drainage and flood control” and revised in 1973. The master plan was modification of the van Breen plan. The principle of flood control of Jakarta city is developing 3 canal system i.e. flood “banjir” canals, urban drainage canals and polder system.

The urban drainage system is to drain-off the rainwater from the area enclosed the protection of Banjir Canal trough gravity flow channels including rivers and canals. Other areas where mostly located in the low-lying area and affected by the high tidal intrusion, the rainwater will be store in the polder and them pumping-out to the sea.

In fact, Eastern Banjir Canal can not be realised. Mean while, Western Banjir Canal can be realised with some modification due to the increase of cost of land.

4. The Flood in 2002

Every January and February Jakarta is always flooding (flood months). The recent floods have been lasted from January 25th to February 10th with the climax on February 4th, 2002. It has been raining the entire of January. On January 30th a very large amount of rain has fallen. That was the highest daily rainfall of the month January 2002. It has been raining from February 1st to 15th with large rainfall on February 2nd. This amount rainfall is too much in a short time.

Meanwhile, due to rainfall before January 25th, the ground was saturated which caused no more water can infiltrate into the soil, rivers were "full" but not really overtopping. When heavy rainfall on January 30th, rivers started to overtop, depression areas flooded, water logging and springtide at sea.

Rainfall continuously fallen in Jakarta and vicinity with large rainfall on February 2nd, it caused a large flood on February 4th, 2002. Consequently, about 10,000 hectares or 100 km^2 were flooded which is $^1/_6$ of the Jakarta City (in total 661.51 km^2).

Many factors caused such severed flood. The following factors are reported:

a. Rapid urbanisation which brought about by the loss of natural flood reduction functions of land use practices. In 1983, green area was about 60% of the total area of DKI Jakarta. In 1995, the green area was about 17%. And in 2000, the green area was about 2%.
b. Poor operation and maintenance (O&M) over the past six years. If O&M are not done, then the sluices, gates pumps, weirs and trash racks will not function smoothly and flow is restricted.
c. Encroachment by man and structures. All 13 rivers have been narrowed more and more. The flood way is narrowed. As a consequence, the water level will become higher and higher.
d. Bottlenecks by bridges, electrical, gas and water lines. Bridges are too low and too short. Electrical, gas and water lines often placed too low. The flow has been obstructed by the bottlenecks, causing backwater effect.
e. Sedimentation of rivers outlets. All river outlets to the sea silted up because the velocity of flow at the outlet is very low. Sand and silt are deposited on the riverbed. The riverbed comes up, and so does the water level.
f. Insufficient size of (old) flood gates. Many floodgates are old and rusty. Due to poor O&M these do not work well. Once, these were designed for a required peak flow. Nowadays these must be enlarged for larger flow in the future.
g. Building in the river. Some small buildings have been built in the river center. This disturbs the flow of water and sediment.
h. Dumping solid waste into the rivers. Solid waste is accumulated on riverbanks. When water overtops, all solid waste is washed into the river and gets floating. Debris, heavy parts like metals, broken glass pieces remain on the riverbed. Approximately 2,400 m^3/day of solid waste is dumped into the rivers.
i. Clogging of gates by solid waste and floating debris. The water flow normally touches at gates and floating debris is withheld at the front of the gates. When there are trash racks, floating debris will clog the trash racks. Large amounts of debris on the trash rack will reduce the flow section. As a consequence, backwater occurs and results in higher flood levels.
j. Gradual land subsidence. Land subsidence is continuing as long as groundwater is extracted. Since the 1996 floods, the land subsided by about one meter in particular areas. This has worsened the flood risks particularly in those specific areas. Land subsidence rate is about 11.4 cm/year.
k. Clogging of main city drains. Due to poor O&M, solid waste and sediments have been accumulated over the years. When the January – February rainstorms came, these main drains could not conveyed the rain storm waters. Streets and houses were flooded.
l. Absence of city drains. Many areas and streets do not have side drains or ditches. When it rains, water is trapped in the streets. As streets are mostly flat and horizontal, all rain storms water cause a stagnant pool. This can be seen in the northern parts of the city.
m. Absence of early warning system. Forecasting rain storms from the position of cyclones and anti-cyclones is well done by Meteorological office (BMG). But the forecasting floods of 13 rivers are very difficult. The flow in the 13 rivers is very fast and arrives at the sea in a couple of hours. Early warning from the field is necessary to anticipate the operation of floodgates.
n. Confluence of rivers. Where rivers meet each other, we always find problems at their confluence. Encroachment, sedimentation and poor design hamper flow.

Reference

1. A.R. Soehoed, 2002." Banjir Ibukota: Tinjauan Historis & Pandangan ke Depan (*Flood in Capital City: Historical and Future Point of View*", Djambatan, Jakarta.
2. Anonymous, 2002. "Flood Survey in DKI Jakarta", Pacific Consultants International, Jakarta.
3. Subandono Diposaptono, 1983. "Working Report on Jakarta Flood Control Project, Cakung Drain – Cengkareng Drain", Jakarta.
4. Pacific Consultants International, 2002: Flood Survey in Indonesia.

PRECISE NEARSHORE CURRENTS MODEL USING SIGMA COORDINATE SYSTEM

Hisamichi Nobuoka[1] and Nobuo Mimura[2]

1) Dept. Urban & Civil Eng., Faculty of Eng., Ibaraki University, Nakanarusawa, Hitachi, Ibaraki, 316-8511, Japan.
Fax: +81-294-38-5268 E-mail: nobuoka@mx.ibaraki.ac.jp

2) Center for Water Env. Studies, Ibaraki University, Nakanarusawa, Hitachi, Ibaraki, 316-8511, Japan

Background and Purpose

Predicting 3-D nearshore current in wide area is important to management for the coastal sand and the water-quality in coastal area. We need a 3-D nearshore current model being able to predict efficiently and accurately in shallow coastal and deep offshore area. Nobuoka et al. (1998) have shown that the driving force of 3-D nearshore currents is the vertical distribution of radiation stress, and have developed the time-mean current model based on this principle. However, using the level method of the Cartesian coordinates, this model was not practical, i.e., the vertical distribution of the current can not be predicted sufficiently in a shallow region, a simulating CPU time is very long because grid points increase as a water depth becomes deep. If the sigma coordinate is used in the model, an equal number of grid points can be set in shallow coastal and deep offshore areas.

The purpose of this study is to develop the nearshore current model using the sigma coordinate system with the vertical distribution of radiation stress for the driving force.

Government Equations

As shown in Fig.(1), the one of the sigma coordinate is a non-dimensional scale of depth so that each water depth is divided by the all of the local water depth. Equation (1) and (2) show the government equations of nearshore current expressed by sigma coordinate system, that derived from Navier-stokes equation and Continuity equation with the assumption that each velocity can express sum of the three components, i.e. wave, time-mean and turbulent components.

The radiation stress term integrated in all of the local water depth should agree with conventional radiation stress regardless of sigma coordinates. This integrated stress is in agreement with the conventional radiation stress showed by C.C.MEI (1989). From this agreement, it has been proved that radiation stress component of Eq.(2) is appropriate.

As keep a conservation law even if the method of finite differences is adopted, Eq.(1) and Eq.(2) are integrated in the vertical direction in the layer which is divided arbitrarily in the water depth. The integrated equations are government equations of nearshore current model proposed by this study.

Calculation of Radiation Stress

Even if sea bottom is very mild, the effect of the inclination needs to be taken into consideration to the vertical distribution of radiation stress (Nobuoka, 2002). Therefore, the vertical distribution of radiation stress in sigma coordinate system was calculated using Biesel wave theory (1952) which can take the inclination into consideration.

One example of the results for the vertical distribution of radiation stress in a surf zone is shown in Fig.(2). The result of sigma coordinate shows the steep momentum gradient generated by the breaking wave near the water surface can be expressed by the radiation stress. The result except the effect of inclination of sigma-line is also shown in Fig.(2). Although the gradient of sigma-line is

$$\frac{1}{D}\left\{\frac{\partial u_f}{\partial x}+\frac{\partial v_f}{\partial y}+\frac{\partial \omega_f}{\partial \sigma_f}\right\}=0 \qquad \text{--(1)}$$

$$\frac{1}{D_f}\left\{\frac{\partial u_f^2 D_f}{\partial x}+\frac{\partial u_f v_f D_f}{\partial y}+\frac{\partial u_f \omega_f}{\partial \sigma_f}\right\}=-\frac{\partial}{\partial x}\left(-gD_f\sigma_f\right)-\overline{\frac{\partial}{\partial x}\left[-gD_w\sigma_w-\frac{\partial}{\partial x}\int_0^{\sigma w} w_w u_w D_w d\sigma-\frac{\partial}{\partial y}\int_0^{\sigma w} w_w v_w D_w d\sigma - w_w\omega_w\right]}$$
$$-\overline{\left(\frac{\partial \sigma}{\partial x}\right)\left(-gD_w\sigma_w-\frac{\partial}{\partial x}w_w u_w D_w-\frac{\partial}{\partial y}w_w v_w D_w-\frac{\partial}{\partial \sigma_w}w_w\omega_w\right)}-\overline{\frac{1}{D_w}\left\{\frac{\partial u_w^2 D_w}{\partial x}+\frac{\partial u_w v_w D_w}{\partial y}+\frac{\partial u_w \omega_w}{\partial \sigma_w}\right\}}+R_x \qquad \text{--(2)}$$

very mild as less than 1/20, the effectiveness of the gradient affects radiation stress greatly.

Verification of Nearshore Current Model

The solution of the proposed nearshore current numerical model is the same as that of Nobuoka (1998). These calculated results for some wave condition were compared with that of experiment in order to examine the capacity of proposed model. The time-mean water level, calculated by sigma coordinate system, showed good agreement with the results of experiment. As shown in Fig.(3), the calculated time-mean current using sigma coordinate system is in agreement with that of experiment well. On the other hand, the result using the Cartesian coordinate system can not express the distribution because of above mention about the limit of layer numbers. This result confirms us that using sigma coordinate system improved the prediction of a vertical distribution profile.

The proposed model was compared with undertow model of Okayasu's type (1988). Figure (4) shows the root-mean-square error of calculation against experimental results in the surf zone. This result indicates that the predictability by this proposed model is as high as that of undertow model.

The proposed model can reduce the calculation CPU time by 80% compared with that of the former model. The calculation CPU time is short also has advantage being able to calculate in a wide area.

Conclusion

In this study, the equations of 3-D nearshore current using sigma coordinate system were proposed, the model for these equations was developed and the following points were clarified.

1) We must take account of the inclination of sigma line in order to calculate the radiation stress correctly.
2) The prediction of on/off-shore currents in a surf zone by the proposed model is as good as that by undertow model which can not estimate offshore and alongshore currents.
3) The calculation CPU time becomes short by the proposed model.

The above results confirm us that nearshore current model using sigma coordinate system is highly practical and is a part of effective tool to predicting beach change and water-quality distribution.

References

Biesel, F. (1952) : Study of Wave Propagation in Water of Gradually Varying Depth, Gravity Waves, pp.243-253.

Mei, C. C. (1989) : The Applied Dynamics of Ocean Surface Waves, World Scientific, 740p.

Nobuoka H., H. Kato and N. Mimura (1998) : Three dimensional nearshore currents model based on the vertical distribution of radiation stress, 26th ICCE, pp.829-842.

Nobuoka H.and N. Mimura (2002): 3-d nearshore current model focusing on the effect of sloping bottom on radiation stresses, 28th ICCE, (in press)

Okayasu A., Shibayama T. and Horikawa K. (1988) : Vertical Variation of Undertow in the Surfzone, 21th ICCE., 478-491.

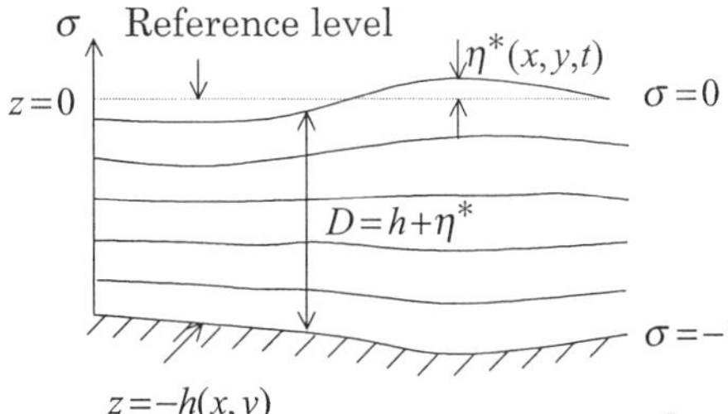

$$\sigma = \frac{z-\eta^*}{D}$$

Figure (1)

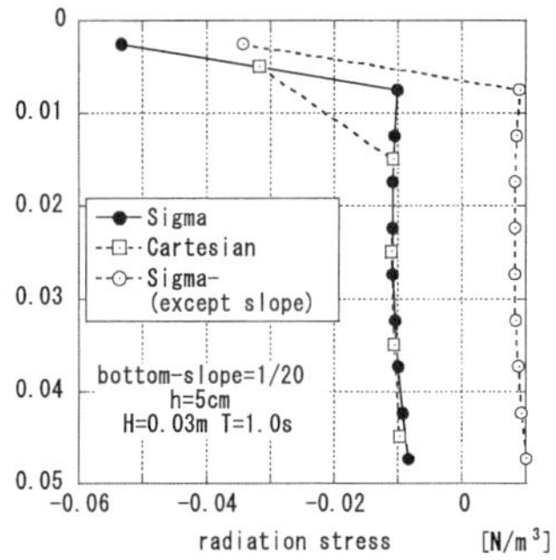

Figure (2)

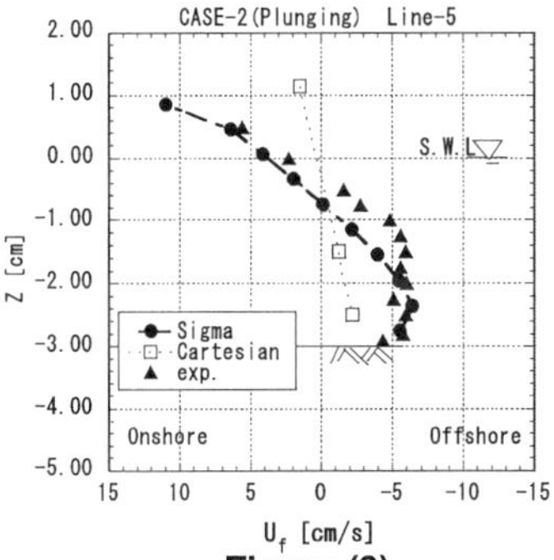

Figure (3)

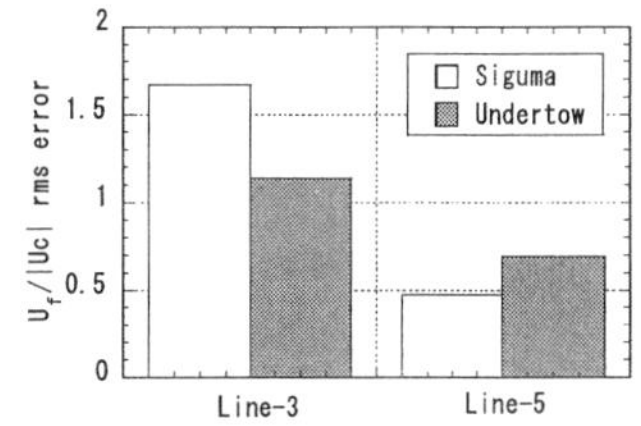

Figure (4)

An application of a nesting procedure to a highly-resolved current simulation in a mangrove area

Yasuo NIHEI[1), *Keita Sato*[1), *Yasunori Aoki*[2), *Tsukasa Nishimura*[1) *and Kazuo NADAOKA*[3)

1) Dept. of Civil Eng., Tokyo University of Science, 2641 Yamazaki, Noda-shi, Chiba 278-8510, Japan, Phone: +81-4-7124-1501, Fax: +81-4-7123-9766, Email: nihei@rs.noda.tus.ac.jp.
2) Design & Eng. Dept., Penta-Ocean Construction Co. Ltd., 2-2-8 Koraku, Bunkyo-ku, Tokyo 112-8576, Japan, Phone: +81-3-3817-7804, FAX: +81-3-3817-7805, E-mail: Yasunori.Aoki@mail.penta-ocean.co.jp.
3) Graduate School of Information Science and Engineering, Tokyo Institute of Technology, 2-12-1 O-okayama, Meguro-ku, Tokyo 152-8552, Japan, Phone: +81-3-5734-2589, Fax: +81-3-5734-2650, Email: nadaoka@mei.titech.ac.jp.

Introduction

A riverine-type mangrove (R-type mangal), which is observed in most mangrove forests in Japan, is composed of meandering creeks and fringing mangrove swamps with a densely vegetated mangrove trees and roots. The ratio of the swamp area to the creek area appears to be typically of order 2-10 and then the horizontal scale of the swamp is quite different with that of the creek. To accurately predict water environments and sedimentary processes in a R-type mangal, it is desirable to conduct current simulations in which the morphological feature of R-type mangals as mentioned above are taken into consideration. For this purpose, it may be expected to employ a nesting procedure, in which the computational results in a larger-scale domain are successively reflected on a smaller-scale computation. From these reasons, in the present study, we have attempted to apply a nesting procedure to the highly-resolved current simulation in Fukido River estuary, one of typical R-type mangals, located in the north-west part in Ishigaki Island, Okinawa, Japan.

Outline of a numerical model and computational condition

The fundamental equations in the present computation are based on the 2-D Navier-Stokes equation with the drag force for densely vegetated mangrove trees in the swamp. To realize a highly-resolved computation for flow field in the mangrove area with computational efficiency, we introduce the three computational domains as shown in **Fig.1**. The larger-, intermediate- and smaller-scale domains, referred here as Grid 1, 2 and 3, have the spatial resolutions of 4m, 1m and 0.25m, respectively. As the nesting procedure, we

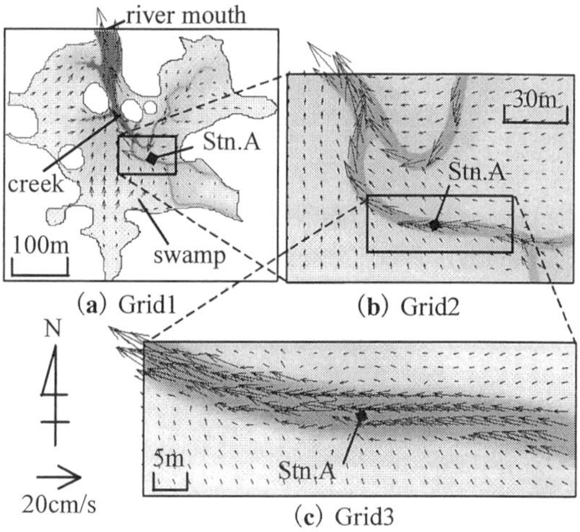

Fig. 1 Computational domains and computed flow structure at ebb tide (8pm, August 16, 2001).

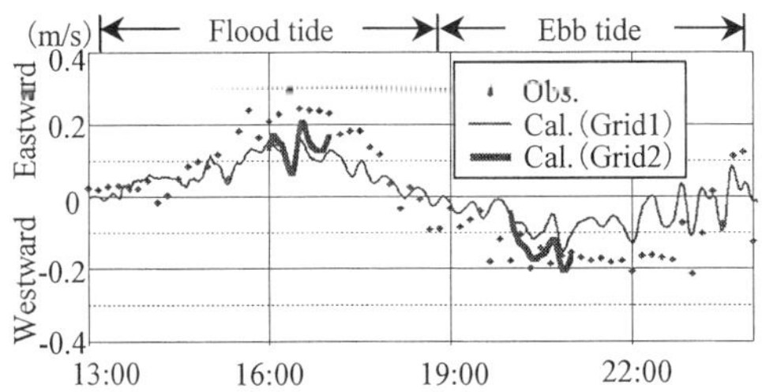

Fig.2 Time sequences of the observed and computed creek velocity at Stn.A (August 16).

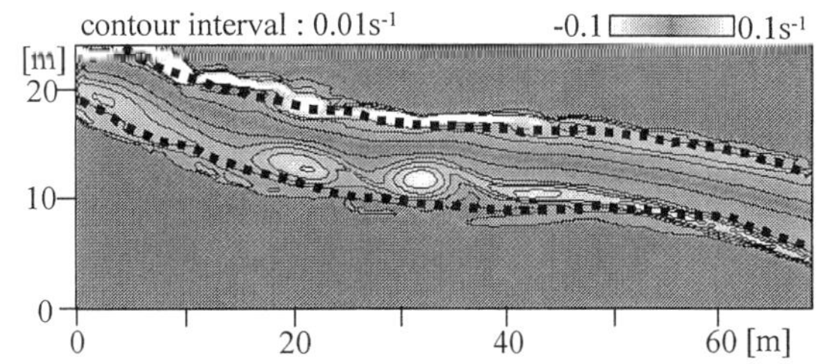

Fig.3 Contour of vorticity around the creek computed in Grid 3 (20:38, August 16).

adopt a one-way nesting approach, in which the computational results in the larger-scale domain are given as the open boundary conditions for the smaller-scale computation. For comparison of the computed results with observed data obtained by the authors, the tidal current simulation is performed for the period from 1pm August 16 to 11am August 17, 2001.

Results and discussion

The horizontal patterns of flow field obtained in Grid 1 show a tidal asymmetry between the flood and ebb water velocities in the swamp; while the swamp current normal to the creek is dominant at the flood tide, the flow in the swamp at the ebb tide goes toward the river mouth as shown in **Fig.1 (a)**. We compare the observed water elevation and horizontal current in the swamp with the computational results in Grid 1, indicating that the computational results give acceptable agreements with the observed data.

As depicted in **Fig.1 (b)**, the appreciable differences of the horizontal velocities in the creek are displayed in the computational results in Grid 2 although those are not clearly shown in Grid1. The lateral distribution of the horizontal velocities in the creek may be formed due to morphological features of the creek and the influence of larger scale flow field in the fringing mangrove swamp. **Figure 2** indicates the time sequences of the observed creek current at Stn.A with the computational results of the velocity in Grid 1 and 2. The comparison of these results shows that the computational results in Grid2 gives better agreements with the observed data than those in Grid 1.

The computational results in Grid 3 with the higher spatial resolution indicate the more complicated flow structure around the creek. **Figure 3** shows the contour of the computed vorticity in Grid 3, indicating that the horizontal large-scale eddies appear around the boundary between the creek and swamp. The evolution of a series of large eddies is found to be different in left and right banks of the creek and also vary in the flood and ebb tides. These large-eddy structures in the creek do not appear in Grid 1 and 2 with relatively lower grid resolutions. These computational results demonstrate that we confirm the fundamental applicability of the nesting procedure to the tidal current simulation in the mangrove area.

NUMERICAL SIMULATION OF WAVE PROPAGATION IN THE WATER AREA OF THE YANGTZE RIVER ESTUARY

Wang Hongchuan[1] Zuo Qihua[2] Pan Junning[3]

The wave conditions in the water aera of Yangtze River Estuary are of significance for the design of deep-water channel regulated works for Yangtze Estuary (Fig1). The regulated structures of the works are laid out in very large scope of the sea area. The wave parameter of the structure must be defined through analysis and calculation depending on the observed information. The parabolic approximation of mild-slope equation is one of the most useful methods in simulating the wave transformation for wave refraction and diffraction. In our research some key parameters are discussed of how to define the friction coefficient, how to use the measuring data, and how the structures influence the propagation of the wave.

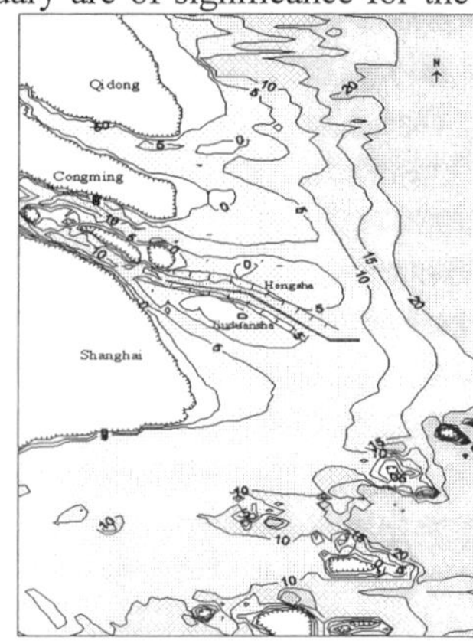

Fig.1 The map of Yangtze Estuary with the 2nd-stage regulated structure

1. Intruduce of the model

Considering the nonlinear parabolic mild-slope equation:

$$\frac{\partial}{\partial x}\{(KCC_g)^{1/2}(1+\frac{P_1M}{K^2CC_g})\}\phi = iK(KCC_g)^{1/2}(1+\frac{P_2M}{K^2CC_g})\phi \qquad (1\text{-}1)$$

in which $M\phi = (-\omega^2k^2D|A|^2 + i\omega F)\phi + (CC_g\phi_y)_y$, $D = \frac{ch4kh + 8 - 2th^2kh}{8sh^4kh}$ (1-2)

when P_1=0.25 , P_2=0.75 , it is a high-order parabolic mild-slope equation. It can be rewrite to the following form:

$$(kCC_g)\phi_x + \{\frac{1}{2}(kCC_g)_x - ik^2CC_g + \frac{i}{2}\omega^2k^2D|A|^2 + \frac{\omega}{2}F\}\phi - (P_1B + iP_2)M'\phi + \frac{P_1}{k}(M'\phi)_x = 0 \qquad (1\text{-}3)$$

where F in Eq.(1-2) may include the dissipation of friction, the inputting energy of wind and the 2nd-order of the wave transmission in y-Direction. F can be expressed: $F = F_r + F_w$, F_r and F_w represent the influence of the friction and the energy of wind respectively.

2. Coefficient of friction

there are several stations on the aera of the Yangtze Estuary such as Niupijiao station , Yinshuichuan station , Dajishan station and so on. At the deep water out of -20m, There exists a P point (122°30′E, 31°15′N) where the wave parameter are obtained during typhoon period. Two set of data during typhoon period between P and Niupijiao stations are participated in investigating the propagation of the wave. when typhoon 9906# period the wave height H_s is 4.1m, and T equals 7.1s in E-direction, and H_s is 3.2m, and T equals 7.2s in NNE-direction at P, several conditions are investigated. The differences of wave height among variable coefficient of friction are listed in Tablet 1. The result of linear model is listed in the tablet either.

Tablet 1 Computed wave height H_s(m) at Niupijiao Station in period of Typhoon for different coefficient of friction

f=	9906# (E)		9912# (NNE)	
	Non-linear	Linear model	Non-linear	Linear model
0.005	3.16	2.62		2.21
0.0075	2.99	1.93	2.36	2.14
0.01	2.87		2.27	1.76
0.02	2.54		1.98	
0.03	2.26		1.86	

According to the Niupijiao station's wave data, the wave height is 2.7m in 9906# period and 2.3m in 9912# period. When we set the coefficient f 0.015 in E-direction and 0.01 in NNE-direction respectively, the result of nonlinear model is agreement with the station's data.

3. the influence of Wind velocity

When the waves propagate for a long distance, the wave energy dissipates and the wave height reduces. The inputting wind energy become significant. According to the permanent measuring station the wave height of return period of 50 yrs is obtained at the stations, as well as the deep wave height out of P point (-20m) is defined. Adding the wind velocity into the term of wind energy, the investigated wave height is obtained (tablet 2). It shows that the wave height increase as the Wind velocity is added. When U reaches 20m/s in NNE-direction, the wave height at Dajishan is in agreement with the value of the return period of 50 years, and when U is 0m/s the wave height at yinshuichuan has good agreement with the value of E-direction. So we define the wind velocity as the parameter in NNE and E-direction respectively.

Tablet 2 Computed wave height H_s(m) at the Stations in return period of 50 yrs for different wind velocity

Station	Measured H_s(m)	Computed H_s(m)		
		U=0m/s	U=10m/s	U=20m/s
Yinshuichuan (E)	4.4	4.34	4.59	4.78
Dajishan (NNE)	6.13	4.83	5.21	6.03

4. Transmission of the breakwater

The transmission and reflection coefficient of submerged structure is studied by many researcher[2][3][4]. The coefficient is related to the type of the structure. From the result of the predecessor, the transmission coefficient over submerged breakwater become small as the types queue semi-circular breakwater, sloping breakwater and vertical breakwater. The transmission coefficient of the submerged semi-circular breakwater are given based on physical model tests in the paper [2], which can be described as $K_t = 0.258[\tanh(-1.55h_c / H_s)]^{1.15} + 0.56$.

5. Wave field

The wave distribution in the area of Yangtze Estuary is computed. Result shows that the wave height behind the leading jetties is about 0.6~0.8 times of the value in front of the jetties.

Reference:

[1] James T. Kirby, Higher-order approximations in the parabolic equation method for water waves, Journal of Geophysical Research, Vol. 91 , No. C1, p933-952., Jan. 1986.

[2] Xie Shileng & Cai Yanjun, Analysis of Wave Height behind a Submerged Semi-Circular Breakwater, Port engineering technology, 1999, No.4.

[3] Beji S. and J. A. Battjes, 1994, Numerical Simulation of Nonlinear Wave Propagation over a Bar , Coastal Engineering , 23:1-16.

[4] Yoshimi GODA, Re-analysis of Laboratory Data on Wave Transmission over Breakwaters, Report of the Port and Harbour Research Institute, Vol. 8 No. 3, 1969 .

1 Doctor, Research Institute of Coastal and Ocean Engineering, Hohai University, Nanjing, China. Senior Engineer, River & Harbour Department, Nanjing Hydraulic Research Institute, Nanjing, 210024, China, E-mail:hcwang@njhri.edu.cn. Tel:+86 25 5829341. Fax:+86 25 5829333.

2 Professor, Nanjing Hydraulic Research Institute, Nanjing, 210024, China.

3 Engineer, River and Harbour Department, Nanjing Hydraulic Research Institute, Nanjing, 210024.

Visualization of Tidal Oscillation in the Taiwan Strait

JUANG Wen-Jye[1], LIN Ming-Chung[2] and CHIANG Chung-Chiuan[3]

1 Researcher, Center of Harbor and Marine Technology, Institute of Transportation.

No.2, Chung-Heng 10th Rd., Wu-Chi, Taichung 435, Taiwan.

Tel.: +886-4-26587185, Fax: +886-4-26564415, jye@mail.ihmt.gov.tw

2 Professor, Department of Engineering Science and Ocean Engineering, National Taiwan University.

No.73, Tzou-San Rd., Taipei 106, Taiwan

Tel.: +886-2-23625470Ext.240, Fax: +886-2-23929885, mclin@mail.na.ntu.edu.tw

3 Assistant Researcher, Center of Harbor and Marine Technology, Institute of Transportation.

No.2, Chung-Heng 10th Rd., Wu-Chi, Taichung 435, Taiwan.

Tel.: +886-4-26587132, Fax: +886-4-26564415, river@mail.ihmt.gov.tw

Abstract

Taiwan situates right on the rim of the Asia Continental Shelf. A typical profile of the shelf bathymetry around the coast of Taiwan is shown as Figure 1. Due to the topographical effect of the shelf (Lin, et al., 2001), the oscillation patterns of a partially standing tides appeared in the Taiwan Strait are investigated firstly by using the theory of co-oscillation, Figure 2. To further identify the oscillation patterns, a two-dimensional depth-averaged shallow water wave model: MIKE21_HD, developed by Danish Hydraulic Institute (DHI, 1996), is applied to compute the regional tidal characteristics (Juang, et al., 2001). By extracting the computed variations of water surface elevation from sections along the coast of China and around Taiwan, Figure 3, the visualizations of tides oscillating in the Taiwan Strait are finally produced as animations and shown statically as Figures 4 and 5, respectively. From Figure 4 of the tidal oscillation along the coast of China, one can visualize a Kelvin wave equipped with nearly constant amplitude propagating from north to south, and accompanied with a partially standing tides oscillating adjacent to the southern end. From Figure 5 of the tidal oscillation around the coast of Taiwan, it is easily to identify that a partially standing tides oscillating on a topographical shelf with the anti-node appeared almost at the central part of the shelf and the nodes located adjacent to the rims. With the aids of the visualizations associated with the animations, the oscillation appearances as well as the characteristics of tides in the Taiwan Strait are illustrated completely.

Keywords

The Shelf, Tidal Oscillation, Partially Standing Tides, Visualization

References

DHI (1996): User guide and reference manual of MIKE 21- coastal hydraulics and oceanography hydrodynamic module. Danish Hydraulic Institute.

Lin, M. C., W. J. Juang and T. K. Tsay (2001): Anomalous amplification of semidiurnal tides along the western coast of Taiwan. *Ocean Engineering*, 28(9), pp.1171-1198.

Juang, W. J., C. C. Chiang and M. C. Lin (2001): Tidal Current Simulations on Seas Surrounding Taiwan. Proc. 1st Conf. Asian and Pacific Coastal Engineering (APACE2001), Dalian, China, Vol.1, pp.86-95.

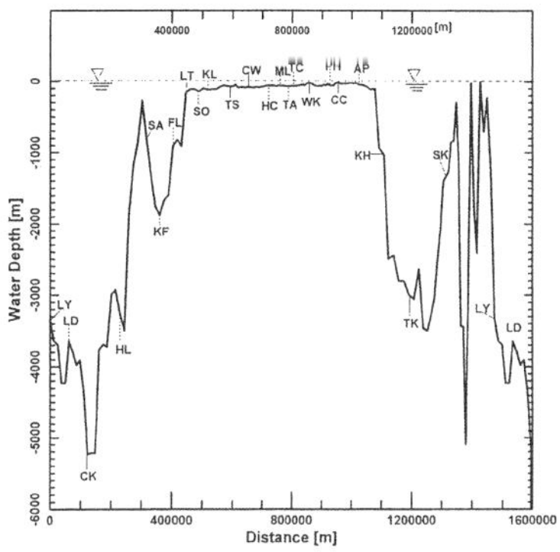

Fig.1 Typical profile of the shelf bathymetry around the coast of Taiwan

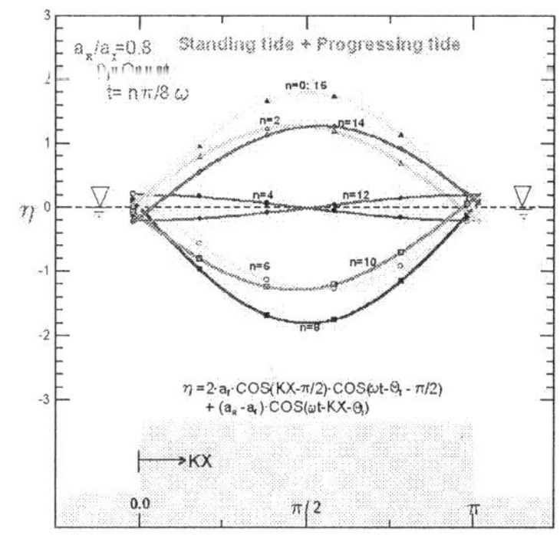

Fig.2 Oscillation of a partially standing tides on an idealized shelf

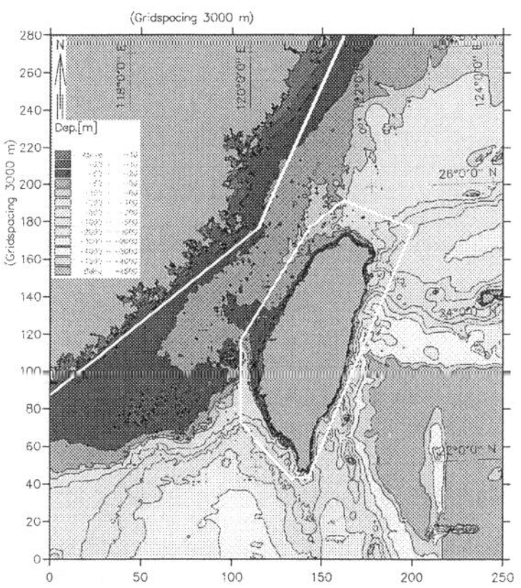

Fig.3 Bathymetry around Taiwan and the topographical sections (white color line) for extracting the tidal oscillation data

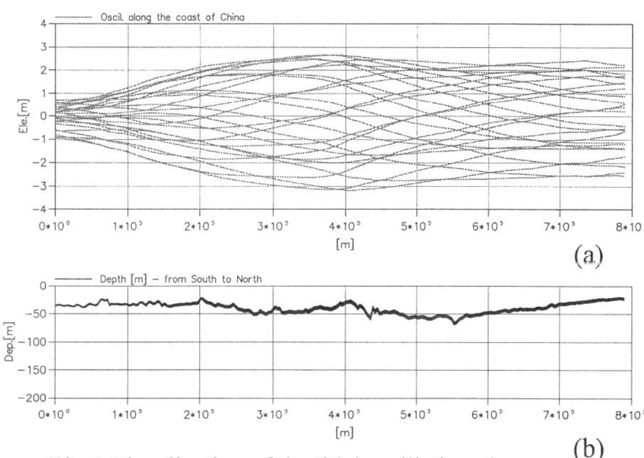

Fig.4 Visualization of the tidal oscillation along the coast of China (a) associates with the bathymetry profile (b)

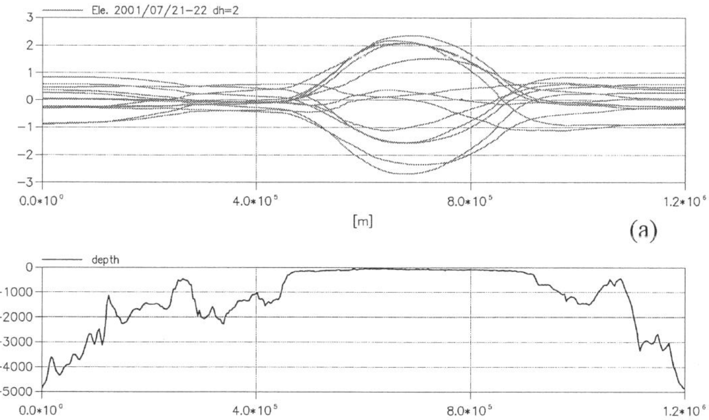

Fig.5 Visualization of the tidal oscillation around the coast of Taiwan (a) associates with the bathymetry profile of a topographical shelf (b)

A NONLINEAR COUPLED NUMERICAL MODEL FOR ASTRONOMICAL TIDES AND STORM SURGE – NUMERICAL SIMULATION OF THE STORM SURGE IN RIVER CHANGJIANG'S ESTUARY

Tan Ya, Zhang Junlun

College of Traffic and Coastal Engineering , Hohai Univ., Nanjing 210098, P.R.C

Abstract

If the extraordinary sea-level rising caused by strong wind or sudden change of air pressure meets the high water of astronomical spring tides, the interaction of the two will probably result in the sudden and sharp rising of water level as well as some damages such as dams, lands, ports and villages being flooded. This kind of damage is usually called damages caused by storm surge. Chinese coastal areas, especially the areas south to the River Changjiang's Mouth in East China Sea, are one of the high frequency storm surge occurrence areas in the world. Because the water of our offshore areas is often relatively shallow, when the strong wind power of typhoon acts on sea surface, the water body will be fully mixed vertically and be transported horizontally in a great extension. Therefore it can be assumed that the upper and lower layers of seawater are symmetrical so that it is proper to adopt the depth-concerned integral 2-D overall storm surge numerical model.

The tide level of a storm surge caused by typhoon, when observed from the shore, includes the astronomical tidal wave from open seas and the additional elevation (decreasing). Likewise, the current occurring in the water area not only includes the runoff from the upper stream and the periodical tidal current, but also includes the additional flow velocity related with storm-surge-caused elevation or decreasing, that is the flow velocity caused by elevation or decreasing. The coupling of the storm surge elevation (decreasing) with the normal astronomical tides will cause the flood-and-ebb of the sea level and the to-and-fro change of the water flow. When the numerical model is applied for the simulation of storm surge, it is necessary to take the non-linear effect of elevation (decreasing) caused both by periodical astronomical tides and by weather abnormity into account. Besides the runoff from upper stream and the periodical tidal current, the water flow in the River Changjiang's Mouth also includes the additional flow velocity which is related with storm surge elevation (decreasing), viz. the flow velocity caused by elevation or decreasing. Every year when there are summer and autumn typhoons, extraordinary high water level and relatively large flow velocity caused by elevation or decreasing occur with high frequency in the river stretch from Tiansheng Port to Yanglin, Qinglong Port in Yangtze River.

In this article a 2-D astronomical tides and storm surge coupled numerical model is used to simulate the storm surge process in River Changjiang's Mouth and to testify the 9711 typhoon. In the numerical simulating process of the storm surge in River Changjiang's Mouth, water boundary controlling is the key factor that will influence the accuracy of the simulation results. In this article the water boundary controlling of the storm surge in River Changjiang's Mouth is provided by the East China sea storm surge and tidal wave coupled model. The computational field of the East China sea storm surge and tidal wave coupled model is from Nan'ao of Shantou, Guangdong in the south, via Penghu islands to Budai at the west bank of Taiwan, and next from Su'ao at the east bank of Taiwan via Japanese Liuqiu archipelago to Jiuzhou, through Duima Channel to Fushan in Korea. The rationality and accuracy of this model has been validated. In this article the integrative water level process line during 9711 typhoon period in Wusong station in River Changjiang's Mouth is computed and the result is in good consistent with the field data. The rationality and accuracy of the computational results of this model will serve as a basis of the further improvement and development of the storm surge inundation numerical forecasting model with a high differentiating rate.

Wave Propagation over Slowly Varying Depth with the Presence of Weak Currents

Ming-Chung Lin[1] Chao-Min Hsu[2] Chao-Lung Ting[3]

Abstract

As waves enter into open shallow water, shoaling and refraction occur due to water depth variations. The presence of various currents in the ocean makes wave motion much more complicated. Isobe (1994) considered variation method to derive a set of equations, which can be simplified to different physical equations according different water depth distributions; however, he did not include the current effects. In this paper, we consider a much more realistic problem, which includes wave condition and current effects simultaneously. A set of governing equations was derived based on dynamic boundary condition with the method of calculus of variation, that the order of current velocity is much smaller than the order of wave phase velocity and group velocity, i.e. $O(|\boldsymbol{U}|^2/CC_g) << 1$.

Firstly we, according to Isobe (1994), use the method of calculus of variation to handle the dynamic free surface boundary condition.

$$L[\phi,\eta] = \int_{t_1}^{t_2} \iint_A \int_{-h}^{\eta} [\frac{\partial\phi}{\partial t} + \frac{1}{2}(\nabla\phi)^2 + \frac{1}{2}(\frac{\partial\phi}{\partial z})^2 + gz] dzdAdt \qquad (1)$$

The relative symbols of the equation and boundary conditions described above can be represented below; $\phi(x,y,z,t)$ is velocity potential of the field, ∇ is the horizontal gradient operator, define as $(\partial/\partial x, \partial/\partial y)$; x and y are the horizontal coordinates, and $\boldsymbol{x}$ is represented as (x, y); t represents time; the gravitational acceleration is g; the surface elevation is $\eta(x,y,t)$, and $h(x,y)$ represents the depth of the water. The horizontal velocity of pure current is represented as $\boldsymbol{U}$–(U, V), and $\boldsymbol{u}$=(u, v) is the wave particle velocity. Let the velocity potential of the field, as shown in equation (2), be represented as the sum of the pure current velocity potential and wave velocity potential after the wave-current interactions, and the pure current velocity potential in vertical distributions is assumed uniform.

$$\phi = \phi_U(\boldsymbol{x}) + \phi_w(\boldsymbol{x},z,t) = \phi_U(\boldsymbol{x}) + \sum_{\alpha=1}^{N} Z_\alpha(z,h(\boldsymbol{x})) f_\alpha(\boldsymbol{x},t) - \boldsymbol{U}\cdot\boldsymbol{x} + Z_\alpha f_\alpha \qquad (2)$$

From the Euler equation, we get the following two equations (3) and (4).

$$Z_\alpha^\eta \frac{\partial\eta}{\partial t} + \nabla[A_{\alpha\beta}\nabla f_\beta] - B_{\alpha\beta} f_\beta + (C_{\beta\alpha} - C_{\alpha\beta})\nabla f_\beta \nabla h + \frac{\partial Z_\beta^\eta}{\partial h} Z_\alpha^\eta f_\beta \nabla\eta\nabla h + \boldsymbol{U}\cdot\left(Z_a^\eta \nabla\eta + Z_a^{,h}\nabla h\right) = 0 \qquad (3)$$

---

[1] Prof., Department of Engrg. Science and Ocean Engrg, National Taiwan University, No.1, Sec. 4, Roosevelt Rd. Taipei, Taiwan 106, Fax:886-2-23929885, Tel:886-2-23625470-240, Email: mclin@mail.na.ntu.edu.tw

[2] Ph.D. Student, Department of Engrg. Science and Ocean Engrg., National Taiwan University, No.1, Sec. 4, Roosevelt Rd. Taipei, Taiwan 106, E-mail: deepblue@ocean221.na.ntu.edu.tw

[3] Assoc. Prof., Department of Engrg. Science and Ocean Engrg, National Taiwan University, No.1, Sec. 4, Roosevelt Rd. Taipei, Taiwan 106, E-mail:chaoting@ccms.ntu.edu.tw

$$g\eta + Z_\alpha^\eta \frac{\partial f_\alpha}{\partial t} + \frac{1}{2}(Z_\alpha^\eta Z_\beta^\eta \nabla f_\alpha \nabla f_\beta + \frac{\partial Z_\alpha^\eta}{\partial z}\frac{\partial Z_\beta^\eta}{\partial z} f_\alpha f_\beta) + \frac{\partial Z_\alpha^\eta}{\partial h} Z_\beta^\eta f_\alpha \nabla f_\beta \nabla h + f_\alpha \frac{\partial Z_\alpha^\eta}{\partial h} \boldsymbol{U}\cdot\nabla h + \tfrac{1}{2}|\boldsymbol{U}|^2 + Z_\alpha^\eta \boldsymbol{U}\cdot\nabla f_\alpha = 0 \quad (4)$$

From these equations, we can get different physical equations with different water depth distribution assumptions. For example, with $z=\cosh k(z+h)/\cosh kh$; $z_1=1$, $z_2=(h+z)^2/h^2$; and $z=1$ respectively, it yields modified Boussinesq equation in equations (5) and (6), and shallow wave equation (7) and (8), including current effects as follows:

$$\frac{\partial\eta}{\partial t} + \nabla[(h+\eta)\bar{u}] + \boldsymbol{U}\cdot(\nabla\eta + \nabla h) = 0 \quad (5)$$

$$g\nabla\eta + \frac{\partial\bar{u}}{\partial t} + \bar{u}(\nabla\cdot\bar{u}) + |\boldsymbol{U}|\nabla U + \nabla(\boldsymbol{U}\cdot\bar{u}) = \frac{h}{2}\nabla[\nabla\cdot(h\bar{u}_t)] - \frac{h^2}{6}\nabla(\nabla\cdot\bar{u}_t) \quad (6)$$

$$\frac{\partial\eta}{\partial t} + \nabla[(h+\eta)\boldsymbol{u}] + \boldsymbol{U}\cdot(\nabla\eta + \nabla h) = 0 \quad (7)$$

$$\frac{\partial\boldsymbol{u}}{\partial t} + (\boldsymbol{u}\nabla)\boldsymbol{u} + g\nabla\eta + |\boldsymbol{U}|\nabla U + \nabla(\boldsymbol{U}\cdot\boldsymbol{u}) = 0 \quad (8)$$

In the abstract, we only show the results of the flow fields of shallow wave and currents around a circular island.

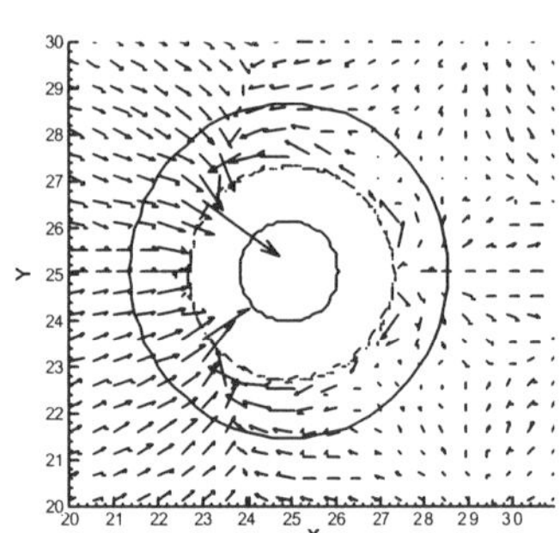

(a) U=0cm/sec

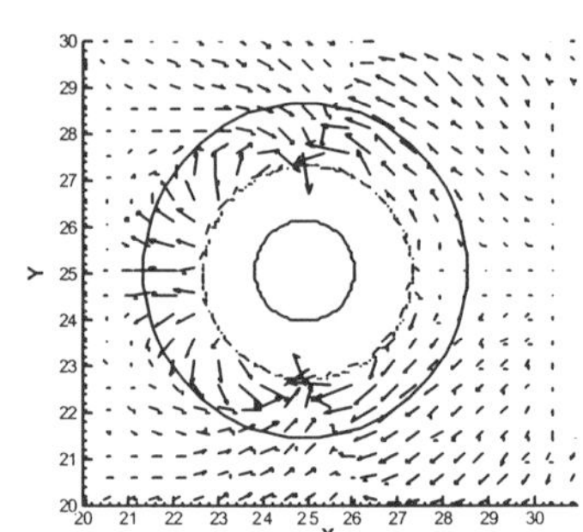

(b) U=5cm/sec

Figure 1. The velocity distributions of the wave interactions with favorable flow around a circular island (T = 6 sec).

The positions of maximum particle velocity move to larger x positions than in the case of no current. Therefore, the current influences the phase and magnitude of particle velocity.

References

1. Isobe, M. (1994), "Comparative Study of Equations for Analyzing Coastal Wave Transformation," *Journal of Hydraulic, Coastal and Environmental Engineering*, Japan Society of Civil Engineers, No.491, pp.1-14. (in Japanese)
2. Liu, P. F.-F, Yong-Sik Cho, M. J. Briggs, U. Kanoglu and C. E. Synolakis (1995), "Runup of Solitary Waves on a Circular Island," *J. Fluid Mech.*, Vol. 302, pp.259-285.

REGIONAL OCEAN TIDE SIMULATOR: SIMULATION OF BARRIER EFFECTS

Byung Ho Choi[1], Jin Hee Yuk[2] and Han Soo Lee[2]

[1]Professor, Department of Civil and Environmental Engineering, Sunkyunkwan University,
Suwon 440-746, Korea
Email: bhchoi@yurim.skku.ac.kr

[2]Graduate Student, Department of Civil and Environmental Engineering, Sunkyunkwan Univeristy,
Suwon 440-746, Korea
Email: jhyuk@skku.edu; ban8303@skku.edu

In view of practical concerns for pollutants, oil spill dispersal, search and rescue operations at sea, and navigation, demands for accurate tidal predictions in both time and space are increasing. The current tide prediction system for regional seas neighboring to Korean peninsula comprised of the modeling systems with finite element mesh based model, addresses the necessity of resolving the detailed estuarine and continental shelf tidal system and timely prediction for operational forecast. Design strategy of predictive modeling system is to provide accurate water level prediction including tidal prediction and meteorologically induced surge prediction. To meet this requirement, fine-resolution relocatable FEM based model is developed during recent years and presented in this paper. Application of the modeling system for estimating the effects on tides due to construction of barriers at Saemangeum area, west coast of Korea and Isahaya Bay in the Ariake Sea, Japan are described Some of results showing the changes in tidal regime including the detailed flow patterns and variations in drying-boundaries are presented and discussions are also make for further modeling studies.

Regional Tidal Simulator

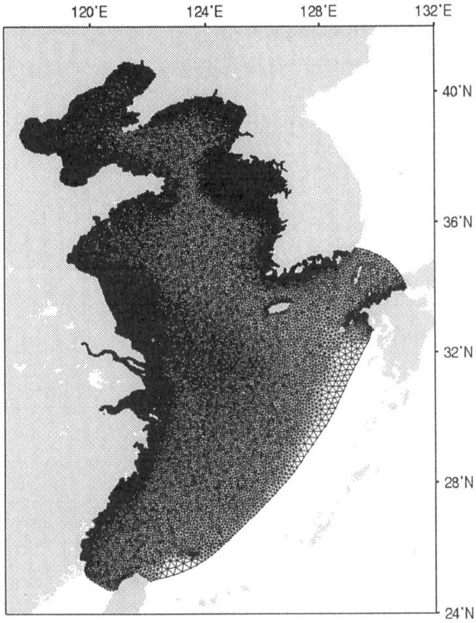

Base Model : FEM Mesh of the Saemangeum area.

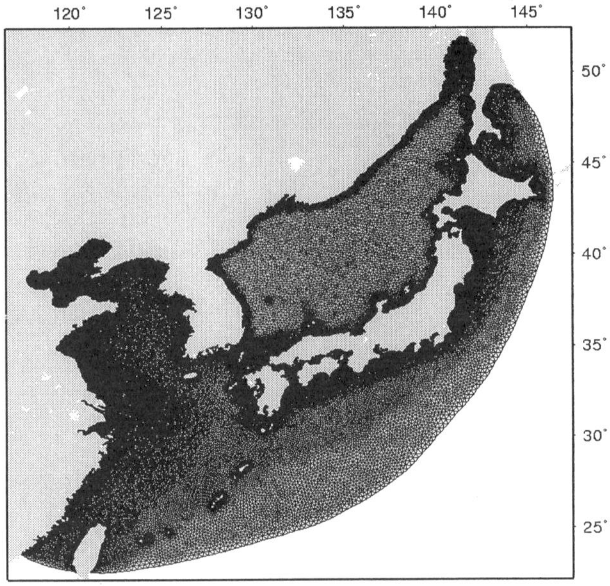

Base Model : FEM Mesh of the Isahaya Bay

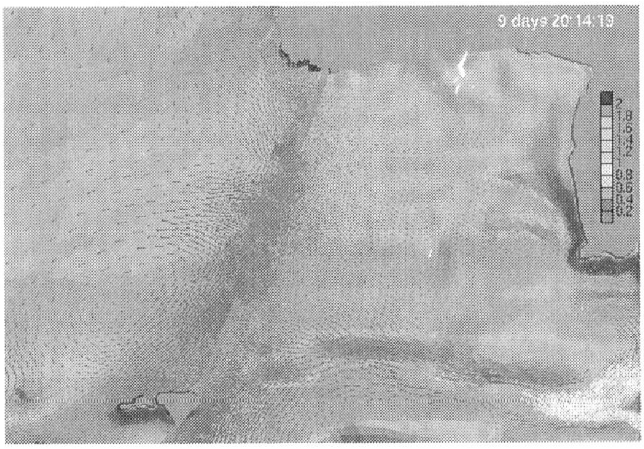

The vector current chart of the Saemangeum area

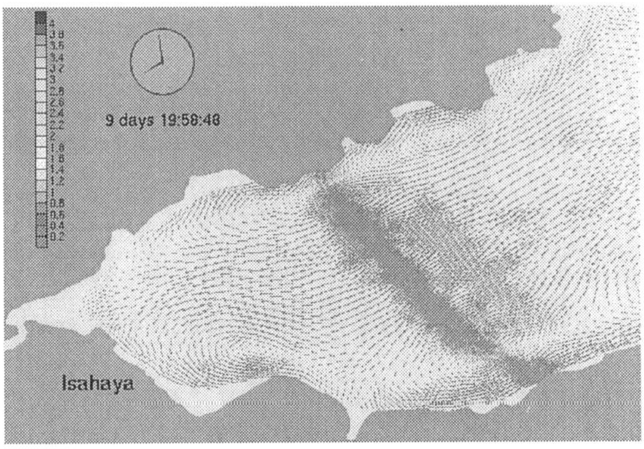

The vector current chart of the Isahaya Bay

FIELD OBSERVATION OF WAVES NEAR THE SHORELINE AND ITS ANALYSIS

K. Seki* and M. Mizuguchi**

*Graduate student, Dept. Civil Eng., Chuo Univ., Kasuga, 1-13-27, Bunkyo-ku, Tokyo Japan
Tel.81-3-3817-1818, Fax.81-3-3817-1803, E-mail, seki@civil.chuo-u.ac.jp

**Professor, Dr. Eng., Dept. Civil Eng., Chuo Univ. Kasuga, 1-13-27, Bunkyo-ku, Tokyo Japan
Tel.81-3-3817-1818, Fax.81-3-3817-1803, E-mail, mm@civil.chuo-u.ac.jp

1. Introduction

Breaking waves are dominant phenomena in the surf zone. However long waves with periods of a few minutes may be more important near the shoreline, as the wave steepness of the long waves is too small to break.

In this study, we carried out an elaborate field observation in the inner surf zone by employing many Ultra Sonic Wave Gages (USWG) and an Electro-Magnetic Current Meter (EMCM). This paper presents both the data obtained and some analysis on the behavior of short waves (wind waves) and long waves.

2. Field Observation

A set of twenty six USWG and an EMCM are installed at Hasaki Observation Pier (operated by PARI, Japan) with spatial interval of about 3m as shown in Fig.1. The data analyzed in this study were sampled continuously for 2 hours around the maximum tide with time interval of 0.2s. Fig.1 also shows the beach profile. Mean shoreline lay at x=4m. Variation of significant wave height indicates that ch.17 was located at the mean secondary breaking point. Ch.8 was located at the nearest point to the beach, where the beach face was not exposed during the period analyzed.

3. Data analysis

Spectra were calculated by FFT with 2^{15} points (about 109 minutes). In time series analysis, short and long waves are separated by numerical low pass filter with cut-off frequency 1/25 Hz. For ch.17 the incident and reflected waves were evaluated, before separating the short and long waves, by a linear long wave theory method using the measured surface elevation and cross-shore velocity. Fig.2 shows examples of surface elevation in the breaking process.

4. Results

Main features to be noted are listed in the following.

Time series analysis

1) Amplitudes of short and long waves become of the same order near the shoreline, as the energy of short waves dissipate by wave breaking as seen in Fig.2 for ch.8.

2) Unification of waves occurs in short waves, and decreases the number of waves. Large waves at

ch.8 were observed only for those unified waves.

Spectral analysis

1) For reflected waves power exists only in the range of f<0.02 Hz, showing complete energy dissipation of short waves near the shoreline.

2) Some peaks are observed in the range of long waves of power spectra. These peak frequencies agree with those of local anti-node calculated by multi-slope standing wave theory.

3) Cross spectra between onshore and offshore components at ch.17 indicates that the long waves are almost perfectly reflected at the shoreline.

4) Close to the shoreline, the shape of power spectra of wind waves changes from f^{-4} decrease, in the high frequency range, to f^{-2} decrease. This may correspond to the fact that the wave shape changes from cnoidal type to triangular one of shock waves.

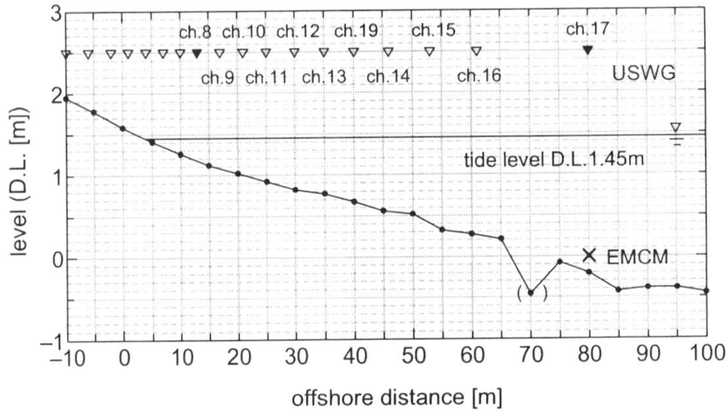

Fig.1 Beach profile and experimental set-up

(The beach profile data at 70m is affected by scouring of the pier.)

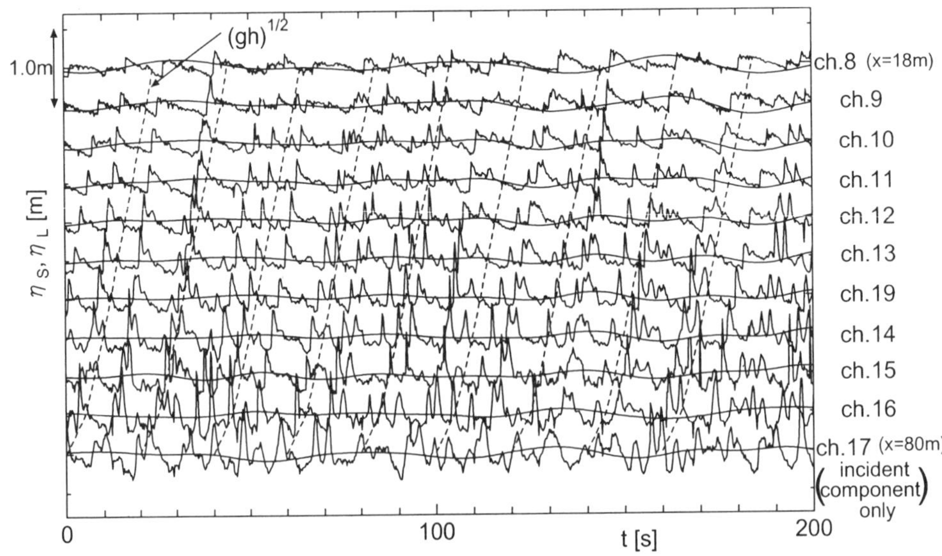

Fig.2 Example of time series data of short and long waves

EXPERIMENTAL STUDY ON BOTTOM SHEAR STRESS UNDER SAWTOOTH WAVE

Suntoyo*), Hitoshi Tanaka and Hiroto Yamaji

* Corresponding author

Department of Civil Engineering, Tohoku University, 06 Aoba, Sendai 980-8579, Japan

Phone & Fax: (+81)-22-217-7451, e-mail: suntoyo@kasen1.civil.tohoku.ac.jp

INTRODUCTION

The bottom shear stress estimation under saw-tooth wave is the crucial step, which is required as an input to most of sediment transport model under rapid acceleration and the relative contribution of pre-suspended sediment to the onshore sediment transport in swash zone. Suntoyo et al. (2002) had shown that the bottom shear stress variation under saw-tooth wave was dependent on high acceleration and deceleration. However, investigation of a more reliable estimation method to estimate the time-variation bottom shear stress under saw-tooth wave over rough bed has not deeply done yet. Turbulent bottom boundary layer flow under saw-tooth wave is investigated with the experiments in oscillating wind tunnel over rough bed by Laser Doppler Velocimeter (LDV) to measure velocity distribution. Bottom shear stress will be estimated by fitting the logarithmic velocity profile to the measured velocity. Furthermore, the experimental results will be examined with the bottom shear stress estimation method based on both the consideration of the friction coefficient for sinusoidal wave motion and that of the acceleration effect for saw-tooth wave.

ESTIMATION METHODS OF BOTTOM SHEAR STRESS AND DISCUSSION

Bottom shear stress of the experimental results are estimated by fitting the logarithmic velocity distribution to the measured data, which is given in Eq.(1),

$$u = \frac{U_*}{\kappa} \ln\left(\frac{z}{z_0}\right) \tag{1}$$

where, u: the flow velocity in the boundary layer, κ : the von Karman's constant (=0.4), U_*: the shear velocity, z: the cross-stream distance from theoretical bed level and z_o: roughness height.

Whereas, the bottom shear stress estimation method based on first consideration for friction coefficient under sinusoidal wave motion, that can be expressed by an equation, as proposed by Samad and Tanaka (1999) for an arbitrary variation of $U(t)$, as follows:

$$\tau_o\left(t - \frac{\varphi}{\omega}\right) = \frac{\rho}{2} f_w U(t)|U(t)| \tag{2}$$

where τ_0: the bottom shear stress, φ: the phase difference, ω: the angular frequency, f_w: the wave friction coefficient. Furthermore, the friction coefficient formula is given in Eqs. (3) and (4), respectively:

$$f_w = \exp\left\{-7.53 + 8.07\left(\frac{a_m}{z_o}\right)^{-0.100}\right\} \tag{3}$$

and

$$f_w = \exp\left\{5.5\left(\frac{a_m}{k_s}\right)^{-0.2} - 6.3\right\} \tag{4}$$

Second consideration, the instantaneous bottom shear stress is calculated from the variation of the free stream velocity as proposed by Nielsen (2002), the acceleration effect can be obtained through with the instantaneous friction velocity, as expressed by

$$\tau_o(t) = \rho|U_*(t)|U_*(t) \tag{5}$$

$$U_*(t)=\sqrt{f_w/2}\left[\cos\varphi\ U(t)+\sin\varphi\left\{\frac{U(t+\delta t)-U(t-\delta t)}{2\omega\delta t}\right\}\right] \tag{6}$$

where δt: the time step of the velocity measurement. Furthermore, it is based on the above equations, namely, from Eq.(2) to Eq.(6) will be examined several methods as given in **Table 1**. **Fig. 1** shows the comparison of the bottom shear stress between the experimental result and the calculation methods for a case at Reynolds number, Re = 1.35 x 10^5 under saw-tooth wave with the roughness parameter, a_m/k_s = 13.5 and the wave skewness parameter, α = 0.35. Method 1 and 2 have given a good agreement with the experimental result, while Method 3 gave the underestimated value due to exclude the acceleration effect in the calculation, and also the phase difference still remain appear though the phase difference value has been included in the calculation in Method 3. The wave friction factor given in Eq.(3) fits to estimate the bottom shear stress with the consideration of the sinusoidal wave motion for that case, though Method 1 exclude the acceleration effect. A correlation between bottom shear stress of measurement and calculation for Method 1 and 2 is shown in **Fig. 2** and **3**, respectively that give a good agreement as previously mentioned. However, those methods need to be applied to other cases for higher Reynolds number and a different wave skewness parameter. Furthermore, it can be obtained which method can overcome the bottom shear stress characteristic under saw-tooth wave in which at the peak part is larger than the trough part.

Table 1 Summary of proposed methods to compute the instantaneous bottom shear stress.

Method	$\tau_o(t)$	f_w	$U_*(t)$
1	Eq. (2)	Eq. (3)	-----
2	Eq. (5)	Eq. (4)	Eq. (6)
3	Eq. (5)	Eq. (4)	$=\sqrt{\frac{f_w}{2}}\cos\varphi\ U(t)$

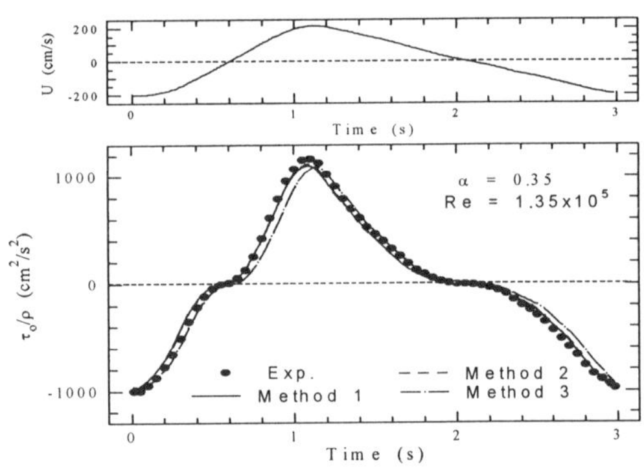

Fig. 1 Comparison of experimental result bottom shear stress and calculation methods.

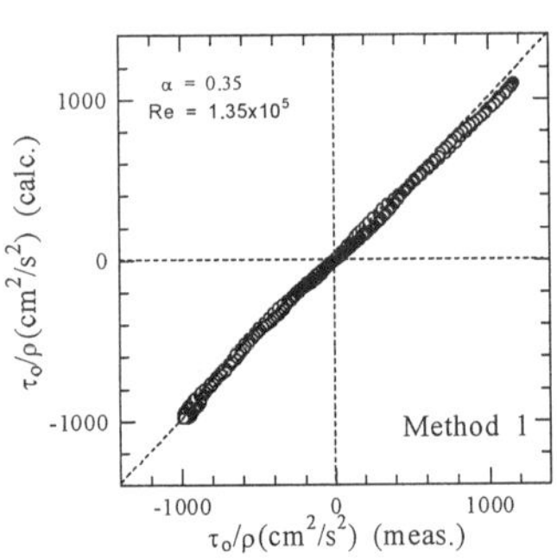

Fig. 2 Correlation between bottom shear stress of measurement and calculation for Method 1.

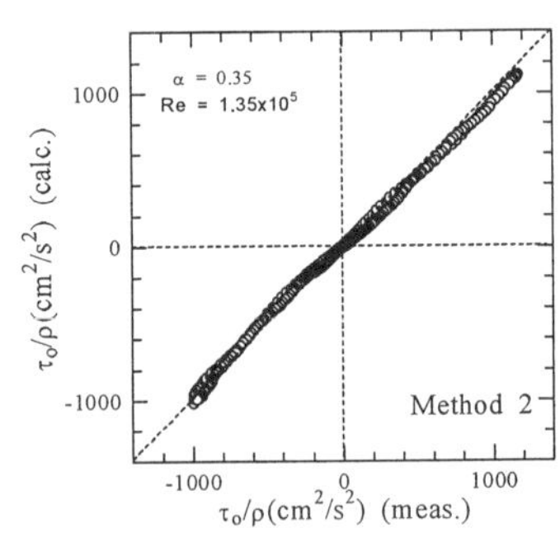

Fig. 3 Correlation between bottom shear stress of measurement and calculation for Method 2.

REFERENCES

Nielsen, P.: Shear stress estimation for sediment transport modeling under waves of arbitrary shape, *Abstracts, 28th ICCE*, paper no. 154, 2002.

Samad, M. A. and Tanaka, H.: Estimating instantaneous turbulent bottom shear stress under irregular waves, *Journal of Hydroscience and Hydraulic Engineering*, Vol. 17 No. 2, pp. 107-126, 1999.

Suntoyo, Tanaka, H. and Yamaji, H.: Investigation of turbulent flow in an oscillatory boundary layer under saw-tooth wave, *Proceedings of the 4th International Summer Symposium*, pp. 171-174, 2002.

A High-Resolution Numerical Scheme for Boussinesq Equations

Xiping Yu
Department of Hydraulic Engineering, Tsinghua University, Beijing, China
Email: yuxiping@tsinghua.edu.cn

ABSTRACT

The Boussinesq equations have been developed as one of the most useful mathematical models for coastal hydrodynamics in the last decade. The distinguished work by Madsen and his colleagues (e.g., Masen et al, 1991) and by Nowgu (1993) extended the classical Boussinesq equations (Peregrine, 1967) to be valid beyond the deepwater limit of water waves. Enhancements by other researchers further generalized the applicability of the mathematical model to various coastal hydrodynamic phenomena including wave breaking (e.g., Madsen et al., 1997, Skotner, 1999). When compared with the extensive efforts in the development of the Boussinesq equation theory, study on the numerical method, however, does not seem to have been enough. The present paper is to present a high-resolution numerical scheme for accurate solution of the Boussinesq equations.

Development of a numerical scheme for the Boussinesq equations is known to be rather delicate. This is not only because the equations involve high-order derivatives which usually cause difficulties in discretization near a boundary, but also because the equations describe a number of important secondary phenomena which can not be distinguished by an arbitrary numerical algorithm. On the other hand, the Boussinesq equations reduce to the classical shallow water wave equations when the ratio of water depth to local wavelength vanishes. One of the most important characteristics of the shallow water wave equations is that they adopt solutions with discontinuities or breakers. An effective numerical method for the Boussinesq equations should then take this fact into consideration. Conceptual application of a shock-capturing scheme such as the TVD, TVB or ENO scheme should thus have advantages.

The Boussinesq equations employed in the present study are those derived by Nowgu's (1993). Considering only one horizontal dimension the equations can be written into the following conservative form:

$$\frac{\partial h}{\partial t} + \frac{\partial E}{\partial x} = 0 \tag{1}$$

$$\frac{\partial P}{\partial t} + \frac{\partial F}{\partial x} = 0 \tag{2}$$

where x is the horizontal coordinate, t is the time, h is the water surface elevation from the still water level, and P is a variable related to the horizontal velocity at the reference level. P, E and F are defined by

$$P = u + B_1 h^2 \frac{\partial^2 u}{\partial x^2} + B_2 h \frac{\partial^2 hu}{\partial x^2} \tag{3}$$

$$E = (h + h)u + A_1 h^3 \frac{\partial^2 u}{\partial x^2} + A_2 h^2 \frac{\partial^2 hu}{\partial x^2} \tag{4}$$

$$F = gh + \frac{1}{2}u^2 \tag{5}$$

in which, A_1, A_2, B_1, and B_2 are constants depending on the relative position of the reference level R through

$$A_1 = \frac{R^2}{2} - \frac{1}{6};\ A_2 = R + \frac{1}{2};\ B_1 = \frac{R^2}{2};\ B_2 = R \tag{6}$$

The value of R can be determined so that the linear dispersion relation resulted from Eqs. (1) and (2) has the best agreement with that from the small amplitude wave theory. This leads to $R = -0.533$ according to Nowgu (1993). Note that the terms with high-order derivatives tend to vanish as the depth-to-wavelength ratio decreases to zero, Eqs. (1) and (2) then reduce to the shallow water wave equations under such a condition.

The finite difference algorithm proposed in the present study is based on a two-step Runge-Kutta method for Eqs. (1) and (2) with an ENO scheme for the numerical fluxes proposed by Shu (1988), which can be easily implemented. Cubic splines are employed to approximate the spatial variation of variables when E is evaluated. High-order method is also utilized for the inversion of u from P according to Eq. (3).

A critical problem for verification of numerical schemes for the Boussinesq equations is known to be waves propagating over a submerged dike with obvious generation of high frequency harmonic modes. The present method is also tested by applying to this problem. The computational results are shown to agree with experimental data very well.

Modified Boussinesq Model on Uneven Bottom

Benlong WANG, Hua LIU*, Leiping XUE and Yousheng HE
School of Civil Engineering and Mechanics, Shanghai Jiao Tong University
Shanghai 200240, P.R. China
*hliu@sjtu.edu.cn

Abstract

The ability to predict wave transformation from deep to shallow water accurately is vital to the numerical models of water wave in coastal waters. During the wave transformation, shoaling is one of the most important processes on the changing of wave profile in intermediate and shallow water zone. In the estuary, wave-current interaction will change the wave significantly as well. The moving shorelines generate complicated wave phenomena on beaches. It is our common interest to develop an integrated mathematical model for simulation of all these water wave processes. In the present work, some aspects mentioned above are discussed in the scope of the higher order Boussinesq equations.

An infinite-order, Boussinesq-type equations for wave propagation over variable bottom is derived. With finite order truncation of the Taylor operator and optimization, the application range of the present model is extended to $h_x \leq 1$ for the bottom slope bottom. The main goal of the present work is to extend the application range of Boussinesq-type wave model from mild slope bottom to rapidly varying bottom. This work completes the models given by Madsen et al [2002]. In their work, it was found that the linear shoaling characteristic is irrelevant to the bottom slope terms h_x in the series expansion for the specified model. This interesting character is only limited to the linear shoaling. Although the nonlinear effects become significant when the wave travels from deep water to shallow water, a linear solution is still very useful. The highlight of the present work is that we take the bottom slope terms into account in the wave models without deterioration the excellent shoaling characteristic. Numerical tests, such as the linear shoaling and the reflection from slope beach, are carried out to investigate the effects of the modification. From these numerical simulations we found the overall capacities on the problems with rapidly varying bottom are improved significantly.

The propagation of wave over current is another interesting topic for the research of the wave dynamics. To accomplish the goal of the study the wave-current interaction in the numerical wave flume, a new kind of combined wave-current generation method has been developed with the conception of the sponge layer. It is demonstrated that this kind of wave-current generator works reliably and is easy to be implemented in the numerical wave flume. Although Chen et al[1999] and Kristensen et al[1995] extended the Boussinesq equations to improve the accuracy of the dispersion relation, the inherent properties of traditional Boussinesq equations could not give a satisfied dispersion relation for large kh number, need not to say the combination of shoaling and the interaction with current. Although the model discussed by Chen et al [1999] could simulate the wave blocking over a submerged bar, strong friction need to be added to stabilize the flow simulation. Another shortcoming of the traditional Boussinesq models is that the model is not fully nonlinear. Consequently, discussion the completeness of the nonlinear terms is another main concern by most study on the topic of wave-current interaction. All these disadvantages could be overcome in the present model. Comparing with the linear theory, excellent agreement is obtained.

In the final part of the present work, modeling the moving shoreline is discussed. Mild slope beach is a typical bathymetry in the coastal zone. Simulation the moving shoreline is another important task for the numerical model. Unfortunately, the process of run-up and run-down is not well simulated and understood in the framework of the Boussinesq equations. The slot method of Tao[1983] is followed and extended in the study of the moving shoreline associated with the Boussinesq equations by Madsen et al [1997] and Kennedy et al [2000]. Comparison their models with the analytical solution of wave run-up on a slope beach shows that the slot method tended to under predict the maximum run-up height, although this error has been partially eliminated in the work of Kennedy et al [2000]. Inspired by the work of Lynett et al [2002], the run-up model with extrapolation through the wet-dry boundary and into the dry region has been implemented associated with the present Boussinesq model. Different from the other Boussinesq models, there is no explicit dispersive terms in the present full nonlinear full dispersive model. Consequently, no need to switch off the dispersion terms near the shoreline to ensure the numerical stability. In addition, no artificial parameters, such as the porosity, shape factor, need to be introduced in this kind of run-up model. The extrapolation needs to be implemented only for the free surface time-stepping problem. It is straightforward to extend to the cases in two horizontal dimensions. The numerical results show that this kind of run-up model works rather well comparing with the analytical solution for the test case.

References

[1] Chen, Qin, Madsen, P.A. and Basco, D.R. 1999 Current effects on nonlinear interactions of shallow-water waves. Journal of Waterway, Port, Coastal, and Ocean Engineering, 125(4): 176-186, July/August

[2] Chen, Qin, Madsen, P.A. Schaffer, H.A. and Basco, D.R. 1998 Wave-current interaction based on an enhanced Boussinesq approach. Coastal Engineering 33 11-39

[3] Kennedy, A.B. Chen, Qin, Kirby, J.T. and Dalrymple, R.A. 2000 Boussinesq Modeling on wave transformation, breaking, and run-up. I:1D. Journal of Waterway, Port, Coastal, and Ocean Engineering 126(1) 39-47

[4] Kristensen, M.K. 1995 Boussinesq equations and wave-current interaction. Master's thesis, International Research Center for Computed Hydrodynamics (ICCH) at Danish Hydraulic Institute, Denmark and ISVA, Technical University of Denmark

[5] Lynett, P.J., Wu, Tso-Ren and Liu, Philip 2002 Modeling wave run-up with depth-integrated equations Coasal Engineering 46 89-107

[6] Madsen, P.A. Bingham, H.B. and Liu, Hua. 2002 A new Boussinesq method for fully nonlinear waves from shallow to deep water. J.Fluid Mech. 462 1-30

[7] Madsen, P.A., Sorensen, O.R. and Schaffer, H.A. 1997 Surf zone dynamics simulated by a Boussinesq type model. Part I. Model description and cross-shore motion of regular waves Coastal Engineering 32 255-287

[8] Suh Kyung Dong, Lee Changhoog and Park W.S. 1997 Time-dependent equations for wave propagation on rapidly vering topography Coastal Engineering 32 91-117

[9] Tao, J. 1983 Computation of wave run-up and wave breaking. Internal Report, Danish Hydraulic Institute

SOME CONSIDERATIONS OF REFINEMENT OF A BOUSSINESQ EQUATION AND ITS VERIFICATION

Shinichiro ONDA[1], Takashi HOSODA[2] and Ichiro KIMURA[3]

[1]Graduate School of Civil Engineering (River Engrg.), Kyoto University,
Yoshida Sakyo-ku, Kyoto 606-8501, Japan
sonda@river4.kuciv.kyoto-u.ac.jp

[2]Department of Civil Engineering, Kyoto University, Japan
hosoda@river4.kuciv.kyoto-u.ac.jp

[3]Faculty of Environmental and Information Sciences, Yokkaichi University,
1200 Kayo-cho, Yokkaichi 512-8512, Japan
kimura@yokkaichi-u.ac.jp

I. INTRODUCTION

A Boussinesq equation is practical depth-averaged flow model. The original one, however, has some problems, for example, the dispersion relation of water waves does not agree with the Airy waves theory in high wave number range.

Recently a number of attempts have been made to improve the performance of the original one. In most of modified models in previous works, a velocity distribution is expressed implicitly as a complicated relation between a local velocity at an arbitrary depth and a depth-averaged velocity. Although the simple models that express the velocity distribution explicitly with a depth-averaged velocity are more useful, such models are not general.

In this study, a refined type equation of depth-averaged flow model is first derived. In this model, the velocity distribution is expressed explicitly with a depth-averaged velocity and a basic equation can be described as a simple form. To evaluate the model performance, the equation is applied to the free surface oscillation in a water tank and the linear and non-linear analyses are compared with the Airy waves theory and the numerical solution by vertical 2-D CFD model.

II. DERIVATION PROCESS OF BASIC EQUATION

The bed is assumed to be horizontal and the vertical 2D Cartesian coordinate is adopted. Assuming that a vertical velocity distribution is uniform as an initial condition($u=U(x)$), the procedure for deriving the basic equation is summarized as follows. 1) Integrating a continuity equation, the distribution of vertical velocity($=w$) is derived. 2) We substitute w into the irrotational condition and the distribution of u is derived by integrating the irrotational condition. 3) We repeat the stage 1), using u at the stage 2). 4) u is derived by repeating 2). 5) w is derived by repeating 1). 6) A pressure distribution is obtained by substituting the above velocity distributions into momentum equation in z and integrating the momentum equation. 7) The depth-integrated momentum equation is derived by integrating the mo mentum equation in x after substitution of the modified vertical velocity and pressure distribution.

$$u = U + \frac{h^2}{2} B\left(\frac{z^2}{h^2} - \frac{1}{3}\right) + \frac{h^4}{12}\frac{\partial^2 A}{\partial x^2}\left(\frac{z^2}{h^2} - \frac{1}{3}\right) - \frac{h^4}{24}\frac{\partial^2 A}{\partial x^2}\left(\frac{z^4}{h^4} - \frac{1}{5}\right), \quad A = B = \frac{\partial}{\partial x}\left\{\frac{1}{h}\left(\frac{\partial h}{\partial t} + U\frac{\partial h}{\partial x}\right)\right\} \tag{1}$$

$$\frac{\partial q}{\partial t} + \frac{\partial qU}{\partial x} + \frac{\partial}{\partial x}\left(\frac{Gh^2}{2}\right) + \frac{\partial}{\partial x}\left\{-\frac{1}{3}h^2\frac{\partial^2 q}{\partial x \partial t} + \frac{1}{3}h^2U^2\frac{\partial^2 h}{\partial x^2} + \frac{2}{3}h^2U\frac{\partial^2 h}{\partial x\partial t} - \frac{1}{45}h^4\frac{\partial^4 q}{\partial x^3\partial t} + \frac{1}{45}h^4U^2\frac{\partial^4 h}{\partial x^4} + \frac{2}{45}h^4U\frac{\partial^4 h}{\partial x^3\partial t}\right.$$
$$\left. - \frac{2}{945}h^6\frac{\partial^6 q}{\partial x^5\partial t} + \frac{2}{945}h^6U^2\frac{\partial^6 h}{\partial x^6} + \frac{4}{945}h^6U\frac{\partial^6 h}{\partial x^5\partial t}\right\} + \cdots = 0 \tag{2}$$

Where, h; depth, U; depth-averaged velocity, q; depth-integrated velocity component and G; gravity acceleration.

As Madsen et al [3] pointed out that the singular point where a dominator in the dispersion relation becomes zero exists, even if only the irrotational condition is considered. Then, to reduce the order of the derivatives in (2) and dissolve the singularity, we consider the modification of the dispersion terms derived by integrating the pressure distribution in the depth direction. Eq. (3) is derived by employing the similar method by Madsen et al [3] to modify the Eq. (2). ξ is a parameter introduced to satisfy the dispersion relation approximately.

$$\frac{\partial q}{\partial t} + \frac{\partial qU}{\partial x} + \frac{\partial}{\partial x}\left(\frac{Gh^2}{2}\right) + \frac{1}{3}h^2U^2\frac{\partial^3 h}{\partial x^3} + \frac{2}{3}h^2U\frac{\partial^3 h}{\partial x^2\partial t} - \frac{1}{3}h^2\left(1+\frac{2}{7}\xi\right)\frac{\partial^3 q}{\partial x^2\partial t} - \frac{2}{21}\xi h^2\frac{\partial}{\partial x^3}\left(\frac{Gh^2}{2}\right)$$
$$-\frac{1}{45}h^4\left(1-\frac{10}{7}\xi\right)\frac{\partial^5 q}{\partial x^4\partial t} + \frac{1}{45}h^4U^2\left(1-\frac{10}{7}\xi\right)\frac{\partial^5 h}{\partial x^5} + \frac{2}{45}h^4U\left(1-\frac{10}{7}\xi\right)\frac{\partial^5 h}{\partial x^4\partial t} = 0 \tag{3}$$

Furthermore, we derive the second revision of a velocity distribution by feed backing the modification of the dispersion terms. In this study, we consider the case where a depth-averaged velocity is assumed to be small. Then, examining the derivation process of the modification of dispersion terms dimensionally, the expression about A can be described by using time integration.

$$\frac{\partial^2 A}{\partial x^2} = -\frac{45\xi}{h^5}\left\{\int_0^t \frac{\partial q}{\partial t}dt + Gh\int_0^t \frac{\partial h}{\partial x}dt - \frac{1}{3}h^2\int_0^t \frac{\partial^3 q}{\partial x^2\partial t}dt\right\} \tag{4}$$

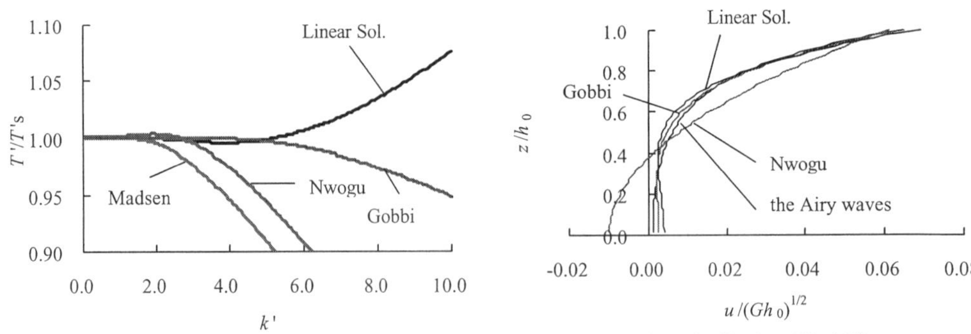

(a) Linear Dispersion Relation (b) Velocity Distribution (k'=4.71)

Fig. 1 Comparison the linear solutions with the Airy waves theory

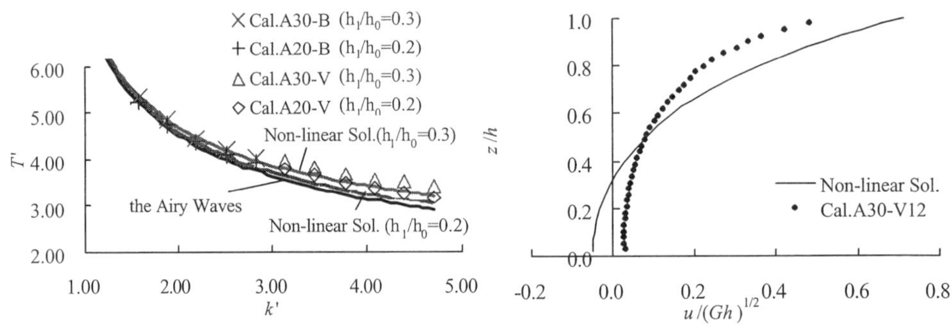

(a) Dispersion Relation (b) Velocity Distribution (k'=3.77)

Fig. 2 Comparison between non-linear analysis and numerical solution
(B: A Generalized Curvilinear Moving Coordinate System, V: VOF method)

III. COMPARISON THE LINEAR SOLUTIONS OF THE FREE SURFACE OSCILLATION IN A WATER TANK WITH THE AIRY WAVES THEORY

The free surface oscillation in a rectangular water tank is employed as a test case to evaluate the model performance. Fig.1 shows the relation between T'/T'_s and k'. T' and T'_s mean non-dimensional period of this model and the Airy waves theory respectively and k' indicates non-dimensional wave number. Fig.2 shows the comparison about velocity distribution between various models. The linear solution of this model is in better agreement with the Airy waves than that by Madsen et al or Nwogu. But, the range of the wave number in the present model is smaller than that of Gobbi's model. The applicable range of wave number in this model will be enlarged if the iteration number using continuity equation and irrotational condition is increased. The basic equations in this model are expressed more simply with the depth-averaged velocity compared with the models with a local velocity at an arbitrary depth.

IV. COMPARISON THE NON-LINEAR SOLUTIONS WITH NUMERICAL SOLUTION

To derive the approximated solution of the surface oscillation in non-linear range, the expressions of h and U with second order are substituted into a continuity equation and Eq. (3), then, the dispersion relation and velocity distribution in non-linear range are obtained by arranging in time and spatial mode. As for the numerical simulation of the free surface oscillation, a generalized curvilinear moving coordinate system and VOF method is adopted to consider the surface elevation. The length of tank L is fixed to 0.1m and still water depth h_0 is varied from 0.05 m to 0.15 m in increments of 0.01m. The initial amplitudes (=h_1/h_0) are 0.03, 0.2 and 0.3. Fig.2 shows the comparison between the non-linear analysis and numerical solution. Since the approximate solution in the refined type equation is in close agreement with the numerical results, it is concluded that this model can be applicable to the range with non-linear effects.

V. CONCLUSION

In this paper, we proposed the refinement of the Boussinesq equations. It is seen that the linear solutions are improved to the high wave number range. Furthermore, the non-linear approximate solutions ware derived. Comparing with the numerical results by vertical 2-D CFD model, the applicability of the present model was tested.

REFERENCES

[1] Madsen, P.A., Murry, R. and Sorensen, O.R.(1991): *Coastal Eng.*, Vol.15, pp.374-388. [2] Nwogu, O.(1993): *J. Water, Port, Coastaland Ocean Eng.*, ASCE Vol.119, No.6, pp.618-638. [3] Madsen, P.A. and Schaffer, H.A.(1998): *Phil. Trans. R. Soc. Lond.*, A356, pp.3123-3184. [4] Madsen, P.A., Bingham, H. and Liu, H.(2000): *Proc. Coastal Eng.* Vol.1, pp.176-189. [5] Agnon, Y., Madsen. P.A. and Schaffer, H.A.(1999): *J. Fluid Mech.* Vol.399, pp.319-333.[6] Gobbi, M.F., Kirby, J.T. and Wei, G.(2000): *J. Fluid Mech.*, Vol.405, pp.181-210.[7] Hosoda, T. et al (1997): *Journal of Hydraulic, Coastal and Environmental Engineering*, No.558/II-38, pp.81-89 (in Japanese).

A quasi-3D numerical shallow water model based on FVM

(Abstract)

Li Shaowu[1] Lu Lifeng[2]

In the prediction of shallow water tidal current field, numerical method becomes the most effective tool. As a practical numerical model, the governing equations used in the model should be as more general as possible to cover various practical engineering situations under the limit of computational capacity. The numerical method should also be good in terms of numerical effectiveness, accuracy and flexibility to various boundary shape. Verification of the model by field data is essential. Although quite a lot of research work has been carried out on the development of numerical shallow water flow model, our purpose is to use the most effective numerical method to solve the quasi-3d shallow water equations to establish a practical numerical model.

Among different numerical methods, the FVM (Finite Volume Method) attracts quite a lot of interest in solving fluid dynamic equations, firstly in aero-dynamic area and then in water dynamic area, because of its standing numerical property in terms of conservative property for physical quantities, good applicability to arbitrary shape of boundary supposing nonstructural griding system is used and has high accuracy of discretization. The key point in constructing a numerical scheme by using FVM is to use a proper method to calculate the flux through the interface between two neighboring control volumes. Among various schemes, the Osher scheme provides good upwind property in calculation of horizontal flux through the interface of two neighboring control volumes with uniform distribution of physical quantities in a control volume. It is also capable of giving high accuracy for simulation of shock without any numerical oscillation around the shock point.

In this research, a numerical tidal current model is developed based on a set of quasi-3d shallow water equations. The equations are transformed into σ coordinate system vertically for simplicity to handle the bottom boundary condition. FVM is used for the numerical calculation. The control volume is of the shape of a prism with triangular shape in horizontal direction. The $k-\varepsilon$ turbulent model is used as the closure. By using this numerical model, a uniform flow over a rectangular domain with slope of 1‰ is simulated(Fig.1). The roughness used in the calculation is $k_s = 5$mm. The model is also applied to a harmonic oscillatory flow over a flatted rectangular domain. The calculated result of water surface elevation is shown in Fig.3 and that of the eddy viscosity in the vertical direction is shown in Fig.2.

The calculated results of the distribution of the horizontal velocity shown above seems reasonable, which indicates that the numerical model can give reasonable result of vertical distribution of velocity.

[1] Assoc. Prof., Dep. of Harb. and Coast. Eng., Civil Eng. School, Tianjin Univ., China. 300072 lishaowu@tju.edu.cn

[2] Post Grad. Student of Harb. and Coast. Eng., Civil Eng. School, Tianjin Univ. 300072.

References

[1]Tan Weiyan, Hu Siyi(1991): A general applicable high quality scheme for two-dimensional shallow water flow——a FVM Osher scheme. Advances in water science, vol. 2(3), pp.154-161.

[2]Wang Xiaojian, Zhang Tingfang(1997): The application of a non-orthogonal FVM in 3d tidal current. Hydrodynamics, 12(1), pp. 56-61.

[3]Peter K. Stansby, Semi-Implicit Finite Volume Shallow-Water Flow and Solute Transport Solver with $k-\varepsilon$ Turbulence, Int. J. Numer. Meth. Fluids, 25(1997):285~313.

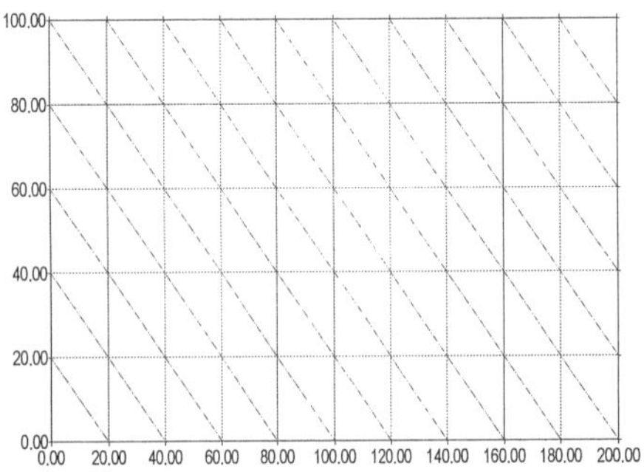

Figure 1

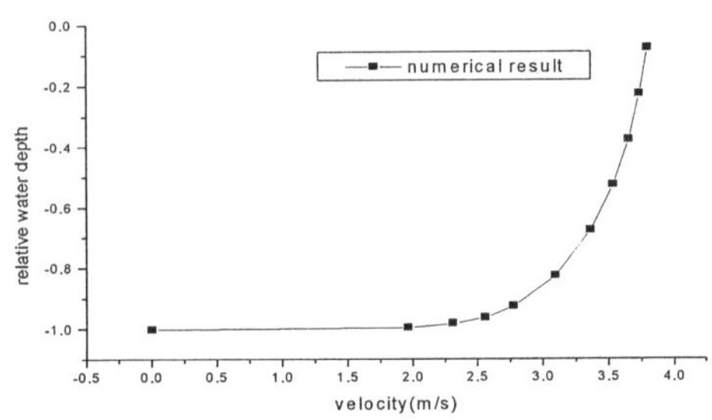

Figure 2

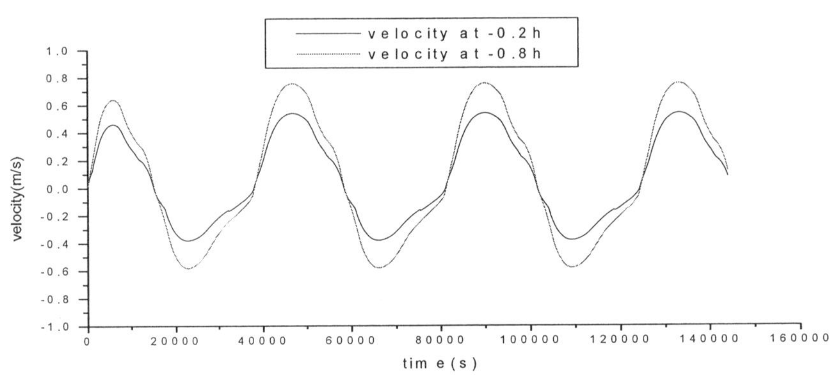

Figure 3

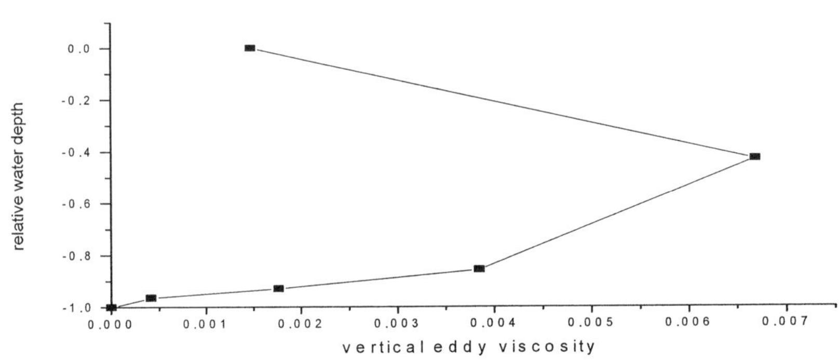

Figure 4 Numerical result of vertical eddy viscosity

An analytic solution of the mild-slope equation for scattering by a truncated conical shoal

Huan-Wen Liu[a,b*], Peng-Zhi Lin[a] and N.Jothi Shankar[a]

[a]Department of Civil Engineering, National University of Singapore, Singapore 119260

[b] Department of Mathematics and Computer Science, Guangxi University for Nationalities, Nanning, Guangxi 530006, P.R. China

Email: cveliuhw@nus.edu.sg, cvelinpz@nus.edu.sg, cvejothi@nus.edu.sg

Since the mild-slope equation was originally derived by Berkhoff [1], various numerical models have been developed to solve it, see [8]. However, an analytic solution is very difficult to obtain although the mild-slope equation is only a simple linear equation. Up to date, only a few analytic solutions have been obtained in two limiting cases, i.e., for long waves and short waves. Examples include the solution of the Helmholtz equation for a cylindrical island standing in an open ocean with constant water depth [4], solutions of the linear shallow-water equation for a circular cylinder mounted on a parabolic or conical shoal [2, 5, 7] and the solution of the linear shallow-water equation for a conical island [6].

The principal difficulty in solving the mild-slope equation analytically comes from the fact that the wave dispersion relationship is implicit for waves on variable depth water, thus the governing equation with its coefficients being transcendental functions seems impossible to be solved exactly. Recently, employing a Padé approximation to the wave dispersion, Liu *et al.* presented an approximate analytical solution of the mild-slope equation in terms of combined Fourier series and Taylor series for a circular cylindrical island mounted on a parabolic shoal [3]. In this paper, the problem of the scattering of plane monochromatic waves by a submerged truncated conical shoal (see Fig.1(a)) is considered and an analytic solution to the mild-slope equation is constructed. Using the current analytic model, wave amplifications along the direction of wave propagation for four wave periods being 120s, 180s, 240s and 480s are calculated and displayed in Fig.1(b). From Fig.1(b), wave focusing on or behind the shoal is clearly observed. It is also seen that the focal point moves downwave with decreasing wave period, probably because the refractive effect over the shoal becomes insignificant as the period decreased.

References

[1] Berkhoff, J.C.W. 1972 Computation of combined refraction-diffraction. In *13th Int. Conf. Coastal Engng*, Vancouver, pp.471-490.

*Corresponding author.

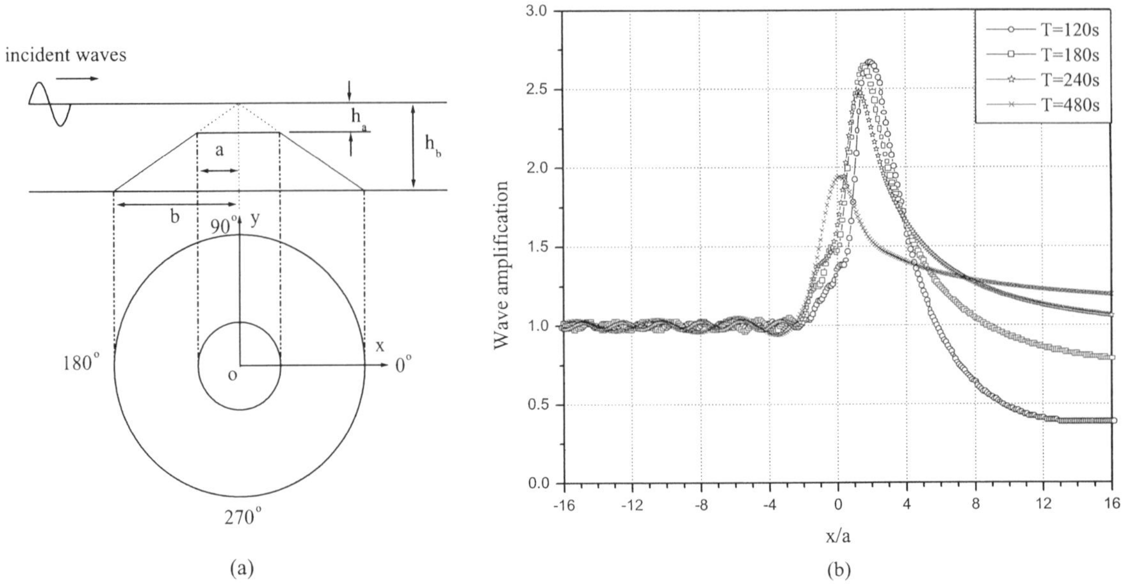

Figure 1: (a) A definition sketch of a truncated conical shoal with $a = 10km$, $b = 30km$, $h_b = 4km$ and $h_a = h_b/3$; (b) Wave amplification along the x-axis for four cases of $T = 120s$,$180s$,$240s$ and $480s$, respectively.

[2] Homma, S. 1950 On the behaviour of seismic sea waves around circular island. *Geophysical Magazine* **21**, 199-208.

[3] Liu, H.-W., Lin, P.-Z. & Shankar, N.J. 2002 An analytical solution for combined refraction and diffraction based on the mild-slope equation. *Proc R Soc Lond A*, submitted.

[4] MacCamy, R.C. & Fuchs, R.A. 1954 Wave forces on piles: a diffraction theory. *US Army Corps of Engng, Beach Erosion Board*, Washington, DC, Tech. Mem. 69.

[5] Yu, X.P. & Chen, Y.M. 2001 A new analytic solution of combined wave diffraction and refraction. *Proc 1st Asian Pacific Coast Engng Conf*, Dalian, PRC, pp.145-154.

[6] Zhang, Y.L. & Zhu, S.-P. 1994 New solutions for the propagation of long water waves over variable depth. *J. Fluid Mech* **278**, 391-406.

[7] Zhu, S.-P. & Zhang, Y.L. 1996 Scattering of long waves around a circular island mounted on a conical shoal. *Wave Motion* **23**, 353-362.

[8] Zhu, S.-P., Liu, H.-W. & Chen, K. 2000 A general DRBEM model for wave refraction and diffraction. *Engng Anal Boundary Elements* **24**, 377-390.

Transient Free-surface Waves due to a Suddenly Stopping Body

Dong-Qiang Lu[1] and Allen T. Chwang[2]

Abstract

The evolution of free-surface waves due to a suddenly stopping body submerged in an incompressible fluid of infinite depth is analytically investigated. The body is assumed to stop suddenly from a steady translational motion. In order to investigate the effect of viscosity on the evolution of waves, the linearized Oseen equations are employed for the disturbed flow. The kinematic and dynamic boundary conditions are linearized for small-amplitude waves, and the initial values of the flow are taken to be those of the corresponding steady-state solutions. By means of integral transforms, exact solutions for the transient waves are obtained. The asymptotic representations for the far-field transient waves are derived by applying Lighthill's two-stage scheme. The results show that although the inviscid waves will keep their profiles due to the conservation of energy, the viscous waves will eventually die out due to the presence of viscosity, which is consistent with the physical reality. The energy dissipation rates for the viscous ship waves are analytically expressed and the evolution of the transient viscous waves is explicitly shown.

---

1 PhD candidate, Department of Mechanical Engineering, The University of Hong Kong, Pokfulam Road, Hong Kong. Email: *dqlu@mail.shu.edu.cn*

2 *Corresponding author*, Sir Robert Ho Tung Chair Professor, Department of Mechanical Engineering, The University of Hong Kong, Pokfulam Road, Hong Kong.
Tel: (+852) 2859 2634; Fax: (+852) 2858 5415; Email: *atchwang@hkucc.hku.hk*

Comparison and Characterization of Bottom Mounted Wave Directional System

Toshihiko NAGAI [1], Noriaki HASHIMOTO [2], Atle LOHRMANN [3],
Masao MITSUI [4], Shoichiro KONASHI [5]

ABSTRACT

1. Purpose of Study

Methods for wave height and direction measurements vary throughout the world, depending on wave climate and local traditions. In Japan, bottom mounted systems have long been the standard for coastal areas (water depth less than 50m) and extensive studies in the 80s and 90s refined their systems to a level where the full wave directional spectrum could be measured. These systems take a multi-parameter approach to the problem by taking into account both pressure fluctuation, orbital particle velocity, and surface elevation to generate a spectrum that covers the full range from 1 second wind waves to 1000 seconds Tsunami. In this study, we are looking at the results from 3 different bottom mounted systems, all using acoustic techniques in combination with pressure. The detailed implementation is different in each system and the purpose is to analyze how the differences impact the characteristics of the wave spectrum.

2. Method of the study

Field comparison test was conducted for one month in September 2001 off the Omaezaki Port faced to the Pacific Ocean. During the test term, high wave conditions due to the Typhoon 0115 were observed. Following three types of the bottom mounted wave gauges were installed and compared at a point of water depth 22m.

1) DWDM (produced by Kaijo Co.)

As one of NOWPHAS coastal wave observation stations, Doppler-type Directional Wave Meter (DWDM) has been installed in this area since 1997. DWDM was connected to the on-land wave monitoring station with a permanent seabed cable. As DWDM observes five components wave actions, acoustic water surface elevation, pressure fluctuation, and three components of oblique directional water particle velocity in orbital depth layer by applying the Acoustic Doppler Principle.

2) AWAC (produced by Nortek Co.)

AWAC's directional wave observation principle is almost the same to the DWDM. As AWAC is a portable

---

[1] **Head, Marine Information Div., Port and Airport Research Institute**
3-1-1 Nagase Yokosuka, 239-0826, JAPAN
tel:+81-468-44-5048 fax:+81-468-42-5246 e-mail: nagai@pari.go.jp

[2] Head, Hydrodynamics Div., Port and Airport Research Institute
3-1-1 Nagase Yokosuka, 239-0826, JAPAN
tel:+81-468-44-5049 fax:+81-468-44-1274 e-mail: hashimoto@pari.go.jp

[3] Managing Director, NORTEK AS
Industriveien 33, N-1337 Sandvika, NORWAY
tel:+47-67-55-6200 fax:+47-67-54-6150 e-mail: inquiry@nortek.no

[4] Team Leader, Measuring and Control Systems Div., Kaijo Co.
3-1-5 Sakae, Hamura, 205-8607, Japan
tel:+81-42-555-6080, fax:+81-42-579-5171 e-mail: m.mitsui@kaijo.co.jp

[5] Managing Director, Sales Div., ALEC Electronics Co.
7-2-3 Ikubukidai-Higashimachi, Nishi-ku, Kobe, 651-2242, Japan
tel: 078-997-8686, fax:078-997-8609, e-mail: konashi@yacht.ocn.ne.jp

self-recording type wave meter, observed wave data were acquired when withdrawing the sensor from the seabed after one month field test.

3)DL2 (produced by Kyowa-shoko Co.)

DL2 is a portable self-recording type bottom mounted wave meter with four wave components observation, acoustic water surface elevation, two components of horizontal water particle velocity, and pressure fluctuation.

3. Results and Conclusions

Significant wave heights and periods obtained by each type of wave gauges showed good agreement during the test term, in both high wave conditions with significant wave heights bigger than 6m and low wave conditions with wave heights lower than 0.5m. Differences of the observed wave height and periods characteristics were not found among the three sensors (Fig.1).

Nevertheless, directional spectrum obtained by each sensor sometimes showed slight differences, in high wave conditions due to the Typhoon 0115 and in low wave conditions with significant wave heights less than 0.5m. The differences in high wave conditions are supposed due to the AWAC's observation layer setting problem and will be improved in the modified software.

We can have directional spectra observed by above three sensors on September 10th, 2001, when the wave has grown up by center of the Typhoon 0115 located southward from the experimental site and coming on. The shape of these directional spectra showed single peak with high concentration of wave direction at the frequency around 0.07Hz. In the some of these directional spectra, however, the peak direction differs, DWDM and DL2 indicate South-East, while AWAC indicates East. Taking account into the Typhoon location and wind field at this time, DWDM's and DL2's results were more reasonable than AWAC's one.

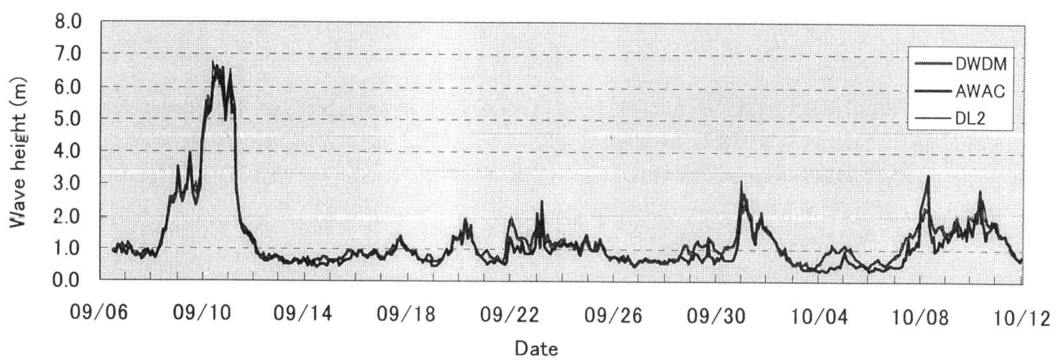

Fig.1 Time series of significant wave height observed by DWDM, AWAC and DL2

Above results proved that observed directional spectrum differed due to observation principle of each sensor and observation layer setting. In order to obtain proper ocean information, proper selection and parameter setting of sensor is very important.

References

Nagai,T., Hashimoto,N., Kawaguchi, K., Yokoi, H., Kudaka, M., and Mitsui, M. (2002). Frequency Banded Wave Climate Description Based on the Observed Directional Spectra, Coastal Engineering Journal (CEJ), Japan Society of Civil Engineers, Vol.44, No.1, 53-65.

Nagai, T., Sugahara, K., Hashimoto, N., Asai, T., Higashiyama, S., and Toda, K.(1994), Introduction of Japanese NOWPHAS System and its Recent Topics, Proceedings of the International Conference on Hydro-Technical Engineering for Port and Harbor Construction (HYDRO-PORT'94), PHRI, 67-82.

Hashimoto, N., Mitsui, M., Goda, Y., Nagai, T., and Takahashi, T. (1996) . Improvement of Submerged Doppler-Type Directional Wave Meter and its Application to Field Observation, Proceedings of 25th International Conference on Coastal Engineering (ICCE'96), Vol.1, 629-642.

Laboratory Experiments on the Effects of Mechanically Generated Waves on the Core Flow under Wind Waves

Shinjiro MIZUNO
Department of Civil Engineering, Hiroshima Institute of Technology
Miyake-2-1-1, Saeki-Ku, Hiroshima, 731-5193, JAPAN
E-mail: mizuno@cc.it-hiroshima.ac.jp

1. Introduction and primary purpose of the study

When a wind speed is higher than 3 m/s, it is well known that numerous streaks nearly parallel to the wind direction sometimes appear on the sea surface of lakes, open oceans and coastal area. Langmuir (1938) showed that these streaks are the visible manifestations of a parallel series of counter-rotating vortices developing below the surface with axes nearly parallel to the wind. Today this kind of flow pattern is well known as Langmuir cells or circulations (LCs). They play an important role in the study of coastal process such as collection of plankton, generation of blue tide in Tokyo Bay, and so on.

Since Craik and Leibovich (1976) have proposed a rational theory on wave-current interaction mechanisms of LCs (CL1 & CL2 mechanism), a large number of theoretical and numerical investigations have focused on application of the CL theory to investigate the three-dimensional structures of wind-driven current in the upper mixed layer of ocean and coastal area. Contrary to CL1 theory, the CL2 theory is more important because it may be applied to mechanically generated waves and/or wind waves usually observed. As far as we know, however, neither laboratory test nor field observation has yet been done to confirm the validity of CL2 mechanism because of the three-dimensional phenomena . The primary purpose of this presentation is to report the experimental results on the effects of mechanical surface waves on Langmuir Circulations in a laboratory tank

We study what role mechanical waves play in the physical process of momentum transfer from the wind to the subsurface flows. There are two possibilities; First, some of previous investigators, who made wind-wave experiments in laboratory tanks, have examined the effects of wave-induced Reynolds shear stress associated with mechanical waves. However, the wave-induced Reynolds stress was found to be not so important to the momentum transfer mechanism; Second, in this study we examine the effect of Lagrangian Stokes wave drift of mechanical waves. It should be noted that there is as yet no research that has investigated this effect on the subsurface structure in a laboratory tank.

We examined response of the pre-existing LCs in a wind-wave tank to two sets of mechanical waves; that is, aligned wind and mechanical waves (aligned-swell case) and opposed wind and mechanical waves (counter-swell case). The momentum transfer in the tank flow was found to be strongly affected by introducing the two sets of mechanical waves. That is, aligned-swell intensified the pre-existing LCs, but markedly weakened the primary circulation in the wind direction, whereas counter-swell acted to the contrary, strengthening the primary circulation more and more toward the wind direction. In this study we report the details of these experimental results, and show that they may be fully explained by the CL2 mechanism based on the Stokes drift of mechanical waves. Our experiments indicate that the primary energy and momentum transfer mechanism from the wind to the subsurface flow in a laboratory tank depends strongly on the CL2 mechanism rather than wind-waves.

2. Experiments and experimental procedure.

The experiments were carried out in a wind-wave tank at Hiroshima Institute of Technology. The tank is 13 m long, $B = 30$ cm wide, 55 cm high. The water depth D was set to 25 cm, and the height of wind-tunnel section H to the remaining 30 cm.. Three experiments were performed at an air free-stream velocity of 7- 8 m/s:

Case PW: Case of pure wind waves(hereafter referred to as PW),

Case CS: Coexisting case of wind waves and mechanically generated counter-waves against the wind (hereafter referred to as CS).

Case AS: Coexisting case of wind waves and mechanically generated aligned-waves propagating to the wind direction (hereafter referred to as AS).

The frequency and height of mechanical waves were selected to $f = 1$ Hz and $Ho = 5$ cm, where Ho denotes the initial height at the mechanical wave machine. A TSI two-component laser-Doppler anemometer (LDA) system was used in the backward scatter mode to measure downwind and vertical current components, u and w.

3. Experimental results

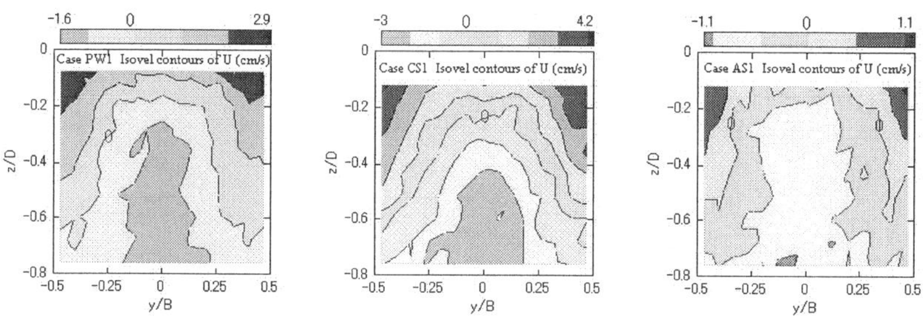

Figure 1 Isovel contour lines of primary flow U (cm/s) for 3 experimental cases. Left, Pure wind waves (PW); middle, Counter-swell (CS); Right, Aligned-swell (AS). Wind speed = 7 - 8 m/s. Figure caption indicates the max. and min. velocity of each current data.

Figure 1 shows response of the primary flow U to 2 sets of mechanical waves in a vertical cross-section, where z/D and y/B denote dimensionless vertical and lateral coordinates, respectively. Case PW(left panel) shows isovels of the primary flow at a constant wind speed of 7 m/s. Cases CS(middle) and AS(right) show the response to counter- and aligned- swells at nearly the same wind speed as case PW, respectively. One of the most interesting effects is that the maximum velocity has become much stronger for case CS but much less for case AS than for case PW, indicating that the response of the subsurface flow to mechanical waves is opposite to each other for two sets of CS and AS.

This is the first experiment for case CS that showed to accelerate the primary circulation toward the wind direction rather than the propagation direction of opposed mechanical waves. This may be considered to be seemingly unreasonable because the Stokes wave drift flows against the wind.

In this presentation we shall discuss that the Stokes drift plays an important role in the modification of the primary circulation, but that its role is not so easy as to be explainable by a sum of the wind-driven current plus the Stokes wave drift. The experimental data show that an organized vertical motion in the center of the tank, i.e., either upwelling or downwelling current (LCs), controls the behavior of the overall kinetic energy of both the horizontal mean flow and turbulence modified by introducing mechanical waves.

Anyway, the response of the subsurface flow in the wind wave tank to mechanical waves is complicated but surprisingly attractive. This presentation may provide some useful hints for the field study of complex coastal process due to the interaction between wind waves, longer waves propagating from deep water, and the low-frequency coastal flow.

Generation of Incident Random Waves in Numerical Extended Mild-Slope Equation Models using a Source Function Method

Gunwoo Kim[1], Chanhoon Lee[2], and Kyung-Duck Suh[3]

A number of random wave propagation models, which can be classified into the mild-slope equation model, have been developed. The models of Kubo et al. (1992) and Lee et al. (2001) were developed by using the Taylor series expansion technique, while the models of Radder and Dingemans (1985) and Suh et al. (1998) were developed by using the Hamiltonian theory of surface waves. In the numerical models for wave propagation, the incident waves along the open boundary should be specified. Recently, for the Boussinesq wave model, Wei et al. (1999) developed a source function method, which can be used for arbitrary grid systems. In this study, we derive two source functions by applying the method of Wei et al. to the mild-slope equation models for random waves: One is for Kubo et al.'s (1992) and Lee et al.'s (2001) models, and the other is for Radder and Dingemans' (1985) and Suh et al.'s (1997) models.

The equations of Suh et al. (1997) including the source function $f_1(x,y,t)$ are given by

$$\frac{\partial\eta}{\partial t}=-\nabla\cdot\left(\frac{\overline{CC_g}}{g}\nabla\widetilde{\phi}\right)+\frac{\overline{\omega}^2-\overline{k}^2\overline{CC_g}}{g}\widetilde{\phi}+\frac{\overline{\omega}^2}{g}\left\{\overline{R}_1(\nabla h)^2+\overline{R}_2\nabla^2 h\right\}\widetilde{\phi}+f_1(x,y,t) \tag{1}$$

$$\frac{\partial\widetilde{\phi}}{\partial t}=-g\eta \tag{2}$$

And the equation of Lee et al. (2001) including the source function $f_2(x,y,t)$ is given by

$$\begin{aligned}&\nabla\cdot\left(\overline{CC_g}\nabla\hat{\eta}\right)+\left\{\overline{k}^2\overline{CC_g}+g\overline{u}_1\nabla^2 h+g\overline{u}_2(\nabla h)^2\right\}\hat{\eta}+i\nabla\cdot\left\{\overline{\frac{\partial(CC_g)}{\partial\omega}}\nabla\frac{\partial\hat{\eta}}{\partial t}\right\}\\&+i\left\{\overline{\frac{\partial(k^2CC_g)}{\partial\omega}}+g\overline{\frac{\partial u_1}{\partial\omega}}\nabla^2 h+g\overline{\frac{\partial u_2}{\partial\omega}}(\nabla h)^2\right\}\frac{\partial\hat{\eta}}{\partial t}=f_2(x,y,t)\end{aligned} \tag{3}$$

The parameters $\overline{\omega}^2\overline{R}_1$ and $\overline{\omega}^2\overline{R}_2$ in Eq. (1) are mathematically equivalent to $-g\overline{u}_2$ and $-g\overline{u}_1$, respectively, in Eq. (3), all of which represent the second-order bottom effects, and the over bar indicates the variables associated with the carrier angular frequency $\overline{\omega}$.

If we assume a constant water depth in the wave generation area, the source function can be obtained by using the Green function method. Because all the above-mentioned equations are reduced to the

[1] Grad. Student, School of Civil, Urban & Geosystem Engineering, Seoul National University, Seoul 151-742, Korea, Tel: +82-2-880-8836, Fax: +82-2-887-0349, E-mail: maossi1@snu.ac.kr

[2] Asst. Professor, Department of Civil and Environmental Engineering, Sejong University, Seoul 143-747, Korea, Tel: +82-2-3408-3294, Fax: +82-2-3408-3332, E-mail: clee@sejong.ac.kr

[3] Professor, School of Civil, Urban & Geosystem Engineering, Seoul National University, Seoul 151-742, Korea, Tel: +82-2-880-8760, Fax: +82-2-887-0349, E-mail: kdsuh@snu.ac.kr

Helmholtz equation in a constant water depth, the procedures of derivation of the source functions are similar to each other. Assuming a Gaussian shape source function, the source functions f_1 and f_2 are obtained as

$$f_1(x,y,t) = \sum_m \sum_n \frac{C_m C_{gm} A_{m,n} \exp(-i\Delta\omega_m t)}{a_{m,n} I_{m,n}} \exp\left[-\beta(x-x_s)^2\right] \exp\left[i(\lambda_{m,n} y + \varepsilon_{m,n})\right] \quad (4)$$

$$f_2(x,y,t) = \sum_m \sum_n -\frac{A_{m,n} \overline{CC}_g \exp(-i\omega_m t)}{\omega_m a_{m,n} I_{m,n}} \exp\left[-\beta(x-x_s)^2\right] \exp\left[i(\lambda_{m,n} y + \varepsilon_{m,n})\right] \quad (5)$$

where $a_{m,n} = -i/2l_{m,n}$, $I_{m,n} = \sqrt{\pi/\beta}\exp(-l_{m,n}^2/4\beta)$, and l and λ are the wave numbers in the x- and y-directions, respectively. In order to generate random waves, the wave number must be calculated by using the dispersion relation, which can be derived by the eikonal equation of each model. Fig. 1 shows numerical solutions of regular waves at 4 different time steps by generating inside the domain, reflecting at the right boundary, and absorbing at the left boundary. Fig. 2 shows a comparison of the spectra of generated random waves (solid line) to the target (dashed line). All the numerical tests were conducted using the Lee et al.'s (2001) model. The results show that waves are generated properly.

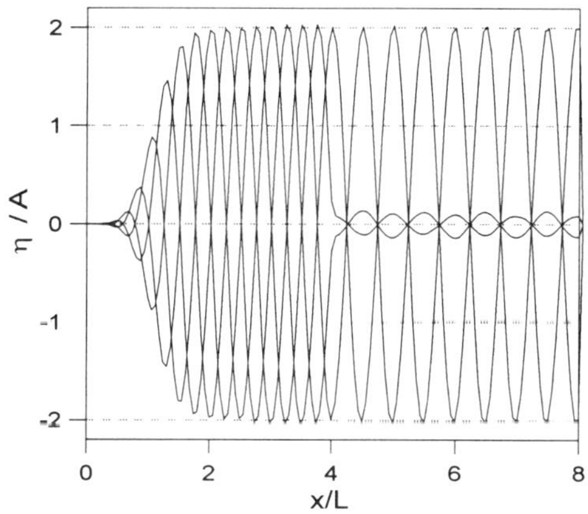

Fig. 1 Wave generation test for regular wave.

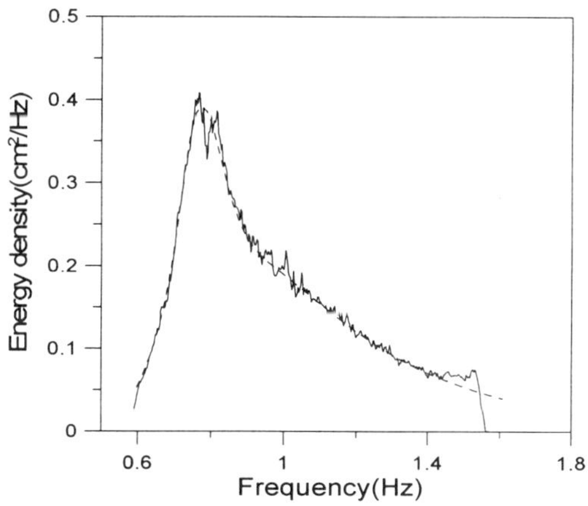

Fig. 2 Wave generation test for random waves.

References

Kubo, Y., Kotake, Y., Isobe, M. and Watanabe, A. (1992) Time-dependent mild slope equation for random waves, *Proc. 23rd Int. Conf. Coast. Engrg.*, ASCE, pp. 419-431.

Lee, C., Kim, G. and Suh, K.D. (2001) Extended mild-slope equation for random waves. *Proc. 4th Int. Symp. on Ocean Wave Measurement and Analysis*, ASCE, pp. 724-733.

Radder, A. C. and Dingemans, M. W. (1985) Canonical equations for almost periodic weakly nonlinear gravity waves, *Wave Motion*, Vol. 7, pp. 473-485.

Suh, K. D., Lee, C. and Park, W. S. (1997) Time-dependent equations for wave propagation on rapidly varying topography, *Coast. Engrg.*, Vol. 32, pp. 91-117.

Wei, G., Kirby, J. T. and Sinha, A. (1999) Generation of waves in Boussinesq models using a source function method, *Coast. Engrg.*, Vol. 36, pp. 271-299.

Simulation of Wave Overtopping on Partially Immersed Breakwater by SPH Model

By Songdong SHAO[1], Hitoshi GOTOH[2] and Tetsu MEMITA[3]

[1] JSPS Postdoc., Dept. of Civil Eng., Kyoto Univ.
Yoshida Honmachi, Sakyo-ku, Kyoto 606-8501, Japan
Tel: 81-75-7535098; FAX: 81-75-7610646; E-mail: shao@coast.kuciv.kyoto-u.ac.jp
[2] Assoc. Prof., Dept. of Civil Eng., Kyoto Univ. (Correspondence)
[3] Senior Res. Officer, Tech. Res. Center, The Kansai Electric Power Co. Inc.

ABSTRACT

The Smoothed Particle Hydrodynamics (SPH) method, coupled with a Large Eddy Simulation (LES) sub-particle scale turbulence model, is proposed to simulate wave overtopping a curtain breakwater. The SPH model solves the Navier-Stokes (N-S) equations by Lagrangian approaches and no grid is needed in the computations. Thus free surfaces can be easily and accurately tracked by particles without numerical diffusion. Classical SPH kernel functions are reviewed and the N-S equation solver is developed based on a semi-implicit algorithm of pressure projection. The accuracy of the SPH model is validated through available data from a wave flume experiment conducted in the Kansai Electric Power Incorporation. The agreement between experimental and numerical wave profiles is quite satisfactory. The efficiency of the incorporated LES model is further demonstrated by the computed velocity and turbulence eddy viscosity distributions of the overtopping flow. The findings in the paper will provide a good understanding on wave mechanics in coastal areas.

Figure Illustrations: Fig.1 gives out the instantaneous experimental wave profiles and computed particle configurations during single wave period at t/T = 0.25, 0.5 and 0.75, respectively. Fig. 2 and Fig.3 are the velocity and turbulence intensity distributions during wave overtopping, showing the robustness of the incorporated sub-particle scale turbulence model.

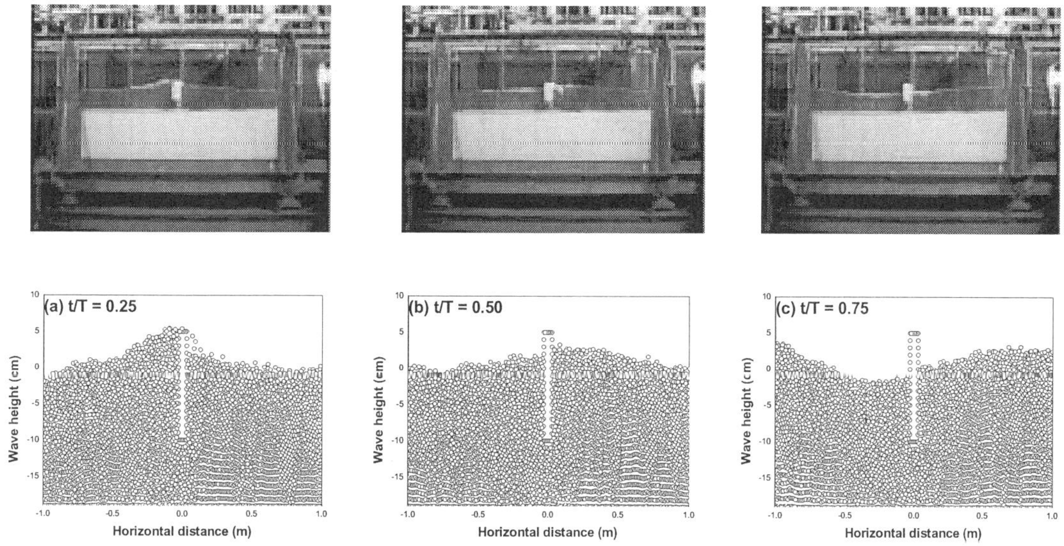

Fig. 1 Experimental and Computational Snapshots by SPH

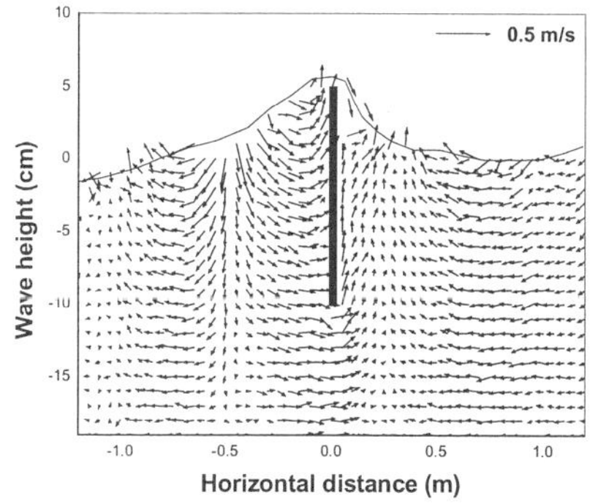

Fig. 2 Velocity Distribution during Wave Overtopping

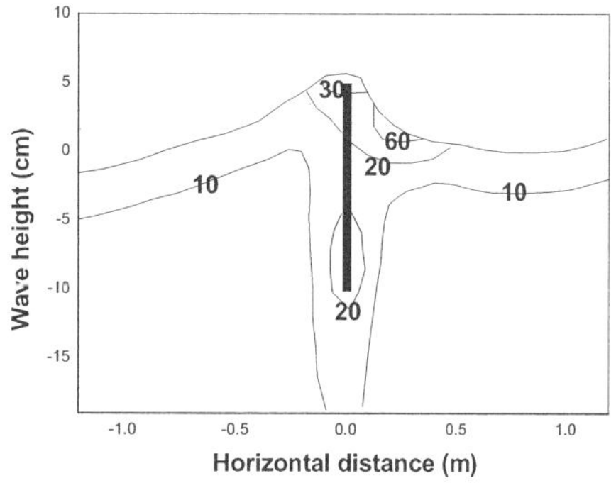

Fig. 3 Turbulence Viscosity Distribution during Wave Overtopping

Numerical Model of Wave Breaking by Lagrangian Particle Method with Sub-Particle-Scale Turbulence Model

Hitoshi GOTOH[1], Minoru HAYASHI[2], Tetsuo SAKAI[3] and Koji ODA[4]

[1] *Associate Professor, Dept. of Civil Engrg., Kyoto Univ.,*
Yoshida Honmachi, Sakyo-ku, Kyoto, 606-8501, Japan
tel:+81-75-753-5098, fax: +81-75-761-0646, e-mail: gotoh@coast.kuciv.kyoto-u.ac.jp (correspondence)
[2] *Engineer, Dept. of Construction, Nishimuro Promotions Bureau,*
Wakayama Prefectural Government, Japan
[3] *Professor, Dept. of Civil Engrg., Kyoto Univ., Japan*
[4] *Graduate Student, School of Civil Engrg., Kyoto Univ., Japan*

Objectives

One of the key factors of computation of wave breaking is the improvement of water surface tracking. The Lagrangian particle method has the advantage in the description of complicated behavior of water surface including a splashing, over the Eulerian grid-base models. Another significant factor of wave breaking is a turbulence. To describe a turbulence, being in the same way as the SGS(=Sub-Grid-Scale) model of turbulence, such as the LES in Eulerian approaches, the SPS(=Sub-Particle-Scale) model is required in the particle method. The analysis of wave breaking by the particle method has just begun, hence there is no studies of numerical simulation by the particle method with the SPS-turbulence model. In this study, the SPS-turbulence model, which had been proposed by the authors, is applied to the wave breaking on the uniform slope.

Simulation Model

The PS(=Particle-Scale) governing equations are derived by the filtering operation, in which the Reynolds stress appears in PS momentum equation. Water is modeled as the assembly of the particles. A behavior of water is calculated by discretizing the governing equations, or continuity equation and momentum equation, in the interaction zone supposed around each individual particle. All of the terms of the momentum equations, such as the convection term, the pressure-gradient term, the viscosity term, the gravity term and the Reynolds stress term are descretized based on the interparticle relations. For the closure of the Reynolds stress term, the Smagorinsky model is employed. The governing equations are discretized without using the computational grids, hence the particle method is free from the numerical diffusion, which brings the difficulties in the ordinary Eulerian approaches.

Results

The calculated domain is vertically two-dimensional wave flume, which is 6.0 m in length and 20.0 cm in mean water depth. By using the absorbing wave maker at the offshore end of the flume, the regular waves, 2.0 s in wave period and 11.0 cm in wave height, are generated. Totally 11,000 numbers of particles, the diameter of which is 1.0 cm, are tracked in this study. Figure 1 shows the time series of breaking wave behavior; while Fig. 2 shows the distribution of the SPS-turbulent intensities in the same moments as waves in Fig. 1. The plunging jet generated at the time t=6.6 s splash down on the water surface at the time t=6.7 s. In the meantime, the high concentration of the SPS-turbulence is found on the on-shore-side wave-front surface. The highly concentrated region of the SPS-turbulence is being stretched by the rapid water flow in the surface on the on-shore-side wave front, consequently, at the time t=6.8 s or 6.9 s, high concentration of the SPS-turbulence is found not only along the wave front surface but also at the foot of wave front. Further detailed description of the breaking wave field will be discussed. As for the accuracy of the numerical model, the selection of appropriate filter size will be also discussed.

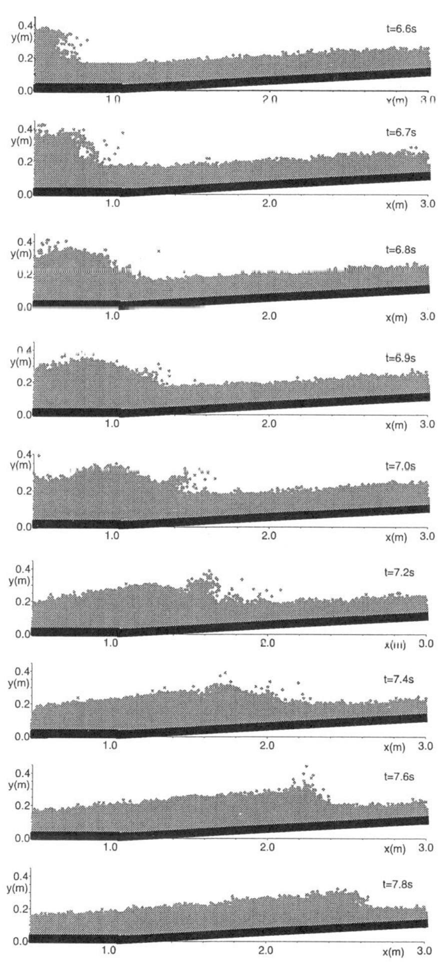

Fig. 1 Typical snapshots of solution

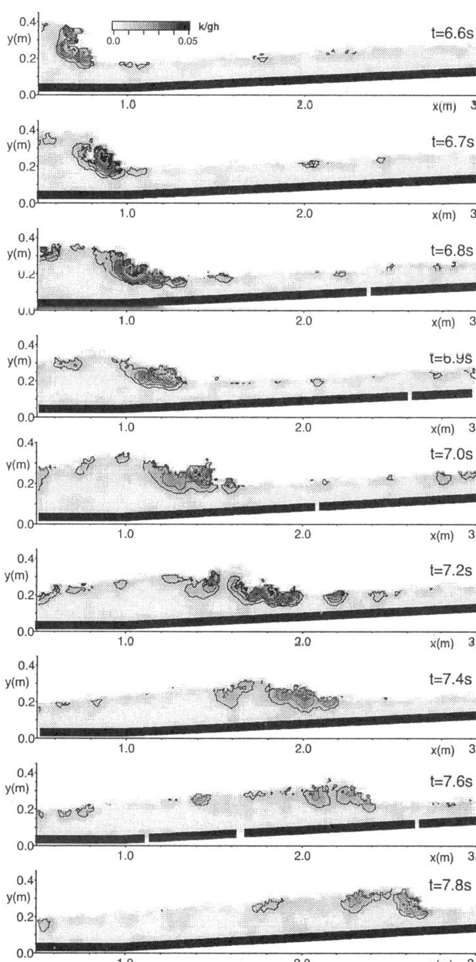

Fig. 2 SPS-turbulent intensities

Influence of incident wave angle on Mach-stem breaking

Tetsu Memita [1] and Tetsuo Sakai [2]

[1] Senior Research Officer, Tech. Res. Center, The Kansai Electric Power Co., Inc.

Nakohji-3 chome, 11-20, Amagasaki, 661-0974, Japan, fax+81-6-6494-9727, e-mail:K423403@kepco.co.jp

[2] Prof., Dept. Civil Eng., Kyoto Univ.

1. OBJECTIVES

When obliquely incident waves with a small incident angle reflect from a vertical wall, a wave propagating along the wall (Mach-stem) can be generated. Such a reflection pattern is called as Mach-reflection. Many researchers have investigated the characteristics of Mach-stem waves experimentally and numerically. To evaluate wave overtopping and the stability of wave dissipating blocks, the influence of Mach-stem waves is significant. There is possibility that Mach-stem neglected at the conventional design is relevant to the damage of coastal structure.

The stem waves formation has much influence on wave breaking along the structure. In the regular and irregular wave fields, the stem waves break more easily than the obliquely standing waves without the Mach-stem effect, for the same incident wave height and period **(Memita et al., 2000)**. To predict the breaking of stem wave is important. There has been no literature dealing with the breaking height of Mach-stem waves.

In this paper, an experiment is carried out to propose the method to predict the wave breaking height of Mach-stem. The development of Mach-stem along the wall with dissipating armor blocks is also examined.

2. EXPERIMENTAL SET-UP

The experiment was carried out in a wave basin (18m×30m) with a wave generator with 60 paddles of 30cm width. The bottom topography is a modeled actual one. A vertical wall having a corner is installed in the wave basin. The wall is high, and does not allow wave overtopping. The regular waves having the wave period of 1.12,1.34,1.73,2.00s are generated. The wave height is chosen to be large enough for the breaking to take place. The angle between the incident wave direction and vertical wall line is 15,25,35 and 55degrees. The total number of incident wave condition amount to about 30. The wave height was measured along the vertical wall at every 50cm interval in parallel to the wall. The breaker point was determined as the point where the wave height becomes maximum.

Photo 1 shows an example of wave field in front of the vertical wall. The angle of incidence is 15 and 35 degrees. The stem waves formation patterns are classified into two main parts. When the angle of incidence is

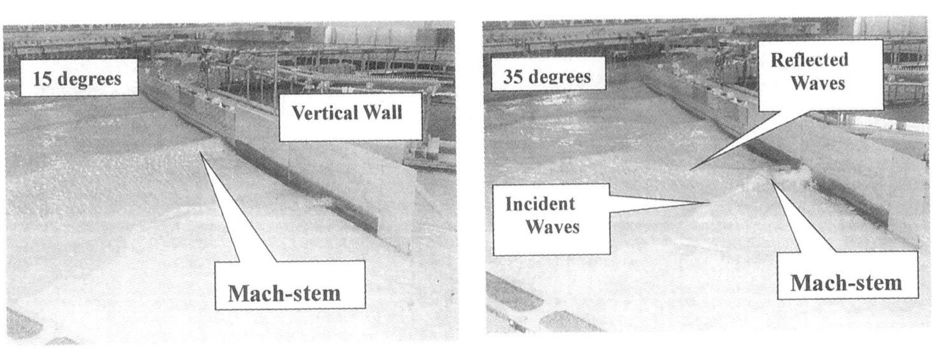

Photo 1. Wave field in front of the vertical wall (incident angle of 15 and 35 degrees)

15-degrees, the wave crest bends, becoming normal to the wall, and no reflected waves appear. When the angle of incidence is 35-degrees, the incident waves, the reflected waves, and Mach-stem waves are present. When the angle of incidence is 55-degrees, the obliquely standing waves are formed.

3. Results and Discussion

Wave breaking of Mach-stem along the wall is quite different from that of standing waves with no Mach-stem effect. The Mach-stem breaking is violent. One can observe that developed stem wave is nearly similar to uni-directional propagating waves in the wave basin. To compare the breaking height of Mach-stem with the wave breaking criterion of progressive waves **(Goda, 1970)** is worthwhile.

Fig.1 compares the experimental results of incident angle of 15 and 25 degrees with the breaker index by **Goda**. The vertical axis is the breaker height to water depth ratio, H_B/h_B, and the horizontal axis is the relative water depth, h_B/L_0. Each dotted point is the average value of 5 waves in each wave condition. The figure shows a good agreement between the experimental data and the breaker index. To predict the breaking of Mach-stem is possible by substituting not incident wave height but Mach-stem wave height in breaker index. In the design of coastal structure, we have to pay attention to this fact. The wave height is getting increased along the wall due to both the diffraction and stem wave development. In this experiment, the wave height ratio normalized by the incident wave height is about 2.0. Even if the incident wave height is about half of wave breaking height by the breaker index, we have to expect the violent breaking of stem waves along structure.

Fig.2 compares the experimental results of incident angle of 35 degrees with the breaker index. The wave crest line of Mach-stem is not so long in the wave basin. The breaking height of Mach-stem wave is higher than that of the progressive waves. It is impossible to predict the breaking of Mach-stem by substituting the Mach-stem wave height in breaker index.

The incident angle has much influence on both stem waves formation patterns and the wave breaking height along structure. It is important to reflect this fact on coastal structure design.

The development of Mach-stem along the vertical wall with dissipating armor blocks is also examined. One can not observe the development of Mach-stem in the wave basin. To set the dissipating armor blocks along coastal structure is useful to prevent the development of Mach-stem.

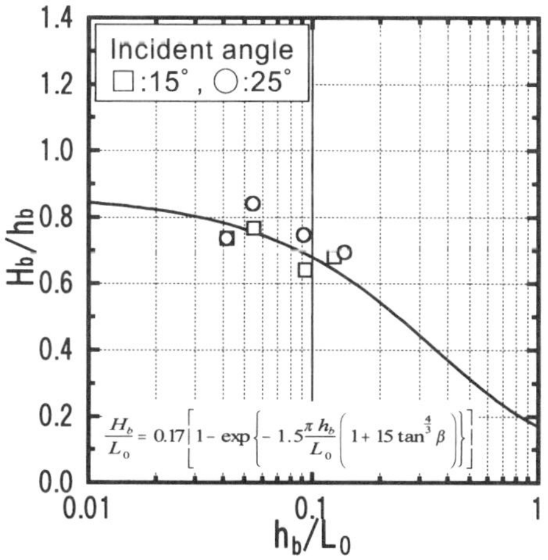

Fig. 1 Comparison with the wave breaking criterion (Incident angle of 15 and 25 degrees)

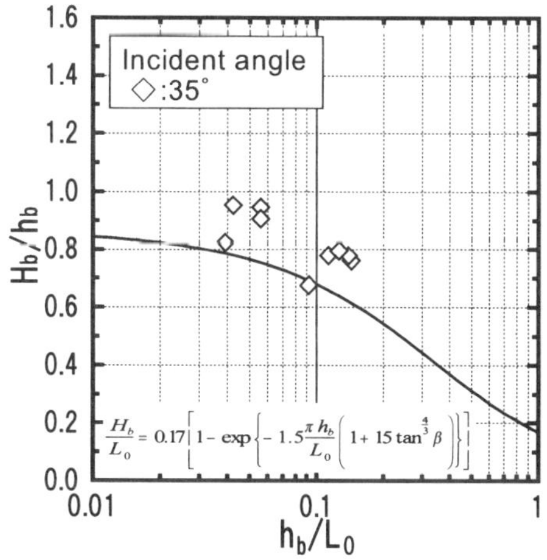

Fig. 2 Comparison with the wave breaking criterion (Incident angle of 35 degrees)

Modeling Near-shore Waves and Surface Rollers

Yoshimitsu Tajima[1] and Ole Secher Madsen[2]
Research Assistant[1], R.M. Parsons Laboratory, Massachusetts Institute of Technology, Cambridge, MA02139, USA; e-mail: yoshitaji@aol.com; phone/fax: +1-617-577-5671
Professor[2], R.M. Parsons Laboratory, Massachusetts Institute of Technology, Cambridge, MA02139, USA

1. INTRODUCTION

This paper will present the development of models for predictions of near-shore wave characteristics and the evolution of surface rollers. The wave model is based on the concept of an equivalent linear wave, but nonlinear wave characteristics may also be reconstructed from the prediction of the equivalent linear wave characteristics. The surface roller model is based on the energy conservation equation which is ideally consistent with the proposed breaking wave energy dissipation model. Because both models are based on simple energy conservation equations, the entire model retains flexibility and computational efficiency for practical applications. The waves may be periodic or narrow-banded random waves, the beach profile may be plane, barred, or movable bed with natural sand grains. Model predictions are compared and show excellent agreement with experimental observations.

2. WAVE MODEL

The wave model (Tajima and Madsen, 2002) consists of the following components:

2.1 Equivalent Linear vs. Non-Linear Wave

Equivalent linear wave is defined as a linear wave which has an identical energy flux to a non-linear actual wave. Based on this concept, the relationship between non-linear and equivalent linear wave characteristics have been explored through numerical experiments using Nwogu's (1993) Boussinesq equations. The numerical simulations are carried out up to values of $H_*/h<0.4$, where H_* is the nonlinear wave height, since the Boussinesq approximation is known to yield unreliable results for bottom orbital velocities when $H_*/h>0.4$. From the experimental results, transform relationships are obtained to express non-linear wave characteristics as functions of deep water wave steepness, H_0/L_0, relative depth, h/L_0 and the bottom slope, β, where all these deep water wave parameters are specified from the local equivalent linear wave conditions. Transform relationships are assumed to be valid for all H_*/h and are applied up to their breaking point by extrapolation. Validity of the proposed non-linear wave model will be presented at the conference though comparison between predicted and measured non-linear near-bottom wave orbital velocity profiles.

2.2 Wave Breaking

Since our wave model is based on linear wave theory, the breaking and broken wave models should also be cast in terms of linear wave theory for compatibility. The breaking point is determined from a modification of Watanabe et al.'s (1984) breaking criterion. Since this criterion is based on linear theory and was established by requiring it to agree with observed breaking water depth, h_b, it is ideally suited for application in conjunction with our equivalent linear wave and leads to the determination of equivalent linear breaking wave height, H_b.

2.3 Broken Waves

Broken wave characteristics are obtained for the equivalent linear wave from the numerical solution of the breaking wave dissipation model

$$\frac{\partial(EC_g)}{\partial x} = -K_b \frac{C_g}{h}(E - E_r) \tag{1}$$

originally proposed by Dally et al. (1985). In Eq.(1), $E = \rho g H^2/8$ is the local wave energy where H is the equivalent linear wave height and E_r is based on the recovery wave height, H_r, of the broken wave if it were to travel on in a constant depth equal to the local depth, h. In contrast to Dally et al., who took K_b=constant, K_b is semi-empirically derived as a function of the bottom slope. It should be emphasized that K_b is theoretically derived so that the deformation of predicted non-linear wave heights, which are transformed from the equivalent linear wave predictions, should agree with the deformation of the observed broken wave heights. In this manner, we can expect that the predicted dissipation of equivalent linear wave energy should be consistent with dissipation of actual wave energy.

3. SURFACE ROLLER MODEL

Under the observational assumption that the surface roller moves with the wave phase velocity, C, time-averaged surface roller energy, E_{sr}, is determined as

$$E_{sr} = \frac{\rho S_{sr} C}{2T} \tag{2}$$

where T is a wave period, and S_{sr} is the area of the surface roller. Evolution of the surface roller energy is then determined from an energy conservation equation

$$\frac{1}{2}\frac{\partial(EC_g)}{\partial x} + \frac{\partial(E_{sr}C)}{\partial x} = -K_{sr}\frac{C}{h}E_{sr} \tag{3}$$

which differs from that originally proposed by Dally and Brown (1995) in the assumption that only the potential wave energy is transferred into the surface roller energy. From observations, the surface roller should disappear, i.e., the surface roller energy becomes zero, when the broken waves stop breaking and recover. In this sense, physical interpretations of Eq.(1) and Eq.(3) are ideally identical. We hence set the decay coefficient, $K_{sr}=K_b$. Figure 1 shows the evolution of the surface roller area in the surf zone for the experiment with periodic waves normally incident on a uniform slope reported by Cox and Kobayashi (1996). As seen in the figure, the model reasonably predicts the growth of the surface roller near the breaking point and the decay well inside the surf zone in a similar manner to empirical expressions proposed by Svendsen (1984) and Okayasu et al. (1988). Moreover, momentum force due to the roller is also expressed in terms of E_{sr} and may be applied to the prediction of wave setup as

$$\rho g h \frac{\partial \bar{\eta}}{\partial x} = -\frac{\partial}{\partial x}\left(S_{xx} + 2E_{sr}\right) \tag{4}$$

where S_{xx} is the wave radiation stress.

3. MODEL EXTENSION TO RANDOM WAVES

Since both wave and surface roller models are governed by simple energy conservation equations based on linear wave theory, the models are readily extended to random wave conditions. Under the assumption of Rayleigh-distributed wave heights with narrow-banded wave frequency, the right hand side of Eq.(1) is represented for random wave conditions as

$$\frac{\partial (EC_g)}{\partial x} = -K_b \frac{C_g}{h} \exp\left(-\xi_b^2\right)\left[E\left(1+\xi_b^2\right) - E_r\right] \tag{5}$$

where $\xi_b = H_b / H_{rms}$. All other model expressions are the same as those for periodic waves with the wave height being replaced by the rms value for the random sea.

4. COMPARISON WITH EXPERIMENTAL RESULTS

Figure 2 shows a comparison of predicted and measured nonlinear wave heights and wave setup for the experiment with periodic waves normally incident on a uniform slope reported by Cox and Kobayashi (1996). In the figure, predicted equivalent linear wave heights are also shown. In order to examine the validity of the proposed surface roller model, the predicted wave setup is also compared with that obtained when only the wave radiation stress is accounted for. As seen in the figure, the surface roller model clearly improves its predictions and reasonably explains the delay of the wave setup near the breaking point. Similarly, Figure 3 shows a comparison of predicted and measured significant wave heights and wave set-up for random waves normally incident on a barred beach reported by Okayasu and Katayama (1992). Other, equally successful comparisons between predictions and experimental observations will be presented at the conference. It should be pointed out that none of the parameters and coefficients in the model was obtained by fitting the model predictions to the data against which the model's capabilities are tested.

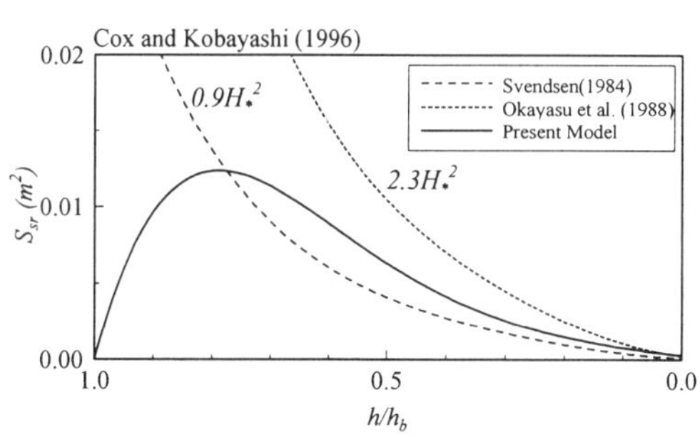

Figure 1: Evolution of the predicted surface roller area with empirical formulae

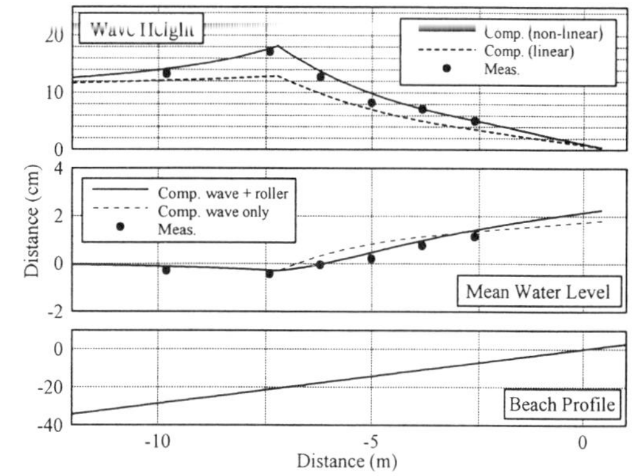

Figure 2: Comparison of predicted and measured wave heights and wave set-up for periodic waves. (Cox and Kobayashi, 1996)

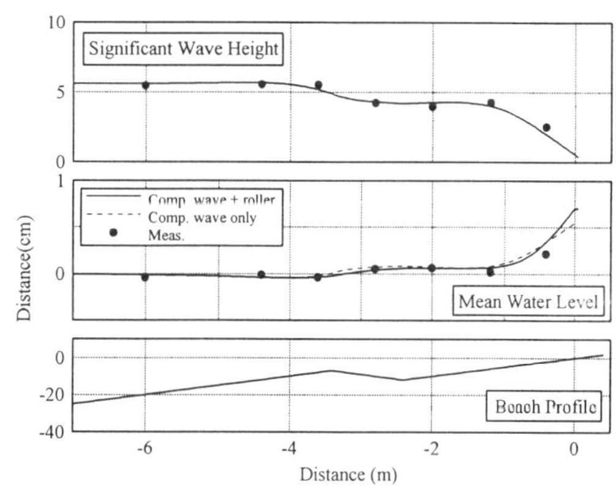

Figure 3: Comparison of predicted and measured significant wave heights, wave set-up, and undertow profiles for random waves. (Okayasu and Katayama, 1992)

REFERENCES

Cox, D.T. and Kobayashi, N., 1996. *Proc. 25th Int. Conf. on Coastal Eng.*

Dally W.R. and C. A. Brown, 1995. *JGR.*

Dally, W.R., R.G. Dean and R.A. Darlymple, 1985. *JGR.*

Nwogu, O., 1993. *Journal of Waterway, Port, Coastal and Ocean Eng.*

Okayasu, A., Shibayama, T., and K. Horikawa, 1988. *Proc.21st Int. Conf. on Coastal Eng.*

Okayasu, A. and H. Katayama, 1992. *Proc. 23rd Int. Conf. on Coastal Eng.*

Svendsen, I. A., 1984. *Coastal Engineering.*

Tajima, Y. and O. S. Madsen, 2002. *Proc. Int. Conf. on Coastal Eng.* (in press).

Watanabe A., T. Hara and K. Horikawa, 1984. *Coastal Eng. in Japan.*

A 3D LES MODEL FOR TURBULENT FREE SURFACE FLOWS IN VEGETATION

Su Xiaohui[1] and Lin Pengzhi[1,*,Θ]

[1]Department of Civil Engineering, The National University of Singapore, Singapore 119260, Republic of Singapore

Abstract

A three-dimensional σ-coordinate LES model is proposed to simulate hydrodynamic behavior of free surface turbulent flows with vegetation. Vegetation is considered as an internal source of resistant force and a phenomenon model is employed to express the performance of vegetation in the flows. The model is modified from the σ-coordinate LES model developed by Lin and Li (2002, International Journal for Numerical Methods in Fluids, Vol 38, Page 1045-1068). The governing equations of the modified model include additional terms for drag force produced by vegetation and the modified model is distinctive in that not only compound channel flow can be solved but also wave motion through vegetation in coastal region can be taken into account. An operator splitting method, which splits the solution procedure into advection, diffusion and pressure correction steps, is employed so that different numerical schemes can be used for the solution of different physical processes. The model has firstly been applied to simulate the hydrodynamic behavior of turbulent flow in compound open channel with partly vegetated region. Then by using the model the attenuation of wave movement due to vegetation in coastal region is

[*] correspondence to Pengzhi Lin, Department of Civil Engineering, The National University of Singapore, 10 Kent Ridge Crescent, Singapore 119260, Republic of Singapore.
[Θ] Email: cvelinpz@nus.edu.sg

simulated. The results reveal that the present model has the capacity of describing three-dimensional structure of large eddy appearing in free surface turbulent flows with vegetation. It is believed that the model will become a useful tool to pursue further study of wave hydrodynamics in coastal region with the presence of aquatic vegetation.

Keyword: σ-coordinate LES model; turbulent flow; vegetation; compound open channel; coastal region

The ExEBEd Model for Multidirectional Random Waves

KAZUYA OKI
Department of Civil Engineering, Kyoto University,
Yoshida-Honmachi, Kyoto, 606-8501, Japan
Tel : +81-75-753-5099 / Fax : +81-75-761-0646
e-mail : oki@coast.kuciv.kyoto-u.ac.jp

HAJIME MASE
Disaster Prevention Research Institute, Kyoto University,
Gokasho, Uji, Kyoto, 611-0011, Japan

and

TERRY HEDGES
Department of Civil Engineering, University of Liverpool,
Brownlow Street, Liverpool, L69 3GQ, United Kingdom

Introduction

Various theories and wave transformation models have been devised to describe nearshore wave fields. Each of the theories and models has certain advantages and limitations with respect to its applicability. In order to estimate the wave conditions, the appropriate model has to be selected depending on the relative importance of the various physical processes and the target coastal area.

Wave prediction models based on the energy balance equation are suitable for applications involving large sea areas because they require little computational power. These models are used for wave forecasting and hindcasting. The calculated quantities are the spectral energy densities, which are phase averaged quantities slowly varying over several wavelengths. Originally, phase averaging wave models did not account for wave diffraction. Recently, however, attempts have been made to introduce wave diffraction effects by Resio (1988), Booij et al. (1997), Rivero et al. (1997) and Mase (2001).

Mase (2001) directly introduced a diffraction term, formulated from a parabolic approximation wave equation, into the energy balance equation and employed the first order upwind difference scheme, which is numerically stable due to numerical diffusion. The numerical diffusion term represented by the second derivative is similar to the wave diffraction term.

In order to suppress numerical diffusion separately from the effect of wave diffraction, quadratic upstream interpolation for convective kinematics (QUICK), proposed by Leonard (1979), is employed in this study. The predictions of the wave model are validated against the Sommerfeld theory for the case of wave transformation through a gap between breakwaters.

ExEBEd Wave Model

For steady-state conditions, the energy balance equation with a wave diffraction term derived by Mase (2001) is written as

$$\frac{\partial(v_x S)}{\partial x}+\frac{\partial(v_y S)}{\partial y}+\frac{\partial(v_\theta S)}{\partial \theta}$$
$$=\frac{\kappa}{2\omega}\left\{\left(CC_g\cos^2\theta\, S_y\right)_y-\frac{1}{2}CC_g\cos^2\theta\, S_{yy}\right\}-\varepsilon_b S \quad (1)$$

where $S=S(f,\theta)$ is the directional wave spectral density, (x, y) are the horizontal coordinates, θ is the wave direction measured counterclockwise from the x axis, ε_b is the coefficient of energy dissipation, and the characteristic velocities,(v_x, v_y, v_θ), are defined as follows:

$$(v_x,\, v_y,\, v_\theta)$$
$$=\left\{C_g\cos\theta,\ C_g\sin\theta,\ \frac{C_g}{C}\left(\sin\theta\frac{\partial C}{\partial x}-\cos\theta\frac{\partial C}{\partial y}\right)\right\} \quad (2)$$

where C is the wave celerity and C_g is the group velocity. In Eq.(1), the expression

$$\left(CC_g\cos^2\theta\, S_y\right)_y-\frac{1}{2}CC_g\cos^2\theta\, S_{yy}\cong 0 \quad (3)$$

has been added to the energy balance equation. The method of adding terms which become identically zero at some condition is similar to the methods employed by Madsen et al. (1991) and Kaihatu and Kirby (1998) to improve the dispersion properties of the Boussinesq equations. Coefficient κ is a free parameter to be optimized in order to change the degree of diffraction. In the previous study of Mase (2001), $\kappa = 2.5$ was adopted.

The first order upwind difference scheme is applied to the first term on the left-hand-side of Eq.(1) and the QUICK scheme is applied to the second and third terms. The central difference scheme is applied to the diffraction term of Eq.(1). Consequently, the finite difference version of Eq.(1) finally becomes:

$$\begin{aligned} &A_1 S_n^{ijk} + A_2 S_n^{i(j-2)k} + A_3 S_n^{i(j-1)k} \\ &\quad + A_4 S_n^{i(j+1)k} + A_5 S_n^{i(j+2)k} + A_6 S_n^{ij(k-2)} \\ &\quad + A_7 S_n^{ij(k-1)} + A_8 S_n^{ij(k+1)} + A_9 S_n^{ij(k+2)} = -B S_n^{(i-1)jk} \quad (4) \end{aligned}$$

The coefficients in Eq.(4), $A_1 \sim A_9$ and B, are functions of v_x, v_y, v_θ, C and C_g.

Since $S_n^{(i-1)jk}$ on the right-hand-side of Eq.(4) are known values, S_n^{ijk} are obtained by solving algebraic equations. In the calculations, conditions such as an open sea boundary, a reflecting wall boundary and a dissipative beach boundary are taken into account. This wave prediction model is called the ExEBEd model (**Ex**tended **E**nergy **B**alance **E**quation model with **d**iffraction effect).

Main Results

In order to check the effect of the QUICK scheme, the wave height distributions behind a breakwater with a gap were calculated and compared with the results from the first order upwind difference model without a diffraction term, called hereafter the EBE model. Monochromatic waves, approaching the breakwater gap at angles of 30° and 45°, were used. The results from the ExEBEd model showed less numerical diffusion than the EBE model.

Obviously, it was necessary to establish the value of κ to be employed in the diffraction term. For this purpose, model results from the ExEBEd model were established for different values of κ and were compared with the Sommerfeld theory for conditions in which the water depth was 12m and the incident significant wave height and period were 1m and 10s, respectively. The energy spectrum was of JONSWAP form with a peak enhancement factor, γ, of 3.3, and the wave directional spreading function was of the Mitsuyasu type with peak values, S_{max}, of 10, 25 and 75. The breakwater gap width B was varied from $B/L = 2$ to $B/L = 8$, where the wavelength $L = 100$m. The grid sizes δx and δy were $B/20$, the number of component waves (of equal energy), N, was 10 and the number of component angles, K, was 36. The value $\kappa = 2.0$ minimized the error between the ExEBEd results and Sommerfeld's theory.

When the gap between breakwaters is wide and the incident random waves have a broad angular spread, the EBE model agrees fairly well with the Sommerfeld solutions. However, for a narrow gap and waves with narrow angular spread, there is poor agreement between the predictions and the Sommerfeld results. Figure 1 compares the calculated significant wave heights obtained from the EBE model with the results of Sommerfeld's theory for the case of $B/L = 2$ and $S_{max} = 75$. The solid lines denote the calculated results and the dotted lines are the Sommerfeld solutions. The prediction from the ExEBEd model for the same case is compared with Sommerfeld's solution in Figure 2. Here, agreement is much better except near the gap and immediately behind the breakwaters where the wave heights are small.

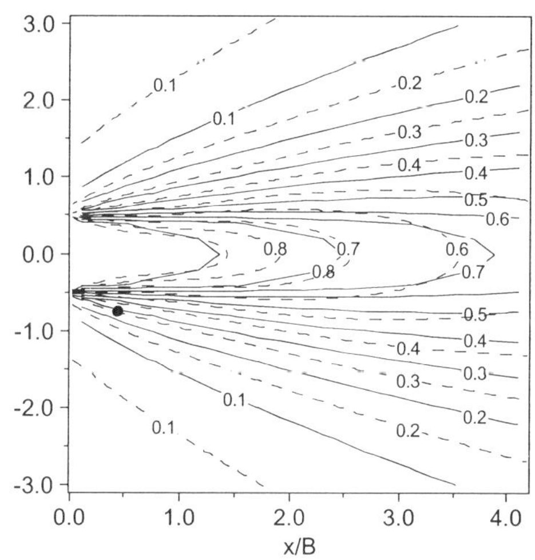

Fig.1 Wave height distribution ;
EBE model and Sommerfeld solution

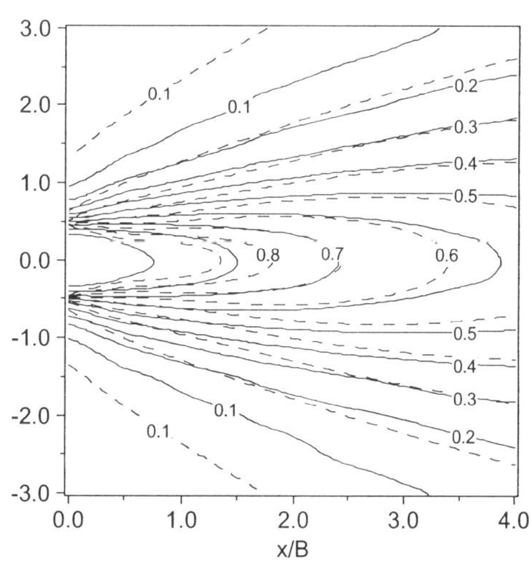

Fig.2 Wave height distribution ;
ExEBEd model and Sommerfeld solution

A METHOD FOR DETERMINING LIMIT STATE DESIGN WAVE USING PERIOD DISTRIBUTION FUNCTION

Atsushi Fujii*(JSCE Member),Yoshifumi Hayashi*,Yoshiharu Shimoda*
Hiroyuki Yamaya**,Eiji Oshita**,Nobutaka Namerikawa***(JSCE Member),Shusaku Miyawaki***

* Nagoya Research and Engneering Office for Port and Airport
Chubu Regional Bureau Ministry of Land,Infrastructure and Transport
1-57-2,Higashimatabei-cho,Minami-ku,Nagoya-shi,Aichi-ken 457-0833 Japan
** Coastal Development Institute Of Technology,3-16,Hayabusa-cho,Chiyoda-ku,Tokyo 102-0092 Japan
*** Japan Port Consultants,Ltd,3-6,Nishi-Gotanda8-chome,Shinagawa-ku, Tokyo 141-0031 Japan
Contact:Dr.Nobutaka Namerikawa, Technological Development Department,Japan Port Consultants,Ltd
Tel:+81.3.5434.5307,Fax:+81.3.5434.5392,E-mail:Nobutaka_Namerikawa@jportc.co.jp

1. INTRODUCTION

All waves acting on a breakwater during its design lifetime (for example, 50 years), represented by their frequency of occurrences according to wave scale, are defined as the fatigue limit state design waves. In Japan, wave heights with a frequency of occurrence equal to 10^4 are adopted as the serviceability limit state design wave for the purpose of controlling cracks in concrete structure.

For determining the limit state design wave for the purpose of designing port structure components, the method proposed by Nagao (1994) employs the average wave period equivalent to 1/1.2 times the significant wave period without considering frequency distributions, although the method takes account of wave height frequency distributions.

From analysis of individual waves contained in raw wave data obtained at the ports of Shimoda and Omaezaki, both located on the Pacific coast of Japan, the authors have verified that the period of large individual waves with a height of 2m or more which is of significance for design purposes corresponds to a significant wave period. Since wave force varies with wave period, it is important to determine an appropriate period for the limit state design wave.

2. WAVE PERIOD DISTRIBUTION FUNCTION BASED ON FIELD WAVE DATA

Period distribution functions for individual waves observed at the ports of Shimoda and Omaezaki were studied with the wave data classified into those having a wave height of 2 m or more and corresponding to significant wave period and those having wave height of less than 2m which can be considered to be the average wave period.

The Longuet-Higgins (1975) equation (1) below was employed with the parameter for use in the formula selected so as to fit the frequency distribution of wave period to be obtained in respect of the field wave data.

$$p(T/Tav) = \nu^2/2\{\nu^2+(T/Tav-1)^2\}^{3/2} \quad \cdots\cdots \quad (1)$$

where p = Probability density of wave period, T = Period, Tav = Average period, ν = Parameter

Fig. 1 shows a curve for the frequency distribution functions in respect of the period of waves with a height of 2 m or more as obtained by using Eq. (1) and a bar graph for the frequency distribution in respect of 15 sets of wave period measurements at Shimoda Port. The good agreement between the values represented by the curve and the bar graph indicates the reproducibility of the frequency distribution of wave period at the above locality by means of Eq. (1). In the case of waves with a height of less than 2 m, similar positive results were obtained by giving other values for Tav and ν.

3. METHOD FOR DETERMINING LIMIT STATE DESIGN WAVE PERIOD

In determining the limit state design wave period several assumptions were made. The first assumption was that the frequency distribution of wave period could be represented by the relevant period distribution function. Secondly, the frequency distribution of wave height was assumed by using the Rayleigh distribution as in the case of the conventional method. Thirdly, it was assumed that wave occurrence characteristics did not change. Based on these assumptions and using the significant wave data covering a period of 11 years obtained at Shimoda Port, the frequency of individual waves over the 50-year design lifetime of the breakwater was tabulated to determine the fatigue limit state design wave.

In Fig. 2, the average period (design period) calculated by the new method according to wave scale is represented by the solid line (▲). Whereas, as seen from Fig. 2, the values calculated by the conventional method (◇) differ widely from those for the average wave period (●) calculated from the actual field measurements, the values obtained by the new method proposed by the authors show a good agreement with those derived from the field measurements in the wave scale above 2 m which is of significance for design purposes. This verifies the appropriateness of the new method.

The new method for determining the limit state design wave period distribution function was standardized so as to be adaptable to Ise Bay where the new offshore Chubu International Airport is under construction, by selecting an appropriate parameter for the equation for determining the wave period frequency distribution function on the basis of field wave data provided by the marine observation station of the Chubu International Airport.

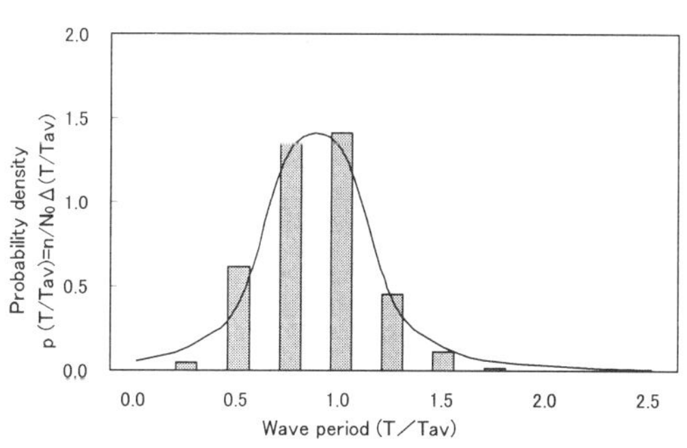

Fig. 1 Wave period distribution on Function based on Shimoda Port Wave Data

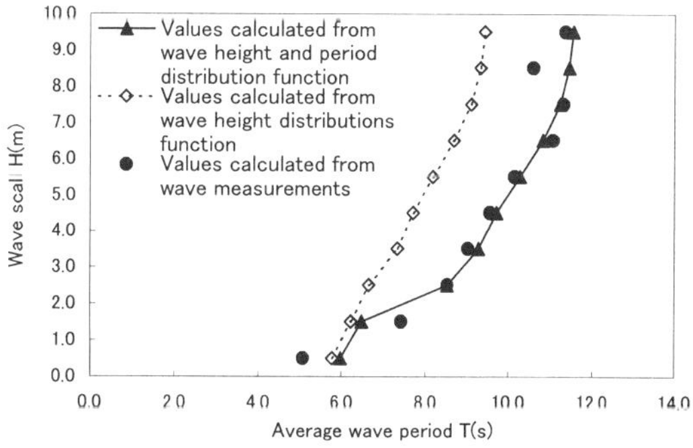

Fig. 2 Average periods of Fatigue limit state design waves

REFERENCES

Takashi Nagao (1994) : Studies on the application of the limit state design method to reinforced concrete port structures,Rept.of PHRI,Vol. 33,No. 4 ,pp.106-107.

Longuet-Higgins, M. S. (1975) : On the joint distribution of the periods and amplitudes of sea waves, J. Geophys. Res. , Vol. 80, No. 18, pp. 2688-269.

Mechanism of Coastal Giant Waves

Guan-Yu Chen*

1.Introduction

In Taiwan, coastal giant waves, nicknamed rabid wave due to their unpredictable and violent nature, are responsible for at least several deaths each year. It is urgent to understand these waves so that human lives can be protected.

These giant waves do not appear often. Related studies on these waves are very rare, and field measurement is still unavailable. To study coastal giant waves, we can only borrow the approach as well as the results of deep-sea freak waves

In this study, all theories that have been proposed to explain deep-water freak waves are surveyed and applied to coastal giant waves, as is given in Table 1.

	Theory	Can it be applied in coastal giant waves?
1	Extreme value in statistics	No. Near-shore waves are from the deep ocean, where the chance of extra-large waves is less than the probability theory has predicted
2	Blockage caustics due to counter current	No. Near a breakwater, the current cannot flow offshore
3	Caustics of wave refraction due to non-uniform current field	No. The phenomena is expected to be quasi-unidirectional
4	Nonlinear superposition for waves of different directions	No. The phenomena is expected to be quasi-unidirectional
5	Unidirectional wave coalescence	Possible, but the lasting time is short
6	Nonlinear evolution of wave groups	Yes (governed by NLS-)
7	Shoaling as well as the concentration of wave energy due to bathymetric refraction and diffraction. Vertical up-shooting when a wave colliding with a vertical wall	No. The bathymetry is constant, while the giant wave is seldom and unexpected. However, these effects increase the wave height and explain why coastal giant waves can only appear in some specific area.

Table 1 Giant wave theories and their applicability in coastal region

Based on wave features provided by coastal giant wave survivors, coastal giant waves are quasi-unidirectional waves that remain large for some duration. Hence, majority of the theories are found to be not feasible except for two theories - the nonlinear evolution of one-dimensional wave groups and the one-dimensional wave-wave coalescence. These two theories are then further investigated with numerical experiments. Their potential of generating extra-large waves is verified. Besides, we also show that the nonlinear Schroedinger (NLS) equation can predict the "freak" phenomenon of shallow water wave groups.

*Associate Researcher, Center of Harbor and Marine Technology, IOT, 2 Chong-Heng Tenth Road, Wuchi 435, Taiwan, TEL: 886-4-26587191,Fax: 886-4-26564415,e-mail: gychen@mail.ihmt.gov.tw

2.Numerical Models and their Verification

The Korteweg-de Vries (KdV) equation is the model equation for shallow water waves and can be applied in the coastal region. In our study, the water surface elevation is integrated by the Fourier Pseusospectral method and this model is verified by comparing with the computation of Zabusky and Kruskal (1966).

3.Wave Coalescence and the Coastal Giant Wave

By utilizing the method of Pelinovsky et al. (2000), we can produce initial conditions that will incur a large hump preceded by a deep trough. The coalescence phenomenon is shown in Figure 1.

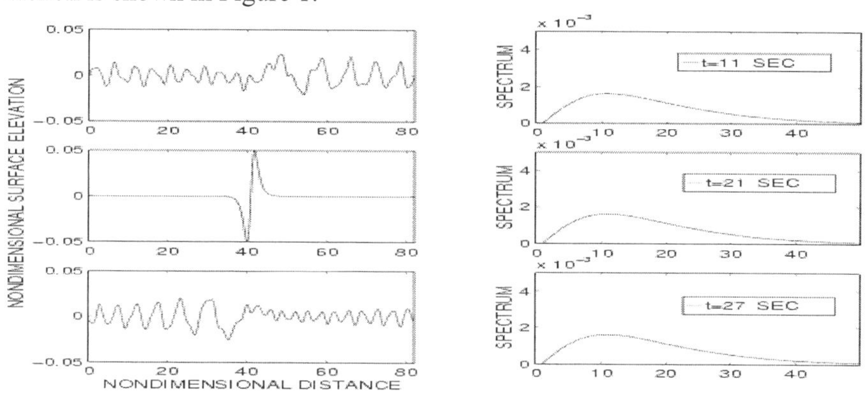

Figure 1. Change of water surface and wave spectra in coalescence.

4. Nonlinear Evolution of Wave Groups

The NLS equation is the general evolution equation for wave packet evolution and can be classified into two categories: the NLS+ and the NLS-. NSL+ is suitable for deep-water periodic waves and is used by some authors as the model equation for deep-sea freak waves. The NLS- for shallow water wave, however, has never been applied in freak wave simulation. We showed that under certain circumstances the maximum amplitude of the NLS- could increase up to 70% in a short time. This "freak" phenomenon is verified by our KdV model, as is given in Figure 2.

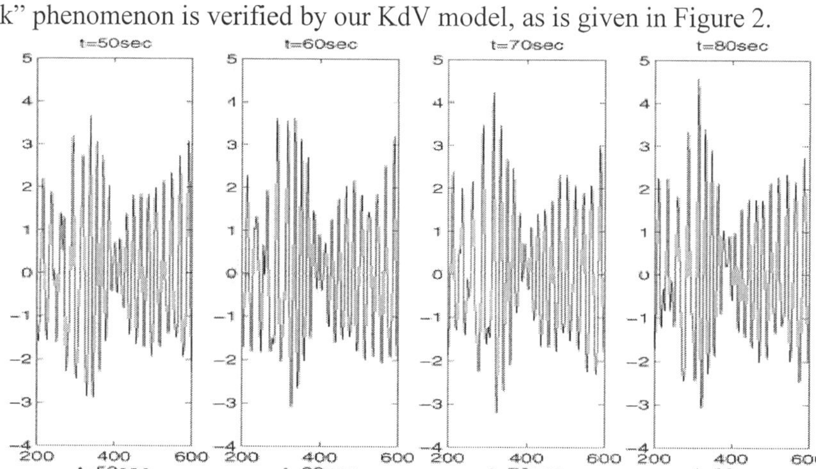

Figure 2. Elevation evolution of a wave group.

REFERENCE

〔1〕Boyd, J. P. and G.-Y. Chen (2001), *Mathematics and Computers in Simulation*, Vol. 55, pp.317.

〔2〕Pelinovsky, E., T. Talipova and C. Kharif (2000) *Physica D*, Vol. 147, pp.83.

〔3〕C.-H. Tsai, et al. (2001,in Chinese), *J. Coastal and Ocean Engineering,* Vol. 1, pp.71.

Application of MEP Method to the Study of Wave Climate Statistical Characteristics

XU Fumin[1] XUE Hongchao[2]

Research Institute of Coastal & Ocean Engineering, Hohai Univ., Nanjing, 210098, P.R. China
Email: xfm67@sohu.com

The probability density distribution of a random event depends on the experimental constraint conditions, there are diversified distributions for the given constraint conditions, the so called MEP distribution is to choose such a distribution whose information entropy is the maximum value among the diversified distributions. In order to find the best fitting method for the wave climate statistical distributions, the Maximum Entropy Principle (MEP) method is introduced, and the corresponding probability density evaluation method for the random fluctuation system is elaborated. According to the MEP method, the probability density evaluation steps can be described as follows: given the random fluctuation signals, then the moments of different orders can be obtained from a sample function, thus the analytical probability density function, then the entropy function, is also obtained according to the MEP concept, in accordance with the constraint conditions, regulating the probability density function in order that the entropy reaches the maximum value, then the corresponding probability distribution function is the MEP distribution. In this article, the constraint conditions are limited to the second order moment, a kind of new MEP distribution expression is obtained for the first time, this expression means that the relationship between the probability density of the random fluctuation signals and the second order moment of the variable has been found. Deferent with all the fitting methods in the past, this kind of MEP distribution can compute the probability distribution of a random fluctuation system conveniently and exactly, so long as the second order moment of the random signal is known, that is, the ratio of the root-mean-square value to the mean value of the random variable is obtained from the random sample, the corresponding MEP distribution expression can be computed according to the deduced expression in this essay. The concept of the wave climate is introduced here, by using the deduced MEP distribution, this article focuses on finding the best method in fitting the significant wave height and spectral peak period probability distributions. Take the Mexico Gulf as an example, three stations at different locations, depths and wind wave strengths are chosen in the half-closed gulf, the significant wave height and spectral peak period distributions at each station are fitted with the MEP distribution, the Weibull distribution and the Log-normal distribution respectively, the fitted results are compared with the field observations, differences between the three distribution fitting methods have been analyzed, the results show that the MEP distribution deduced in this article is the best method, and the Weibull distribution is the worst method when fitting the significant wave height and spectral peak period distributions at different locations, water depths and wind wave strengths in the half closed Gulf, this conclusion shows the reasonability and feasibility of the deduced MEP distribution expression in describing wave climate statistical properties, and it also proves the great potential of applying the MEP method to the study of wave statistical characteristics.

It is widely known that MEP distribution consists of diverse distributions, giving enough terms, the MEP distribution can fit all sorts of distributions effectively, more study needs to be done in the application of the MEP method to the wave statistical properties, the moments discussed here concerns only up to the second order, higher order moments should be considered in order to fit the real distribution better in the future.

Keywords: MEP method, wave climate, significant wave height, spectral peak period, the MEP distribution, the Weibull distribution, the Log-normal distribution.

1.XU Fumin (1967—) associate professor, major in coastal dynamics study.
2 .XUE Hongchao (1928—) professor, major in coastal engineering.

A New Analysis Method for Extreme Wave Statistics by Poisson-GPD Model

Toshikazu Kitano[1], Wataru Kioka[2] and Hajime Mase[3]
1 Assistant Professor, Dept. of Arch. and Civil Engineering, Nagoya Institute of Technology,
Gokiso-cho, Showa-ku, Nagoya, 466-8555, Japan
e-mail: kitano@ace.nitech.ac.jp, tel: +81-52-735-5498, fax: +81-52-735-5503
2 Professor, Dept. of Arch. and Civil Engineering, Nagoya Institute of Technology,
Gokiso-cho, Showa-ku, Nagoya, 466-8555, Japan
e-mail: kioka@ace.nitech.ac.jp, tel: +81-52-735-5487, fax: +81-52-735-5503
3 Associate Professor, Disaster Prevention Research Institute, Kyoto Univ.,
Gokasho, Uji, 611-0011, Japan
e-mail: mase@kaigan.dpri.kyoto-u.ac.jp, tel: +81-774-38-4146, fax: +81-774-38-4321

1. Introduction: In the statistical analysis for extreme wave data proposed by Goda (1988, 2000), the Fisher-Tippett distributions of type I and II (hereafter, abbrebiated as FT-I and FT-II) and the Weibull distribution are employed as candidates for the population function. The reason of employing the Weibull distribution is the flexiblity for fitting to sample. The extreme value distributions are obtained theoretically by the asymptotic process and the extreme value distributions can not be always applied to the samples of wave heights collected by the annual maximum series as well as by the peak-over-threshold series. Besides, the Fisher-Tippett distiobution of type III (hereafter, abbrebiated as FT-III) is not employed as a candidate, even though FT-III is an extreme value distribution as well as FT-I and FT-II. Because the FT-III has the upper bound, it is taken into consideration to avoid the possibility of underestimating the return wave height if the FT-III is selected by mistake as the best-fitting function. In these points of view, it can be said that the conventional method is only a means of fitting the empirically supposed populations and that it is not based on the rigorous extreme value theory. Recently in the field of the theoretical statistics, the asymptotic process to the extreme values are being modeled by employing the generalized Pareto distribution (abbrebiated GPD) for the tail-part of the population functions of the peak-over-threshold series. In the model by GPD, the excess by an appropriate threshold in the sample follows asymptotically the GPD, and the asymptotic process can be practically confirmed by changing thresholds and estimating the slope of mean residual lifes and the shape parameters, which is independent from threshold value, for different thresholds.

In this study we apply the model by GPD to extreme wave data and we show a new method of analyzing extreme wave statistics by demonstrating some examples.

2. Theory: For the sample of "Block Maxima" like as the daily rain fall, the occurence of exceeding a threshold of a rain fall quantity is modeled by the binomial process. The sample of the peak-over-threshold series like as the local maximum of the hourly-measured significant wave heights has no temporal block but the occurence can be modeled by the Poisson process. By introducing the mean occurence period by an appropriate threshold of wave height u, the return value x_R for return period R is given by

$$x_R = u + \frac{\sigma}{\xi}\left\{\left(R/\tau_u\right)^{\xi} - 1\right\} \tag{1}$$

with the GPD's parameters σ and ξ, and the variance in the statistical valiability of return wave height

$Var(x_R)$ is obtained as

$$Var(x_R) = Var(\tau_u)\left(\frac{\partial x_R}{\partial \tau_u}\right)^2 + {}^t(\nabla x_R)\, Var(\sigma, \xi \mid \tau_u)\nabla x_R; \quad \nabla = \left(\frac{\partial}{\partial \sigma}, \frac{\partial}{\partial \xi}\right) \tag{2}$$

by the delta method. In Eq.(2), $Var(\tau_u)$ is the variance of the mean occurence period which is given by regarding the population of individual occurence periods as the exponential distribution, and $Var(\sigma, \xi \mid \tau_u)$ is the variance of σ and ξ with τ_u fixed, which is given in the maximum likelihood estimation of the GPD's parameters. The fittness to GPD is checked by the Probability-Probability Plot and Quantile-Quantile Plot and the fittness for the Poisson process as the temporal occurence model is examined by the chi-square test for the annual or biannual occurence numbers.

3. Results: As an explanatory example, the Kodiak data of hindcasted storm waves are analyzied here, which was examined by the method in Goda (2000). In the mean residual life plot of Fig. 1, the slope of the mean residual life curve seems to take a negative constant and it suggests that the shape parameter is negative and the population will follow asymptotically FT-III distribution, which is not target in the conventional method. This fact is supported also by the result in the change of the shape parameter's estimates by various thresholds, as shown in Fig. 2. Figure 3 show the return value wave heights for various return periods with the confidence intervals. It is noted that the increase of return value by return period is not large which is similar to the properties of FT-III. Figure 4 show the occurence probability in a year toghether with the theoretical probability by the Poisson distribution with the measured mean occurence number per year. The squared-sum of deviations of each annual occurence numbers from the mean number is not rejected as abnormal large value by chi-squre test, that is to say that the extremal wave hights is acceptable to occur by the identical Poisson process.

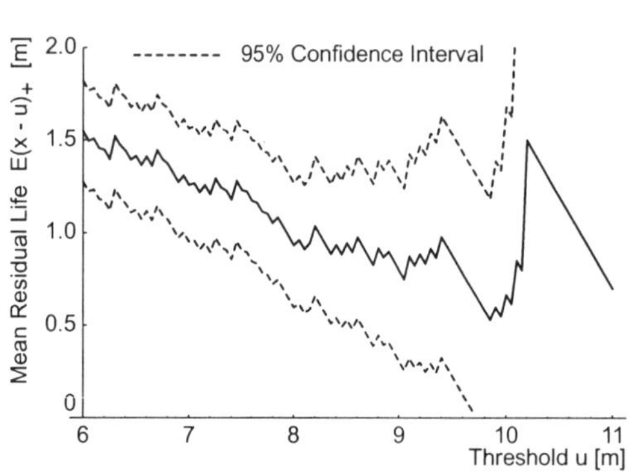

Fig. 1 Mean residual life plot

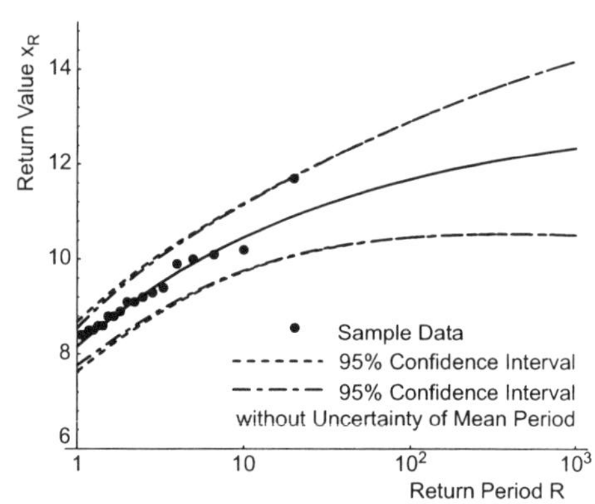

Fig. 3 Return wave heights with confidence interval

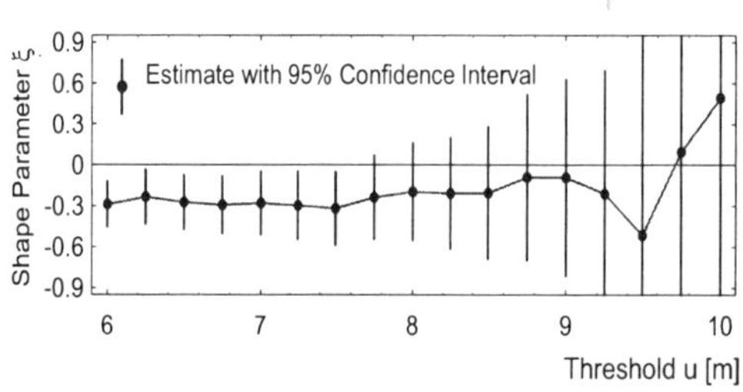

Fig. 2 Change of estimated shape parameter by various thresholds

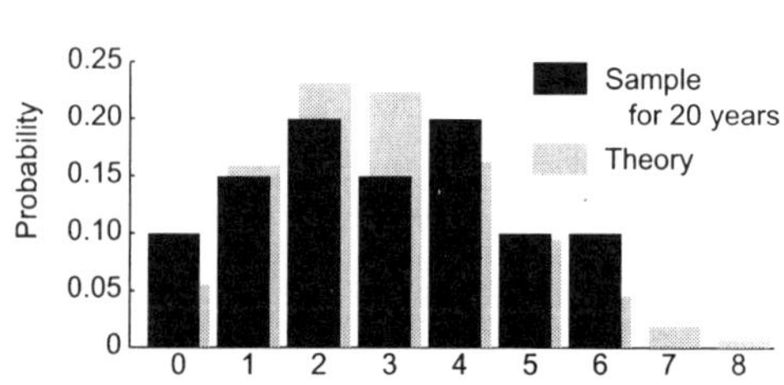

Fig. 4 Occurence probability in a year

STATISTICAL CHARACTERISTICS OF UNUSUAL WAVES OBSERVED AT DANANG, VIETNAM

Nguyen Ngoc Tran,

Director General, Project Management Unit No. 85, Ministry of Transport (MOT), Gov. of S.R. Vietnam, VIETNAM

Kohei Nagai (Corresponding Author),

Director and General Manger, Overseas Project Dept., Japan Port Consultants, Ltd. (JPC),

T.K. Gotanda Bldg., 8-3-6 Nishi Gotanda, Shinagawa-ku, Tokyo 141-0031, JAPAN.

Tel. 81-3-5434-8163, Fax. 81-3-5434-5385, e-mail. Kohei_Nagai@jportc.co.jp

Hajime Kubota,

Project Manager, Danang Port Improvement Project, Japan Port Consultants (JPC), VIETNAM

Nguyen Ngoc Hue,

Vice General Director, Transport Engineering Design Inc. (TEDI), MOT, VIETNAM

Dao Xuan Quang,

Manager, Survey Dept., Port & Waterway Engineering Consultants (POWECO), TEDI, MOT, VIETNAM

ABSTRACT

Almost all the existing ports in Vietnam are located in rivers, inner bays or lagoons. Coastal ports exposed to the open sea are now under construction such as Danang Port and Dung Quat Port. In such coastal ports, sea waves constitute one of the most fundamental and important natural conditions for planning, design and construction of protective facilities such as breakwaters and seawalls.

A JICA study was carried out from 1997 to 1998 to formulate plans to develop new international trade ports on the central coast of Vietnam.[1] At that time, information on the waves in the study area, however, was substantially non-existent, or very limited to those analyzed based on visually observed data, e.g. U.S Navy (1976), Hogben et al. (1986), Nguyen et al. (1995). TEDI (1996) and POWECO (1996) Therefore, the wave measurements in the JICA study from April 1997 to February 1998 were an epoch-making survey, in which actual in-situ wave records were taken at two locations of Danang and Ky Ha by means of ultrasonic-cum-water pressure type wave recorders. Invaluable information on the wave characteristics obtained in the study is that wind waves due to typhoons as high as 9 m in maximum height were successfully recorded at Typhoons No.9721 (*Fritz*) in September 1997 and No.9726 (*Linda*) in November 1997. Another important result is that long period waves up to 16 seconds in significant period were revealed to exist. It is proved that the long period waves are swells from large-scale typhoons remotely located in the west Pacific, and propagated through the Luzon Strait to the central coast of Vietnam. Not only these results of wave measurements at site and their analyses, but also wave hindcast calculations and their statistical analyses in the JICA Study revealed several significant characteristics of the waves at the area. [2]

Two years after the JICA study, wave observation was resumed at the same place, or the mouth of Danang Bay, under the consulting services of the Danang Port Improvement Project.[3] The period of the measurement was selected to cover the most dangerous months due to typhoons and the NE monsoon from August 2000 to April 2001. The results of the wave observation and their statistical analyses are useful in port development projects in Vietnam, and presented in this paper, which can be briefly summarized below:

1) Waves generated by two typhoons, i.e. Typhoons No. 0011 (*Kaemi*) in August and No. 0016 (*Wukong*) in September 2000, were successfully measured.
2) Waves generated by several tropical depressions and the NE monsoon were also recorded from November 2000, to February 2001.
3) The highest wave measured during the 8 months had a magnitude of significant wave height: $H_{1/3} = 5$m, maximum wave height: $H_{max} = 8$m, and wave period: $T = 9$sec, generated by Typhoon No. 0011.
4) Until September 2000, the sea was rather calm except the time of typhoons.
5) From November to February, rough sea repeated frequently by the tropical storms and lows of the NE monsoons. As shown in the figures below, the intervals of the high waves over 2 m in significant wave height were statistically less than one week in November and December, and less than two weeks in January and February. Their durations were shorter than four days.

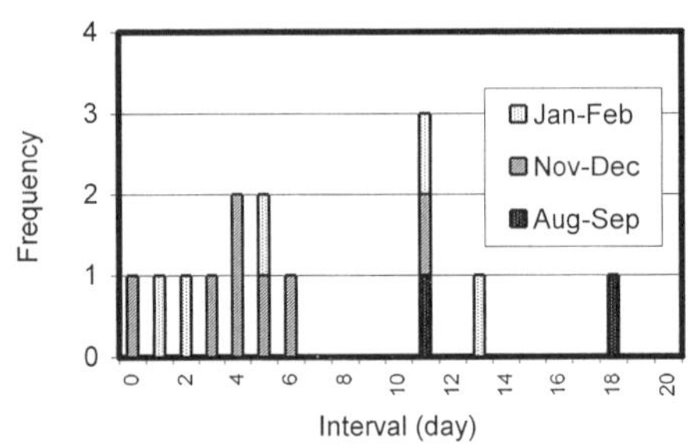

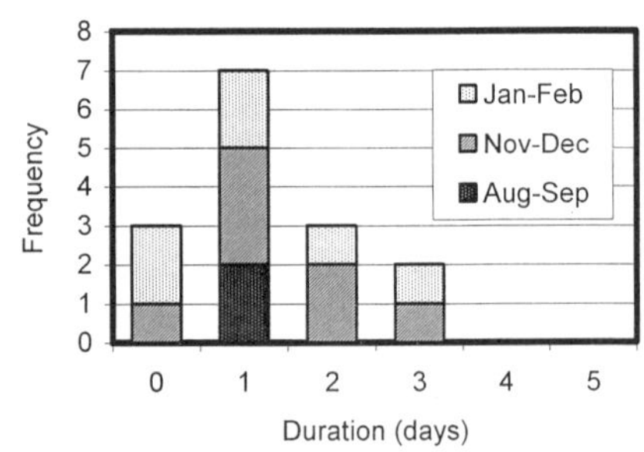

6) In March and April, high wave more than 2m in significant wave height was not recorded.
7) Directional wave spectra [4] of high waves are analyzed for 15 records, and the directional characteristics were clarified.
8) Directional wave spectra of long waves are analyzed for 7 waves, and existence of swells from remote atmospheric disturbances are confirmed.
9) The overall occurrence frequency of waves is analyzed and shown in the form of wave roses. The most frequent wave direction is ENE, followed by NE. If typhoons and tropical depressions pass to the north of Danang, the wave direction can be N in Danang.

This is the first report on statistical analysis of unusual waves quantitatively on the central coast of Vietnam.

References

1) OCDI and JPC: "The Study on the Port Development Plan in the Key Area of the Central Region in the Socialist Republic of Vietnam," JICA, August 1998.

2) Nagai, Kohei; S. Kono; and Dao X. Q.: "Wave Characteristics on the Central Coast of Vietnam in the South China Sea," Coastal Engineering Journal, Vol. 40, No. 4, pp. 347-366, 1998.

3) JPC, TEDI and Maunsell: "Wave Spectral Analysis, Directional Wave Spectra of Unusual Waves Observed at the Mouth of Danang Bay," Technical Rep. No. ANL. 4, Danang Port Consultants Office, 61pp., 2001

4) Hashimoto, N., T. Nagai, T. Asai and K. Sugawara: "Extension of Maximum Entropy Principle Method (MEP) for Estimating Directional Ocean Wave Spectrum", Rep. PHRI Vol. 32, No. 1, pp. 1-51, 1993 (in Japanese)

EMPIRICAL RELATIONSHIP BETWEEN MAXIMUM SUSTAINED WIND AND CENTRAL PRESSURE FOR TYPHOONS IN NORTHEAST ASIAN SEA

See-Whan Kang, Ki-Cheon Jun and Kwang-Soon Park

Coastal and Harbor Engineering Research Laboratory, KORDI
Ansan P.O.Box 29, Seoul 425-600, KOREA

Phone: +82-31-400-6330 Fax: +82-31-408-5827
E-mail: swkang@kordi.re.kr

Abstract

The empirical relationship between minimum central pressures and maximum wind speeds for Northeast Asian tropical cyclones was obtained by analyzing typhoon parameters and sea-surface winds observed at the East China Sea, the South Sea and the East (Japan) Sea for 20 years (1979 - 1999). The resulting empirical equation, $V_{max} = C(P_\infty - P_0)^\gamma$ was compared with various relationships obtained from previous studies. The result shows a very close agreement with SPM's(1984) relation. However, it was much lower in wind intensity compared with other relations, and the differences in the maximum wind speed were increased as the minimum central pressures were decreased. The main cause of these differences may lie in the quality of observation data obtained by either direct measurement or indirect estimation of sea-surface winds.

Empirical parameters obtained from this present and previous studies

	C	γ
Kang *et al*(2002)	5.02	0.5
SPM(1984)	4.86	0.5
Atkinson et al(1977)	3.44	0.644
Dvorak(1975)	3.53	0.671
Fletcher(1955)	8.23	0.5
Schloemer(1954)	4.90	0.5

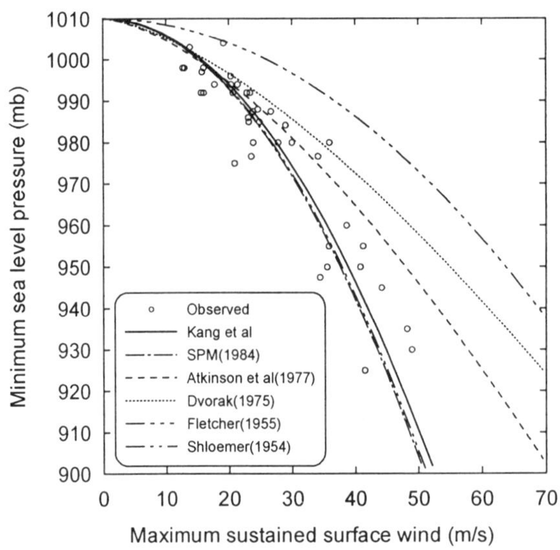

Comparison of the various maximum wind relationships with typhoon wind observations in Northeast Asian Sea.

References

Atkinson, G. D., and Holliday, C. R., 1977. Tropical cyclone minimum sea level pressure-maximum sustained wind relationship for western North Pacific, *Mon. Wea. Rev.*, **105**(2), pp. 421-427.

Dvorak, V. F., 1975. Tropical cyclone intensity analysis and forecasting from satellite imagery, *Mon. Wea. Rev.*, **103**, pp. 420-430.

Fletcher, R. D., 1955. Computation of maximum surface winds in hurricanes. *Bull. Amer. Meteor. Soc.*, **36**, pp. 246-250.

Holland, G. J., 1980. An analytical model of the wind and pressure profiles in hurricanes, *Mon. Wea. Rev.*, **108**, pp. 1212-1218.

Schloemer, R. W., 1954. *Analysis and synthesis of hurricane wind patterns over Lake Okeechobee*, Florida Hydrometeorol. Rep. No. 31, U.S. Weather Bureau.

SPM, 1984. *Shore Protection Manual*, Vol. 1, U.S. Army Corps of Engineers, Washington, D.C.

Wind-Wave Correlation Analysis Based on Field Data and Wind Simulation in Sea Area

Norimi MIZUTANI
Dept. of Civil Engineering, Nagoya University
Nagoya 464-8603, JAPAN
Fax:+81-52-789-1665 E-mail:mizutani@civil.nagoya-u.ac.jp

Tatsuya YOSHIDA
Tokai Branch, Japan Weather Association
1-21-5 Mizukusa-cho, Kitaku, Nagoya 462-0042, JAPAN
Fax:+81-52-914-0171 E-mail:tatsuya@tokai.jwa.or.jp

and

Toyokazu BANBA
Aichi Prefecture
Nagoya 460-8501, JAPAN
Tel:+81-52-961-2111 E-mail:tbanba@pop21.odn.ne.jp

It is difficult to evaluate the deformation of beach profile due to the wave excitation for long duration of time in terms of a single representative wave because the beach deformation is a result of dynamic interaction with unsteady wave excitation. The wave observation stations are not enough to cover whole the coastal line, numerical estimation of waves in time and space domains is necessary to estimate the waves where the observed waves are not available. In the wave generation process, wind plays primary roles on it, and therefore developing a wind simulation technique and analysis of a wind-wave interaction problem seem to be very significant.

In this paper, field data of winds and waves observed at wave station near Shirahama Beach, Wakayama Prefecture, Japan, were analyzed to discuss the wind-wave interaction mechanism. Also, wind simulation method in sea areas is developed based on direct numerical simulation (DNS) method.

First, field data of wind and wave observation have been analyzed using a data set for a duration of one month in September, 1997. In this analysis, the significant wave height, mean wind speed, and power spectra have been obtained.

In case of the significant wave height less than 0.5m, the significant wave heights have been found to have linear relationship with mean wind speed. On the other hand, the significant wave height larger than 0.5m have no clear relationship with wind speed. The investigation of wind-wave relation with atmospheric pressure pattern shows that the wave-wind relation depends on existence of the low pressure in the Pacific Ocean. When the low pressure does not exist in the Pacific Ocean, there is a clear relationship between them as shown in Fig.1. Contrary, if the low pressure exists, swell is developed and propagates to observation field, which makes the relationship unclear.

As far as the wind energy is concerned, main part of the wind energy is found to be stored in lower frequency domain less than 0.02Hz and the power spectrum of wind in lower frequency domain grows up as wind speed increases (see Fig.2), whereas that in the higher frequency domain does not change and always follows -5/3 rule, which is seen in turbulence flow.

When the wind field grows up, the energy of waves around 0.3Hz increases, which makes the spectrum shape twin-peak one (see Fig.3). The major energy of waves also concentrates around 0.1Hz, which is confirmed to be generated by swells from the low pressure area. This indicates the dual behavior of the wind waves, that is the components around 0.3Hz, which caused by the local wind field around observation and the swell around 0.1Hz which is developed around low pressure in far field.

The wind field in sea area is then simulated by the finite difference method based on SOLA scheme. Governing equations are the equation of continuity of incompressive fluid and the Navier-Stokes equations of motion. The initial and boundary conditions are given from GPV (grid point value), which is provided by Japan Weather Association, using the MASCON model. The time variation of wind vector distribution has been computed in this study. Time increment of the computation was chosen as 1.0s, because the high frequency components of wind variation have been confirmed not to be effective for the wave generation or development. The computation results have been compared with observed ones, which shows good agreement between them.

Further results will be presented at the conference and proceedings.

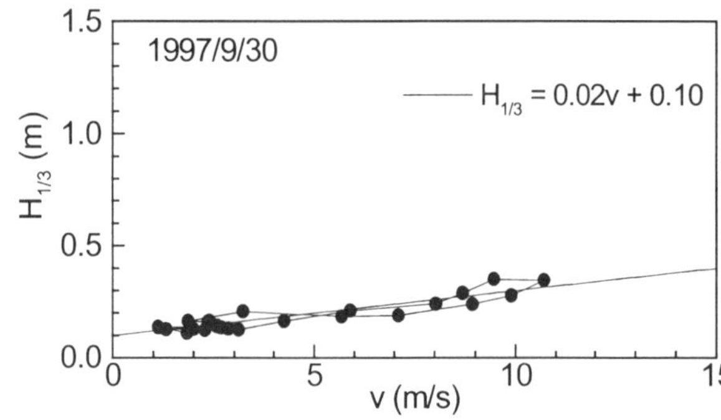

Fig.1 Wind-wave relationship

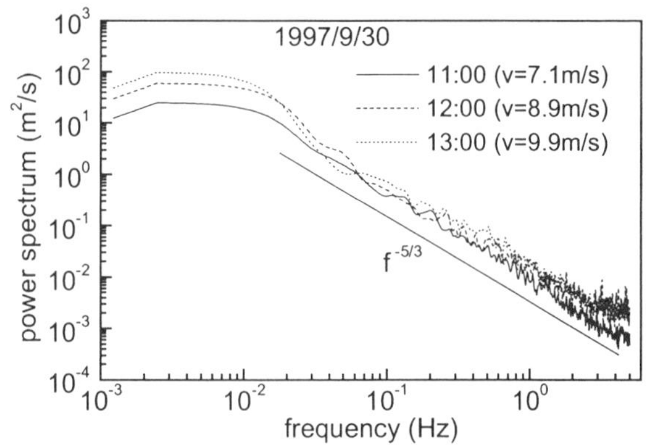

Fig.-2 Power spectrum of winds

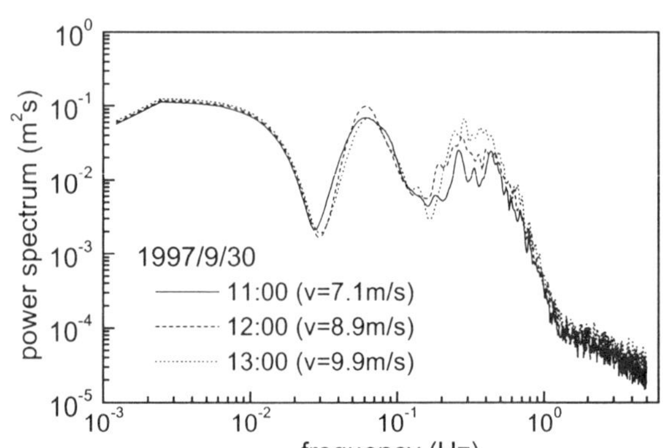

Fig.-3 Power spectrum of waves

New Approaches for Computing Wave Growth Rate due to Wind Induced Shear Instabilities

S. Beji† and K. Nadaoka‡

† *Department of Naval Architecture and Ocean Engineering, Istanbul Technical University, Maslak 80626, Istanbul, Turkey*

‡ *Graduate School of Information Science and Engineering, Tokyo Institute of Technology, 2-12-1 O-okayama, Meguro-ku, Tokyo 152, Japan*

Introduction

In studying the instability of sensitive jets Rayleigh suggested an improvement in the theory by supposing a gradual change in velocity and thus proceeded to derive an equation which is known today as the Rayleigh equation. Miles (1957) proposed a model for the growth of wind waves on the basis of Rayleigh's equation. Later, Conte and Miles (1959) gave accurate computations of wave growth rates by numerical solution of Rayleigh's equation for a *logarithmic* wind profile. This work first considers a different approach of solving the Rayleigh equation for *arbitrary* mean wind profiles by implementing the ideas of Rayleigh, which render the equation analytically solvable in the immediate vicinity of the singular point hence providing the initial values for the numerical integration. Furthermore, the wave growth rate is obtained from the dispersion relation of the air-water interface, which involves the vertical integration of the disturbed vertical velocity. The growth rates obtained are then compared with those of Conte and Miles (1959) and found to be in excellent agreement. In closing, approximations to the vertical profile of the disturbed vertical velocity are suggested as a basis for the development of a vertically integrated model of coupled air-water system.

Rayleigh Equation and Its Approximate Solution

Rayleigh equation for waves moving with a constant celerity $c_0 = \omega/k$ in terms of the $z-$dependent part of the disturbed vertical air velocity $w(x,z,t) = W(z)\exp\imath(kx-\omega t)$ is

$$[U(z) - c_0\,](W'' - k^2W) - U''(z)W = 0 \tag{1}$$

where $U(z)$ is the mean wind velocity and the prime denotes differentiation with respect to z. The equation is obviously singular at the critical height $z = z_c$ where $U(z_c) = c$. In the vicinity of z_c the mean wind profile may be linearized to obtain an approximate equation of the form

$$W'' + \left[\frac{1}{\tilde{z}} - k^2\left[\frac{U'(z_c)}{U''(z_c)}\right]^2\right]W = 0 \tag{2}$$

where $\tilde{z} = -U''(z_c)(z-z_c)/U'(z_c)$. Near the critical point $\tilde{z}$ approaches zero hence equation (2) may be approximated as

$$W'' + \frac{1}{\tilde{z}}W = 0 \tag{3}$$

which is a Riccati equation. The two linearly independent solutions of equation (3) are given in terms of the Bessel functions of the order one

$$W(\tilde{z}) = \sqrt{\tilde{z}}\left[AJ_1(2\sqrt{\tilde{z}}) + BY_1(2\sqrt{\tilde{z}})\right] \tag{4}$$

where A and B are arbitrary constants. Initial values for the numerical integration of equation (1) are computed from the above linearly independent solutions and their derivatives.

Boundary conditions at the interface $z = z_0$ and infinity provide the necessary equations for determining the arbitrary constants of integration. Once the solution of $W(z)$ is obtained the growth rate can be computed from $\gamma = k\Im(c) = \frac{1}{2}skc_0\Im(I_c)$ where the complex integral I_c as obtained from the dispersion relation of the air-water interface is given by

$$I_c = \frac{(k/c_0)}{W(z_0)} \int_{z_0}^{+\infty} [U(z) - c_0]\, W(z)dz \tag{5}$$

The relation between the growth rate γ and the growth rate β as defined by Miles (1957) is $\gamma = \frac{1}{2}(U_1/c)^2 skc_0\beta$ with $s = \rho_a/\rho_w$ being the ratio of the air density to the water density.

Computed Growth Rates and Velocity Profiles

Computations using the above described approach are compared with the results given by Conte and Miles (1959) for the growth rate β in Table 1 with excellent agreement.

c_0/U_1	z_c	Present work	Conte-Miles	z_c	Present work	Conte-Miles
1	.00815	3.533	3.536	.0544	2.744	2.747
3	.00670	3.431	3.433	.0446	2.775	2.778
5	.01780	3.297	3.300	.1190	1.907	1.909
7	.06710	2.438	2.439	.4480	0.677	0.677

Table 1: Comparisons of the dimensionless growth rate β for two different groups of wind parameters $\Omega = gz_0/U_1^2 = 3 \times 10^{-3}$ (left) and $\Omega = 2 \times 10^{-2}$ (right).

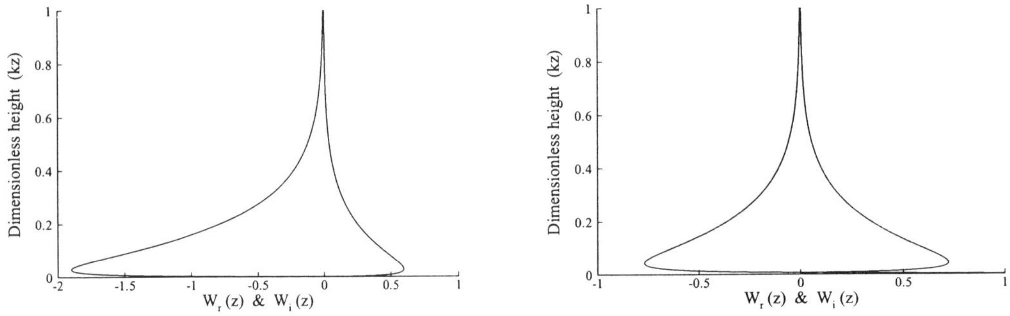

Figures 1a, b: Real and imaginary parts of the vertical air velocity $W(z)$ for $\Omega = 3 \times 10^{-3}$ (left) and $\Omega = 2 \times 10^{-2}$ (right). For both cases $c_0/U_1 = 1$.

To give an idea about the nature of the solution, real and imaginary parts of the vertical air velocity $W(z)$ are shown in Figures 1a and 1b. Continuation of the present work involves the development of a vertically integrated model for wave growth by employing an approximate functional representation of the vertical velocity profiles as obtained from the above analysis.

References

Conte SD, Miles JW, 1959. On the integration of Orr-Sommerfeld Equation. J. Soc. Indus. Appl. Math. 7, 361-369.

Miles JW, 1957. On the generation of surface waves by shear flows. J. Fluid M., 3, 185-204.

STUDY OF THE EFFECT OF HARBOR SHAPES ON WAVE INDUCED OSCILLATIONS

Asur Baris Derun[1] **and Masahiko Isobe**[2]

Dept. of Civil of Engineering, University of Tokyo, 7-3-1 Hongo, Bunkyo-ku,
Tokyo 113-8656 Fax: 03 5841 8503

Abstract

Isobe's (1994) nonlinear mild slope equations have been utilized by employing ADI method in the discretization to build up a numerical model to study the harbor oscillation problem by using regular waves over constant water depth. Line boundary method together with a sponge layer has been used as a wave maker condition to prevent the reflections at the wave generation point. At first, for a rectangular harbor (30*150cm) ("I" shaped) amplification factors have been found then by assembling new square basins (50*50cm) to the existing "I" shaped harbor, "L", "T", "F" and "Y" shaped harbors have been obtained and amplification factors have been calculated for each basin's end point. Coupling basins with dimensions of 25*50cm and 75*50cm have also been used to make up "Y" shaped harbors. Amplification factors have been determined by taking the square root of the ratio of peak frequency spectrum values for the wave height inside the harbor to the incident wave height outside the harbor, thus amplification factor curves can give information about the wave energy phenomenon. Comparison among the obtained amplification factor curves, revealed out that complicated harbors generate less oscillation than simpler models due to their efficiency in energy dissipation.

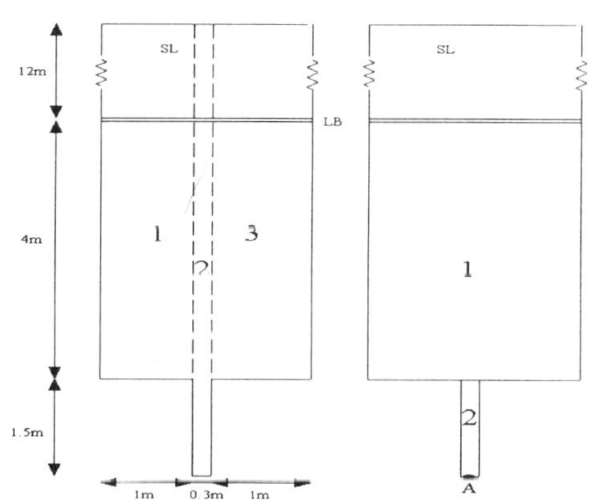

Fig.1. Definition sketch for I shaped harbor (not to scale).

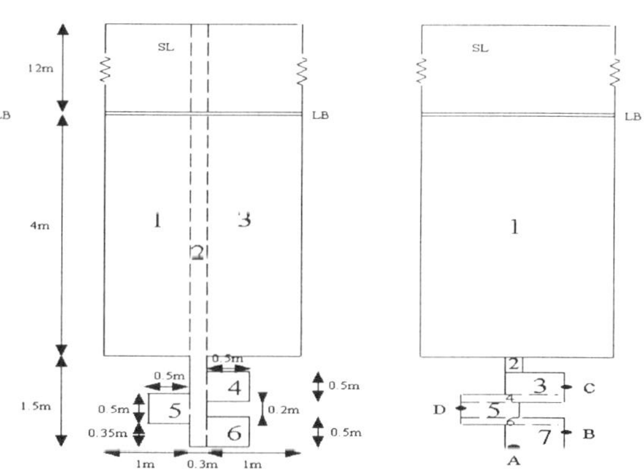

Fig.2. Definition sketch for Y shaped harbor (not to scale).

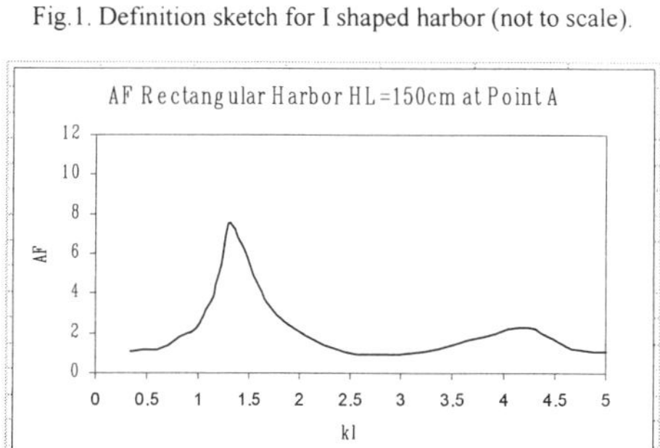

Fig.3. Amplification factors in I shaped harbor at point A

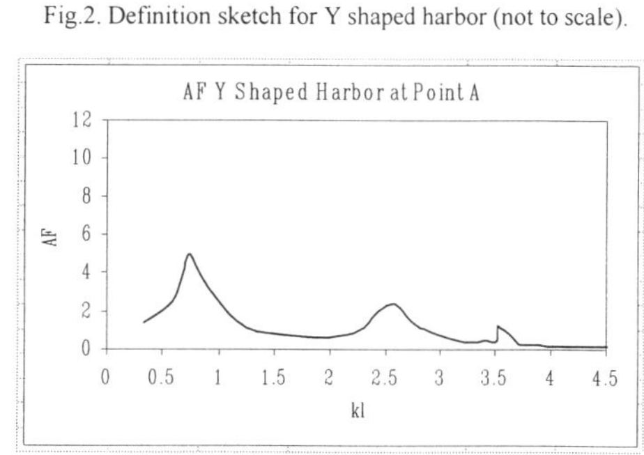

Fig.4. Amplification factors in Y shaped harbor at point A

---

[1] For correspondence: Doctor student, e-mail:derun@coastal.t.u-tokyo.ac.jp

[2] Professor, e-mail:isobe@k.u-tokyo.ac.jp

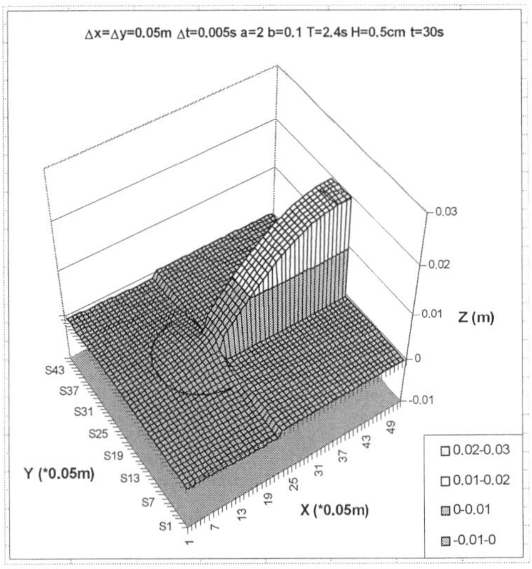

Fig.5. Water surface elevation for I shaped harbor in primary oscillation mode

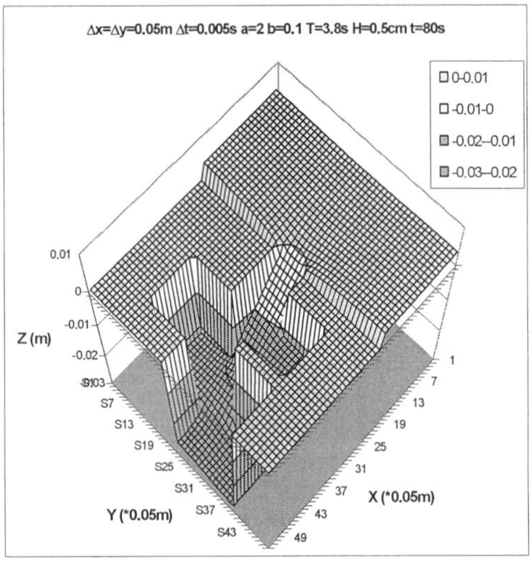

Fig.6. Water surface elevation for Y shaped harbor in primary oscillation mode

References

1. Goda, Y. (1985): *Random Seas and Design of Maritime Structures,* University of Tokyo Press, Tokyo, Japan, 323p.
2. Hwang, L. and Tuck, E. (1970): "On the Oscillations of the Harbors of Arbitrary Shape", J. Fluid Mechanics, V.42, 447-464.
3. Ishii, T., Isobe, M. and Watanabe, A. (1994): "Improved Boundary Conditions to a Time Dependent Mild Slope Equation for Random Waves", Proceedings of the 24th International Coastal Engineering Conference, ASCE, 272-284.
4. Isobe, M. (1994): "Time Dependent Mild slope Equations", Proceedings of the 24th International Coastal Engineering Conference, ASCE, 284-299.
5. Ippen A.T. (Editor) (1966): *Estuary and Coastline Hydrodynamics*, McGraw-Hill, New York, pp.281-340.
6. Larsen, J. (1978): "A Harbor Theory for Wind Generated Waves Based on Ray Methods", J. Fluid Mechanics, V.87, 143-158.
7. Larsen, J. and Dancy, H. (1983): "Open Boundaries in Short Wave Simulations", Coastal Engineering, Elsevier, Vol.7, 285-297.
8. Lee, J.J. (1971): "Wave Induced Oscillations in Harbors of Arbitrary Geometry", J. Fluid Mechanics, V.45, 375-394.
9. Mei, C.C. (1983): *The Applied Dynamics of Ocean Surface Waves*, John Wiley and Sons, New York, pp.183-252.
10. Miles, J. and Munk, W. (1961): "Harbor Paradox", Journal of the Waterways and Harbour Division, ASCE, WW 3, 111-130.
11. Raichlen, F. and Ippen A.T. (1965): "Wave Induced Oscillations in Harbors", Proc. of the ASCE, J. of Hydraulics Division, V.91, 1-26.
12. Unluata, U. and Mei, C.C. (1975): "Effects of Entrance Loss on Harbor Oscillations", J. of the Waterways, Harbors and Coastal Engineering Division, ASCE, WW 2, 161-180.

NUMERICAL MODELLING OF IRREGULAR WAVE PROPAGATION IN HARBOURS

Pan Junning[1], Zuo Qihua[2] and Wang Hongchuan[3]

A numerical irregular wave model based on the Extended Mild-Slope Equation is established to simulate wave propagation in harbours. The main wave phenomena occurred in harbours are involved in the model, such as refraction, diffraction, reflection, dissipation due to bottom friction and energy input due to wind. The time-dependent Extended Mild-Slope Equation can be written as

$$\frac{\partial^2}{\partial t^2}\Phi - \nabla\cdot\left(CC_g\nabla\Phi\right) + \left(\omega^2 - k^2CC_g - \mathrm{i}\omega F - gG + \mu\right)\Phi = 0 \qquad (1)$$

in which C and C_g are wave celerity and wave group velocity respectively, $\Phi(x, y, t)$ is wave potential, k is wave number, ω is wave angular frequency, g is gravity acceleration, F is energy exchange factor, G is the term of $\nabla^2 h$ and $(\nabla h)^2$, h is water depth and μ is the nonlinear term.

Irregular wave is simulated by superposing wave components in different directions and frequencies.

$$\zeta(x, y, t) = \sum_{n=1}^{N}\sum_{m=1}^{M} a_{mn}\cos\left[k_m\left(x\cos\theta_n + y\sin\theta_n\right) - \omega_m t + \varepsilon_{mn}\right] \qquad (2)$$

in which $\zeta(x, y, t)$ is wave surface level, a_{mn} and ε_{mn} are amplitude and initial phase of the component wave with frequency ω_m and direction θ_n respectively.

To solve the extended MSE for each wave component, an evolution equation is drived from Eq. (1)

$$-\frac{2\mathrm{i}\omega}{CC_g}\frac{\partial\phi_{mn}}{\partial t} = \nabla^2\phi_{mn} + \tilde{k}_c^2\phi_{mn} \qquad (3)$$

in which ϕ_{mn} is wave potential with frequency ω_m and direction θ_n, and $\tilde{k}_c^2 = k^2 - \frac{\nabla^2\sqrt{CC_g}}{\sqrt{CC_g}} + \frac{gG - \mu + i\sigma F}{CC_g}$. A modified ADI method is applied to solve equation (3) for each wave component. The finite difference scheme can be expressed as

$$f_{i,j}\frac{\phi_{i,j}^{l+1/2} - \phi_{i,j}^{l}}{\frac{1}{2}\Delta t} = \delta_x^2\phi_{i,j}^{l+1/2} + \frac{1}{2}\left(\tilde{k}_c^2\right)_{i,j}\phi_{i,j}^{l+1/2} + \delta_y^2\phi_{i,j}^{l} + \frac{1}{2}\left(\tilde{k}_c^2\right)_{i,j}\phi_{i,j}^{l} \qquad (4)$$

$$f_{i,j}\frac{\tilde{\phi}_{i,j}^{l+1}-\phi_{i,j}^{l+1/2}}{\frac{1}{2}\Delta t}=\delta_x^2\phi_{i,j}^{l+1/2}+\frac{1}{2}\left(\tilde{k}_c^2\right)_{i,j}\phi_{i,j}^{l+1/2}+\delta_y^2\tilde{\phi}_{i,j}^{l+1}+\frac{1}{2}\left(\tilde{k}_c^2\right)_{i,j}\tilde{\phi}_{i,j}^{l+1} \tag{5}$$

$$\phi_{i,j}^{l+1}=\lambda\tilde{\phi}_{i,j}^{l+1}+(1-\lambda)\phi_{i,j}^{l} \tag{6}$$

in which λ is relaxation factor, $f_{i,j}=-2i\omega/\left(CC_g\right)_{i,j}$, $\delta_x^2\phi_{i,j}^l=\dfrac{\phi_{i-1,j}^l-2\phi_{i,j}^l+\phi_{i+1,j}^l}{\Delta x^2}$, and $\delta_y^2\phi_{i,j}^l=\dfrac{\phi_{i,j-1}^l-2\phi_{i,j}^l+\phi_{i,j+1}^l}{\Delta y^2}$. The model is theoretically unconditionally stable and of high efficiency.

The accuracy of the present model is verified by two typical tests. First, the model is used to calculate irregular wave diffraction and reflection around a semi-infinite breakwater. The results given by the model coincide well with theoretical solution. Second, the model is applied to simulate wave refraction and diffraction on a bottom with an elliptic shoal. The numerical results agree with the experiment data given by Vincent and Briggs (1989).

As an application example, wave field in a harbour is calculated by the present model. The model gives sound wave height distribution in the harbour.

[1]Corresponding author
Engineer, Nanjing Hydraulic Research Institute, Nanjing 210024, P.R. China
(Ph: 86-25-5829340; Fax: 86-25-5829333)
Email: jnpan@njhri.edu.cn

[2]Professor, Nanjing Hydraulic Research Institute, Nanjing 210024, P.R. China
(Ph: 86-25-5828008; Fax: 86-25-5829333)
Email: qhzuo@njhri.edu.cn

[3]Senior Engineer, River and Harbor Department, Hydraulic Research Institute, Nanjing 210024, P.R. China
Email: hcwang@njhri.edu.cn

AN ANALYTICAL STUDY ON HEAVY SILTATION IN THE KEUM RIVER ESTUARY AFTER A DIKE CONSTRUCTION

By Jung L. Lee
Associate Professor, Department of Civil and Environmental Engineering,
Sungkyunkwan University, Suwon, 440-746, Korea
email: jllee6359@hanmail.net
Tel: 82-31-290-7519 Fax: 82-31-290-7549 (Korea-after Feb. 25)
Tel: 1-808-956-9655 Fax: 1-808-956-5014 (Univ. of Hawaii-until Feb. 25)
and
Il-Hyoung Cho
Assistant Professor, Cheju National University, Ara-dong, Jeju 690-756, Korea
email: cho0904@cheju.ac.kr
Tel: 82-64-754-3482

ABSTRACT:
The scattering of water waves for the various configurations of thin barriers has been studied extensively over the past century, despite the mathematical difficulties encountered within the framework of linearized potential theory. In this paper, a mathematical solution for the trapping waves on a circular basin surrounded by a thin vertical barrier with gaps is obtained by using wide-spacing approximate method. In the presence of the barrier, it is expected that part of the incident waves is trapped in the basin, and part scattered to infinity. The approximation is only applied in treating with boundary condition along the barrier: It is assumed that the ratio of basin radius to depth of barrier immersion is large enough so that local effects are negligible. The solution demonstrates the resonant interaction between the barrier and incident waves. Numerical results are also obtained using both an eigenfunction expansion method (Cho, 2002) and the finite difference method by a mild-slope equation as described in Lee and Lee (2001) and are compared with the approximate solution.

Reference

Lee, J. L., Lee, K. J. 2001 Effect of a surface-piercing vertical thin breakwater to harbour tranquility. *1st International Conference on Asian and Pacific Coastal Engineering.*

Cho, I-H, 2002 Wave energy absorption by a circular cylinder oscillating water column device. J. Korean Society of Coastal and Ocean Engrg. 14(1), 8-18.

Hydrodynamic Characters of the Coastal Area of Bohai Bay

Dekui Yuan, Jianhua Tao
Department of Mechanics, Tianjin University, Tianjin 300072, China
Tel: 0086-22-27890726 Fax: 0086-22-27401647 Email: dkyuan@tju.edu.cn

Bohai Bay locates in the west part of Bohai Sea, which is a semi-enclosed internal sea in the North-East China (see Fig.1). Bohai Bay is a shallow sea bay with wide mild slope mud beach and receives large amount of wastewater from Beijing, Tianjin and Hebei province. Due to the development of the economy along the BoHai Sea in recent decades, the water quality and ecological environment in Bohai Bay has been becoming very poor, especially in the coastal area. In this paper, the numerical models for simulating tidal flow, residual tidal current and advection-dispersion of pollutant are setup for Bohai Bay, and verified with the field measured and monitoring data. The models for simulating wave propagation and wave-current interaction are also developed and setup for Bohai Bay. Then the hydrodynamic characters of Bohai Bay are studied by using these numerical models, and the significative results are obtained by analyzing the numerical results.

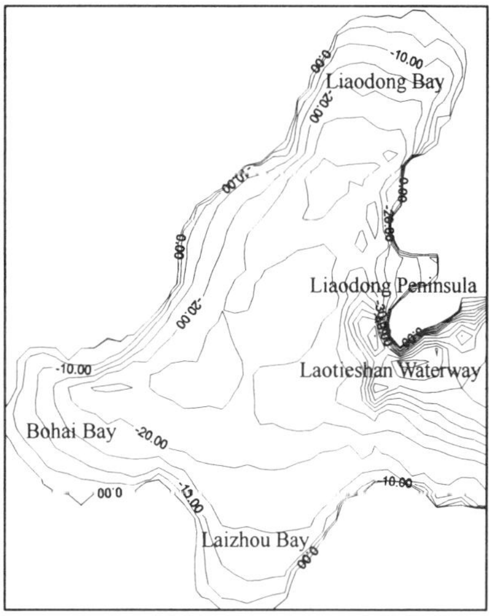

Fig.1 Topography and Location of Bohai Bay

The depth integrated 2-D tidal wave and advection-dispersion equations with moving boundary condition (Wang and Tao, 1998) are setup and applied to simulate the tidal flow and the movement of dissolved or suspended substances respectively. The residual current and the movement of particles in Bohai Bay are studied by using Eulerain and Lagrangian methods, respectively. Based on the study of the wave propagation model, wave breaking model, and the effect of wave on current in the near-shore area with a mild bottom slope, the long-shore current model has been set up (Sun T. and Tao J.H., 2002).

The field measured data (Yuan etc.1999) of tidal current and the concentration of COD in Bohai Bay have been used to verify the numerical models at first (see Fig.2). Then by using the numerical models mentioned above, the hydrodynamic characters of Bohai Bay are studied, and the conclusions are obtained as follows.

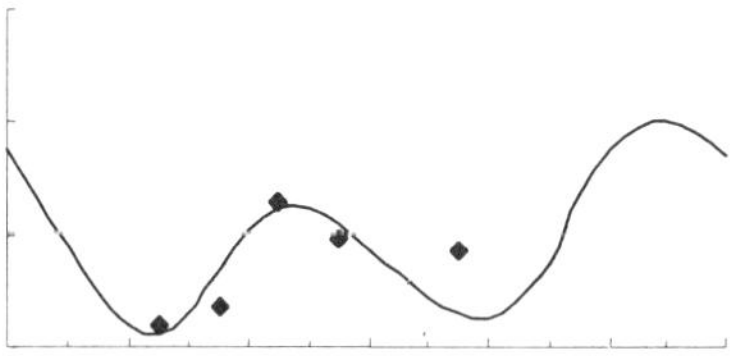

Fig.2 Concentration of COD at F1 station August 3rd 1993

Hydrodynamic characters of the coastal area of Bohai Bay :

1) The numerical results are in good agreement with field measured data (velocity and

concentration of COD) in Bohai Bay, so it is reliable to study the hydrodynamic characters of Bohai Bay by using the numerical models.

2) The movement of the water in Bohai Bay is studied by using Lagrangian method, and the numerical results show that the water exchange between Bohai Bay and outer water is poor.
3) The residual tidal current in Bohai Bay is calculated with Eulerian method, and the results show that the residual current and tidal circulation current in Bohai Bay are weak. This is disadvantageous for the transport of pollutants, and it is one of the main reasons for the accumulation of pollutants and coastal eutriphication.
4) The distribution of COD in Bohai Bay is studied, and the results show that the transport and concentration of the COD is determined mainly by the residual current and the discharge of outfalls.
5) The transport of the pollutant around the main outfalls in Bohai Bay is studied by taking water waves into account. It is found that acting on the wave-induced current, the pollutants will move along the beach due to long-shore current. This phenomenon will increase the pollution on the beach, even far from the outfalls (see Fig.3 and Fig.4).

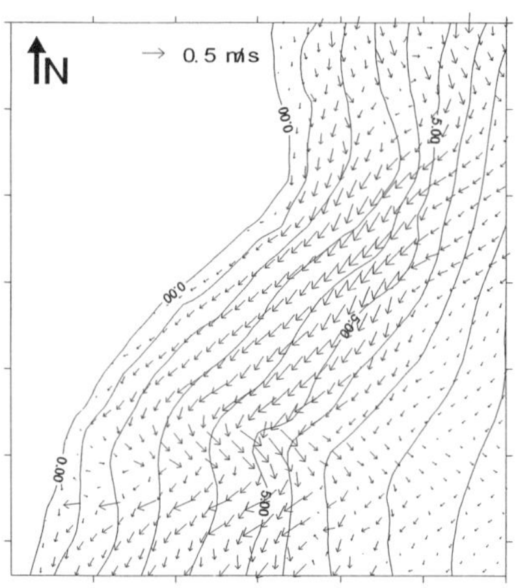

Fig.3 Velocity of long-shore current
Wave height: 2.9m; Period: 7.6s
Wave direction: East

Fig.4 Contour of polluted probability
Wave height: 2.9m; Period: 7.6s
Wave direction: East

References

1 China Ocean Environmental Book 2000. 2001. State Oceanic Administration of China.

2 Sun T, Han G and Tao J H. Numerical Study of Wave-induced Long-shore Currents and Experimental Verification. Journal of Hydraulic Engineering (in press)

3 Sun T, Han G and Tao J H. Pollutant Transport under Waves in the Near-shore Zone. Acta Oceanologica Sinica (in press)

4 Wang Z L and Tao J H. 1998. Numerical Study of Pollutant Transport in the Sea-bay with Mild Slope Beach. Acta Oceanologica Sinica. 17(3), 399-408.

5 Yuan D K, Lin B L, Tao J H and Falconer R A. 1999. Verification of a Numerical Model using Field Monitoring Data for Modelling Bohai Bay. Proceeding of the 28th IAHR Congress. 346.

Comparison of Time-Dependent Extended Mild-Slope Equations for Random Waves

Changhoon Lee[1], Gunwoo Kim[2], and Kyung-Duck Suh[3]

Recently, by including second order bottom variation effects, Suh et al. (1997) extended the time-dependent mild-slope equations of Smith and Sprinks (1975) and Radder and Dingemans (1985), and Lee et al. (2001) also extended Kubo et al.'s (1992) equation so that the extended equations can predict the water wave transformation over a rapidly varying topography. Numerical results of the models showed better accuracy of the extended models compared to the previous models especially for the cases of waves on a steep slope and the Bragg reflection of waves on a ripple patch. In this study, in view of random wave transformations, comparisons are made between the extended mild-slope equations of Suh et al. and Lee et al. by both analytical methods and numerical simulations.

The equations of Suh et al. are given by

$$\frac{\partial \eta}{\partial t} = -\nabla \cdot \left(\frac{\overline{CC_g}}{g} \nabla \tilde{\phi} \right) + \frac{\overline{\omega}^2 - \overline{k}^2 \overline{CC_g}}{g} \tilde{\phi} + \frac{\overline{\omega}^2}{g} \left\{ \overline{R}_1 (\nabla h)^2 + \overline{R}_2 \nabla^2 h \right\} \tilde{\phi} \tag{1}$$

$$\frac{\partial \tilde{\phi}}{\partial t} = -g\eta \tag{2}$$

and the equation of Lee et al. is given by

$$\begin{aligned} &\nabla \cdot \left(\overline{CC_g} \nabla \hat{\eta} \right) + \left\{ \overline{k}^2 \overline{CC_g} + g\overline{u}_1 \nabla^2 h + g\overline{u}_2 (\nabla h)^2 \right\} \hat{\eta} + i\nabla \cdot \left\{ \overline{\frac{\partial (CC_g)}{\partial \omega}} \nabla \frac{\partial \hat{\eta}}{\partial t} \right\} \\ &+ i \left\{ \overline{\frac{\partial (k^2 CC_g)}{\partial \omega}} + g \overline{\frac{\partial u_1}{\partial \omega}} \nabla^2 h + g \overline{\frac{\partial u_2}{\partial \omega}} (\nabla h)^2 \right\} \frac{\partial \hat{\eta}}{\partial t} = 0 \end{aligned} \tag{3}$$

The parameters $\overline{\omega}^2 \overline{R}_1$ and $\overline{\omega}^2 \overline{R}_2$ in Eq. (1) are mathematically equivalent to $-g\overline{u}_2$ and $-g\overline{u}_1$, respectively, in Eq. (3), all of which represent the second-order bottom effects, and the over bar indicates the variables associated with the carrier angular frequency $\overline{\omega}$.

By using the method of geometrical optics, the dispersion relation and the shoaling coefficient of each model can be obtained. For the waves with a carrier frequency, both equations satisfy the linear dispersion relation exactly. For waves with a local frequency different from the carrier one, however, the dispersion relations of the two models become different from the exact linear dispersion relation. This tendency of differences also occurs in the model's shoaling coefficient, which also depends on the model's dispersion

[1] Asst. Professor, Department of Civil and Environmental Engineering, Sejong University, Seoul 143-747, Korea, Tel: +82-2-3408-3294, Fax: +82-2-3408-3332, E-mail: clee@sejong.ac.kr

[2] Grad. Student, School of Civil, Urban & Geosystem Engineering, Seoul National University, Seoul 151-742, Korea, Tel: +82-2-880-8836, Fax: +82-2-887-0349, E-mail: maossi1@snu.ac.kr

[3] Professor, School of Civil, Urban & Geosystem Engineering, Seoul National University, Seoul 151-742, Korea, Tel: +82-2-880-8760, Fax: +82-2-887-0349, E-mail: kdsuh@snu.ac.kr

relation. In shallow water, the two models have wide frequency ranges of accurate dispersion relations. In deep water, the models of Suh et al. and Lee et al. have wide ranges of accuracy for the local frequency higher and lower, respectively, than the carrier one. Therefore, for Suh et al.'s model, the peak frequency, which is lower in whole frequency range, is chosen as a carrier frequency, and, for Lee et al.'s model, the weight-averaged frequency, which is higher than the peak frequency, is chosen as a carrier frequency.

Using the two models with optimally chosen carrier frequencies, we have simulated random waves that propagate over a slightly inclined bed from deep water ($kh = 2\pi$) to shallow water ($kh = 0.1\pi$). The resulting shoaling spectra are shown in Fig. 1. The result of Lee et al.'s model is closer to the exact shoaling spectrum than that of Suh et al.'s model. This is because the shoaling coefficient of Lee et al.'s model considers the random wave characteristics but the shoaling coefficient of Suh et al.'s model does not consider sufficiently.

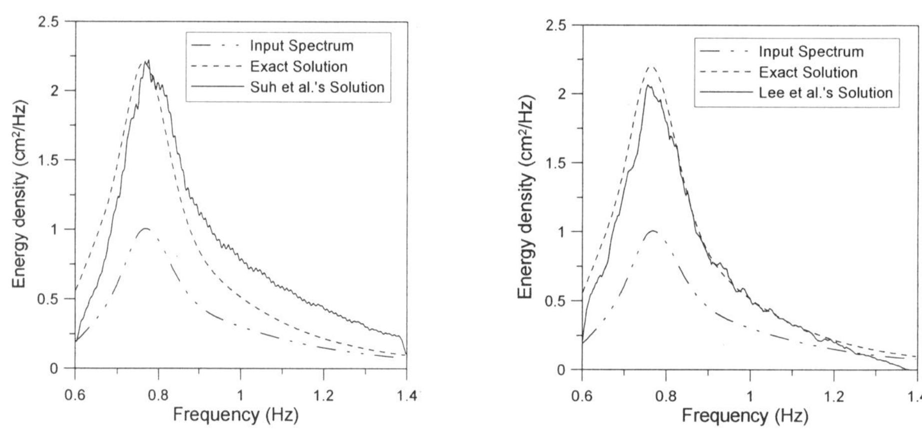

Fig. 1. Input and shoaling spectra of each model

References

Kubo, Y., Kotake, Y., Isobe, M. and Watanabe, A. (1992) Time-dependent mild slope equation for random waves, *Proc. 23rd Int. Conf. Coast. Engrg.*, ASCE, pp. 419-431.

Lee, C., Kim, G. and Suh, K.D. (2001) Extended mild-slope equation for random waves. *Proc. 4th Int. Symp. on Ocean Wave Measurement and Analysis*, ASCE, pp. 724-733.

Radder, A. C. and Dingemans, M. W. (1985) Canonical equations for almost periodic weakly nonlinear gravity waves, *Wave Motion*, Vol. 7, pp. 473-485.

Smith, R. and Sprinks, T. (1975) Scattering of surface waves by a conical island, J. *Fluid Mech.*, Vol. 72, pp. 373-384.

Suh, K. D., Lee, C. and Park, W. S. (1997) Time-dependent equations for wave propagation on rapidly varying topography, *Coast. Engrg.*, Vol. 32, pp. 91-117.

WAVE DAMPING AND A NEW DISPERSION EQUATION DUE TO POROUS SEABED EFFECT

Zhao-Chen SUN   Yong-Bo YUAN
Dalian University of Technology, Dalian, China

Liang CHENG and Mark RANDOLPH
The University of Western Australia, WA

Abstract In most existing wave propagation researches it is usually assumed that the seabed is impervious. But in shallow water area seepage phenomenon in porous seabed has a certain effect on wave propagation. It can cause wave damping. Wave seepage in porous seabed can also induce seabed deformation, sliding, and even liquefaction. When the elasticity of seabed is small and porosity is large, the interaction between waves and seabed is more significant.

In this paper, seabed response, porous water pressure and wave damping are given by analytically solving the governing equation of wave field—Laplacian Equation and The governing equations for seabed response — Biot equation simultaneously.

The interaction between waves and the porous seabed has a certain effect on wave parameters, and can lead to the deformation, sliding ,or even liquefaction of the porous seabed. Wave damping and strength reduction are the direct results of the interaction. From hydrodynamic point of view, when waves propagate over such a porous seabed, part of the wave energy will get into and dissipates inside the seabed, and hence causing the wave height reducing. In context of sediment transport, only horizontal water particle velocity is considered as the incipient factor in classical sediment transport studies. If permeable condition is considered on the interface of sea water and porous seabed, there will be a vertical velocity. Is it big enough to have any influence on the incipient velocity? From geotechnical point of view, the stability of seabed under wave action is a major concern. Big waves may cause instantaneous liquefaction to the seabed.

Researches have been done by Madsen (1978), Yamamoto (1978), Tsai (1995), Lin (1997), Reid (1957). In the existing studies of wave-seabed interaction, different assumptions are made, which include (1) seabed is impervious or rigid; (2) seabed consists of rigid porous medium, and the pore water is incompressible; (3) the seabed is impervious, but the skeleton is elastic; (4) the seabed consists of porous medium, the skeleton is elastic, and the pore water is compressible.

In the researches mentioned above, two demerits exist, which are (1) wave field and seepage field are solved as two separate problems, i.e., the wave field is solved under the assumption that the seabed is impervious. Error induced from this assumption is negligible in some cases. However, significant inaccuracy will arise in situations in that water depth is shallow, waves travel long distance, and the porosity of the seabed is large; (2) wave damping is neglected. Due to the porosity of the seabed, part of the wave energy will get into the porous seabed with wave traveling, which will lead to wave energy loss, and hence decrease of wave height.

The purpose of this paper is to present a method to study the wave damping due to the porous seabed effect. Coupling between wave field and the seepage field in porous seabed is considered. Analytical solution for the case of finite seabed thickness is presented.

Results show that wave height decrease is considerable in some cases, which is of significance to coastal and offshore engineering.

Estimation of Wave Fields with Open Boundaries by Applying an Adjoint Model

Tomonao Kobayashi
Associate Professor, Gifu University
1-1 Yanagito, Gifu-shi, Gifu 501-1193, JAPAN
phone: +-81-58-293-2463, fax: +-81-58-230-1891, e-mail: kobat@cc.gifu-u.ac.jp

Tadayuki Adachi
Applied Technology Co., Ltd.
1-2-23 Minamimorimachi, Kita-ku, Osaka 530-0054, JAPAN
phone: +-81-6-6354-5430, fax: +-81-6-6354-5433, e-mail tadayuki@sea.apptec.co.jp

Takashi Yasuda
Professor, Gifu University
1-1 Yanagito, Gifu-shi, Gifu 501-1193, JAPAN
phone: +-81-58-293-2431, fax: +-81-58-230-1891, e-mail: coyasuda@cc.gifu-u.ac.jp

1. Introduction

Recently, third generation models, *e.g.* WAM (WAMDI Group, 1988) or SWAN (Booij *et al.*, 1996), are used to evaluate wave fields. These models take nonlinear wave-wave interaction into account explicitly, and therefore the wave field evaluation is highly accurate. The boundary conditions for these models are usually open boundaries, which are difficult to evaluate with the existent models. Since the boundary conditions affect the wave field evaluation, they should be accurately computed. In this model we include the data assimilation model in the wave model, and assimilate field data in the wave field evaluations with open boundaries. Because this model computes the wave fields and evaluates open boundary conditions at the same time, an improved wave fields evaluation is obtained.

2. Numerical model

WAM Cycle 4 (Günther *et al.*, 1992) is employed, and an adjoint method is applied for the data assimilation to the wave model. The adjoint method can employ not only the field data observed in offshore but the data at the end of the wave propagation, and modifies the computed wave field with the field data by tracing waves backward.

The fundamental equation of WAM is as follows:

$$\frac{DF}{Dt} \equiv \frac{\partial F}{\partial t} + \dot{\boldsymbol{x}} \cdot \frac{\partial F}{\partial \boldsymbol{x}} + \dot{\boldsymbol{k}} \cdot \frac{\partial F}{\partial \boldsymbol{k}} = S_{in} + S_{nl} + S_{ds} \tag{1}$$

where, F: wave energy spectrum, $\boldsymbol{x}$: position vector, t: time, $\boldsymbol{k}$: wave number vector. Source functions S_{in}, S_{nl}, S_{ds} are energy input from wind, nonlinear wave-wave interaction and energy dissipation due to whitecapping, respectively.

We define the cost function H as the difference between the wave spectra F computed by the numerical model and the observed one F_{obs}.

$$H = \sum (F - F_{obs})^2 \tag{2}$$

The adjoint equation for WAM is derived from Eqs.(1) and (2), and applying the adjoint method.

$$-\frac{\partial \delta' F}{\partial t} + \dot{\boldsymbol{x}} \cdot \frac{\partial \delta' F}{\partial \boldsymbol{x}} + \dot{\boldsymbol{k}} \cdot \frac{\partial \delta' F}{\partial \boldsymbol{k}} = \frac{\partial}{\partial F}(S_{in} + S_{nl} + S_{ds})\,\delta' F + 2\sum (F - F_{obs}) \tag{3}$$

where $\delta' F$ is the modification vector for the target variable F.

In this study, the target variables F are the wave energy spectra at the computational grid points on the open boundaries. In the adjoint method, the target variables F are modified iteratively with the vector $\delta' F$ in Eq.(3) until the cost function H becomes minimize.

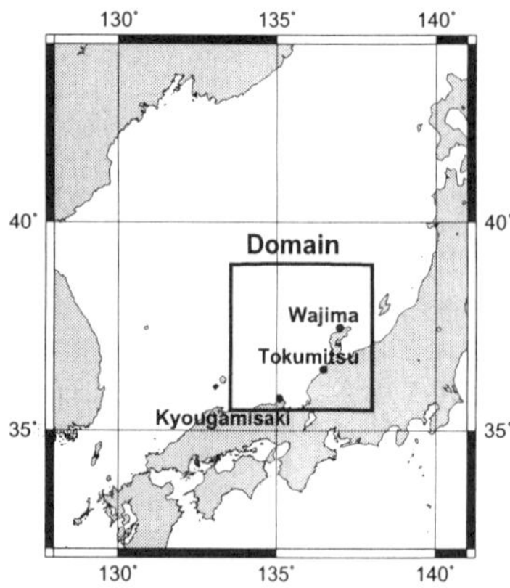

Fig.1: Computational Domain

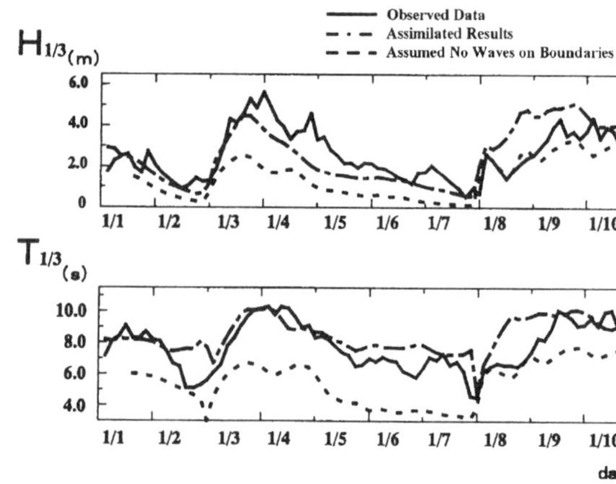

Fig.2: Observed data and Computational Results

3. Numerical results

We apply the adjoint wave model to the west coast of Japan. Fig.1 shows the Sea of Japan and the computational domain in this study. The computational domain is about 400 km × 400 km, and it has open boundaries, where the wave spectra are evaluated with the adjoint model. The wave observation sites at Kyougamisaki Tokumitsu and Wajima are shown in the figure. We compute the wave field in the domain from January 1st 1996 to 10th. In this period, strong seasonal winds blowed from North-West in the Sea of Japan.

Figure 2 shows the significant wave height $H_{1/3}$ and period $T_{1/3}$ observed and computed at Tokumitsu. The solid lines in the figure indicate the observed data. The computed results, indicated with dotted lines, correspond to wave spectra at the points on the open boundaries equal to zero.

The dot-dashed lines in the figure indicate the results computed with the adjoint model, which evaluates the wave spectra at the computational grid points on the open boundaries by assimilating the observed data at Kyougamisaki and Wajima, and computes waves at Tokumitsu. The significant wave height $H_{1/3}$ and period $T_{1/3}$ in the figure shows that the evaluation of the wave height is improved with the appropriate open boundary conditions given by the adjoint model.

4. Conclusions

The proposed data assimilation model is based on the application of an adjoint method to the third generation wave model, WAM. The model is applied to the wave field with open boundaries. Wave fields are accurately computed by simultaneously evaluating the open boundaries. The validity of the adjoint model was examined under several computational conditions.

References

Booij,N.R., L.H.Holthuijsen and R.C.Ris (1996): The SWAN wave model for shallow water, Int. Conf. Coastal Engrg., ASCE, Orland, pp668-676.

Günther,H. *et al.* (1992): The Wamodel Cycle 4 (revised version), Deutsches Klima Rechen Zentrum, Technical Report No.4, 101p.

WAMDI Group (1988): The WAM model — a third generation ocean wave prediction model, Journal of Physical Oceanography, No.18, pp.1775-1810.

Boundary-Fitted Nonlinear Dispersive Wave Model for Applications in Geometrically Complicated Regions

S. Beji[†], B.Barlas[†] and K. Nadaoka[‡]

[†] *Department of Naval Architecture and Ocean Engineering, Istanbul Technical University, Maslak 80626, Istanbul, Turkey*

[‡] *Graduate School of Information Science and Engineering, Tokyo Institute of Technology, 2-12-1 O-okayama, Meguro-ku, Tokyo 152, Japan*

Introduction

Practical applications in coastal engineering usually demand the solutions of governing equations in geometrically complicated areas. Therefore, a circulation or a wave model expressed in boundary-fitted coordinates is much preferable. Earlier attempts were directed to transformation of the governing equations alone, keeping the cartesian velocity components unchanged as dependent variables. Later, it was recognized that transforming both the governing equations and the velocity components would be much better for accuracy and numerical stability (see Muin and Spaulding, 1996). In this work, one-component form of the nonlinear dispersive wave equations of Nadaoka *et al.* (1997) along with the cartesian velocities are transformed to a curvilinear coordinate system, which is not necessarily orthogonal. Thus, the resulting equations are expressed in terms of the contravariant velocity components, which render the specification of the lateral boundary conditions as simple as possible.

Wave Equations in Boundary-Fitted Coordinates

One-component forms of the wave model of Nadaoka *et al.* (1997) in cartesian coordinates can be found in that particular work and are not repeated here. Using the well-known relations (see for example Hoffman and Chiang, 1995) and combining appropriate equations, the transformed forms of these equations in curvilinear coordinates (ξ, η, τ) become

$$\zeta_\tau + J\left[\left(C_p^2/g+\zeta\right)U^*\right]_\xi + J\left[\left(C_p^2/g+\zeta\right)V^*\right]_\eta = 0, \tag{1}$$

$$\begin{aligned} rJU_\tau^* &+ \xi_*^2 Q_\xi + \xi_*\eta_* Q_\eta \\ &= \frac{\xi_*^2}{\omega^2 C_p^2}\left[C_*\left(U_\xi^* + V_\eta^*\right)\right]_{\xi\tau} + \frac{\xi_*\eta_*}{\omega^2 C_p^2}\left[C_*\left(U_\xi^* + V_\eta^*\right)\right]_{\eta\tau}, \end{aligned} \tag{2}$$

$$\begin{aligned} rJV_\tau^* &+ \xi_*\eta_* Q_\xi + \eta_*^2 Q_\eta \\ &= \frac{\xi_*\eta_*}{\omega^2 C_p^2}\left[C_*\left(U_\xi^* + V_\eta^*\right)\right]_{\xi\tau} + \frac{\eta_*^2}{\omega^2 C_p^2}\left[C_*\left(U_\xi^* + V_\eta^*\right)\right]_{\eta\tau}, \end{aligned} \tag{3}$$

with

$$\begin{aligned} Q = g\zeta &+ \frac{1}{2}\left(1 - 3\omega^2 C_p^2/g^2\right)\left[\eta_*^2 U^{*2} + \xi_*^2 V^{*2} - 2\xi_*\eta_* U^* V^*\right] \\ \xi_*^2 &= \xi_x^2 + \xi_y^2, \qquad \eta_*^2 = \eta_x^2 + \eta_y^2, \\ \xi_*\eta_* &= \xi_x\eta_x + \xi_y\eta_y, \qquad C_* = C_p^4(1-r)J. \end{aligned} \tag{4}$$

where ζ is free surface elevation, $U^* = U/J$ and $V^* = V/J$ with U and V being the contravariant components of the horizontal velocity vector at $z = 0$. J is the Jacobian and $\xi_x, \xi_y, \eta_x, \eta_y$ are the metrics of the transformation. $r = C_g/C_p$ with C_p and C_g denoting respectively the phase and group velocities computed according to linear theory for a specified wave frequency ω and a given local depth h.

Sample Computations

Computations involving non-rectangular lateral boundaries are presented to demonstrate the accuracy and versatility of the equations derived. First, linear long wave propagation along a converging channel is simulated and the magnification of the wave amplitude is compared with the theoretical solution of Green. As it is seen from the figures the computed amplitude variations for both cases agree almost perfectly with the theoretical solution depicted by the solid lines.

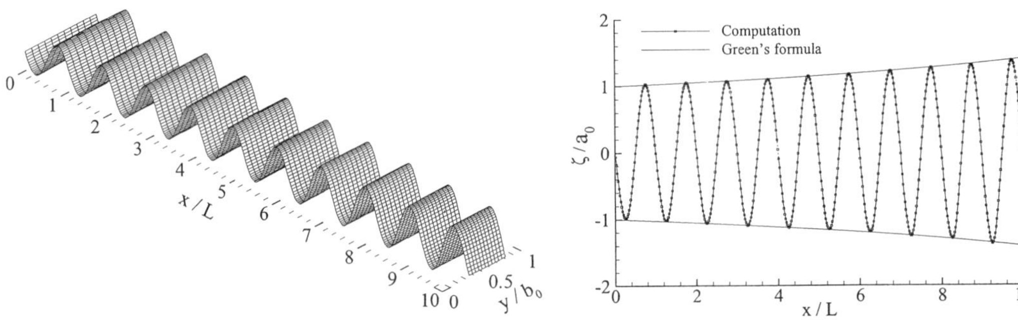

Figures 1a, b: Linear long wave propagation along a converging channel: perspective view and amplitude magnification compared with the theoretical formula of Green.

The second simulation is a hypothetical case showing nonlinear wave propagation across a region enclosed by bending lateral walls which cause diffraction (left wall, bending out) and reflection (right wall, bending in) of waves. Figures 2a and 2b, depicted at two different instances, show that qualitative aspects of the wave deformations are well simulated.

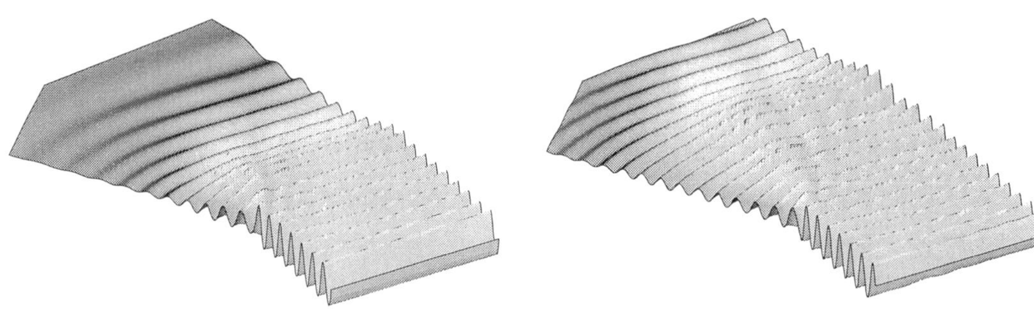

Figures 2a, b: Nonlinear wave propagation across a region of non-uniform cross section.

References

Hoffmann KA, Chiang ST, 1995. *Computational Fluid Dynamics for Engineers* Volume II, 3rd print. Wichita, Kansas: Engineering Education System; p.18-19; 39.

Muin M, Spaulding M, 1996. Two-dimensional boundary-fitted circulation model in spherical coordinates. J. Hydr Eng; Sept.:512-21.

Nadaoka K, Beji S, Nakagawa Y, 1997. A fully dispersive weakly nonlinear model for water waves. Proc R. Soc London A; 453:303-18.

TRUNCATION ERROR OF NUMERICAL SIMULATION OF LINEAR DISPERSIVE WAVE THEORY

Koji FUJIMA[1] and Chiaki GOTO

[1] Department of Civil and Environmental Engineering, National Defense Academy, 1-10-20 Hashirimizu, Yokosuka, 239-8686 Japan
E-mail: fujima@cc.nda.ac.jp

1 Introduction

The dispersion effect plays an important role in wave propagation and transformation. Thus, many studies have been conducted on the characteristics of wave dispersion, and some numerical models are proposed. In the present paper, the truncation error of numerical simulation using linear dispersive wave theory is discussed on a basis of exact solutions of the difference equations.

2 Solutions of the difference equations

The following linear dispersive wave equations are adopted for the governing equations:

$$\frac{\partial \eta}{\partial t} + \frac{\partial M}{\partial x} + \frac{\partial N}{\partial y} = 0 \tag{1}$$

$$\frac{\partial M}{\partial t} + c_0^2 \frac{\partial \eta}{\partial x} = \frac{\partial}{\partial x}\left[\alpha_x h^2 \frac{\partial^2 M}{\partial t \partial x} + \beta_x h^2 \frac{\partial^2 N}{\partial t \partial y} + \gamma_x c_0^2 h^2 \frac{\partial^2 \eta}{\partial x^2} + \delta_x c_0^2 h^2 \frac{\partial^2 \eta}{\partial y^2}\right] \tag{2}$$

$$\frac{\partial N}{\partial t} + c_0^2 \frac{\partial \eta}{\partial y} = \frac{\partial}{\partial y}\left[\alpha_y h^2 \frac{\partial^2 N}{\partial t \partial y} + \beta_y h^2 \frac{\partial^2 M}{\partial t \partial x} + \gamma_y c_0^2 h^2 \frac{\partial^2 \eta}{\partial y^2} + \delta_y c_0^2 h^2 \frac{\partial^2 \eta}{\partial x^2}\right] \tag{3}$$

where $c_0 = \sqrt{gh}$, h = water depth. In the case of $\alpha_x = \alpha_y = \beta_x = \beta_y = 1/3$ and $\gamma_x = \gamma_y = \delta_x = \delta_y = 0$, the above equations coincide with Boussinesq equations. The coefficients of $\alpha_x = \alpha_y = \beta_x = \beta_y = 2/5$ and $\gamma_x = \gamma_y = \delta_x = \delta_y = 1/15$ lead to Madsen and Sorensen equations. The model proposed by Sayama et al. (1987) is expressed as $\alpha_x = \frac{1}{3} - \frac{1}{12}\frac{1-K_x^2}{H_x^2}$, $\alpha_y = \frac{1}{3} - \frac{1}{12}\frac{1-K_y^2}{H_y^2}$, $\beta_x = \frac{1}{3} + \frac{1}{12}\frac{K_x^2}{H_x^2}$, $\beta_y = \frac{1}{3} + \frac{1}{12}\frac{K_y^2}{H_y^2}$ and $\gamma_x = \gamma_y = \delta_x = \delta_y = 0$, and the explicit model proposed by Cho and Yoon (1998) is $\alpha_x = \alpha_y = \beta_x = \beta_y = \gamma_x = \gamma_y = 0$ and $\delta_x = -\frac{1}{12H_y^2}$, $\delta_y = -\frac{1}{12H_x^2}$, where $K_x = c_0\Delta t/\Delta x$, $K_y = c_0\Delta t/\Delta y$, $H_x = h/\Delta x$, $H_y = h/\Delta y$.

On the other hand, the following three schemes are considered for the difference model.

1. The Leap-Frog method. This scheme is called '2/3 implicit scheme' in this paper. In this method, the momentum equations (Eqs.(2) and (3)) are solved implicitly, and the continuity equation (Eq.(1)) is solved explicitly. M and N are obtained simultaneously. This method was used to simulate a transoceanic propagation of tsunami by Goto et al. (1988).
2. The Crank-Nicholson method (3/3 implicit scheme). All equations (Eqs.(1),(2) and (3)) are solved implicitly. η, M and N are obtained simultaneously. Although the computation cost is huge, the computation is always stable.

3. The two-step mixed finite difference method (1/3 implicit scheme). In this method, the momentum equations are solved by a simple implicit procedure separately. Thus, this method can save the computation cost. At present, this method is mainly adopted to simulate a one dimensional propagation of nonlinear dispersive waves by Iwase et al. (2002).

The exact solutions of the difference equations are obtained from the governing equations (Eqs.(1),(2) and (3)) by the above difference models.

3 Characteristics of truncation error

The characteristics of the truncation error are examined on the magnitude of numerical error and the direction-dependency of numerical error.

The truncation error of the Crank-Nicholson method, 3/3 implicit scheme, does not depend on the wave direction, but the magnitude of the numerical error is relatively large. The truncation error of the Leap-Frog method, 2/3 implicit scheme, is small but depend on the wave direction. The truncation error of the two-step mixed finite difference method, 1/3 implicit scheme, is similar as the Leap-Frog method on the magnitude, and the direction-dependency is slightly smaller than the Leap-Frog method. Thus, the two-step mixed finite difference method appears to be most practical. However, by adopting a small grid, the simulation results of Leap-Frog method and Crank-Nicholson method become accurate. On the other hand, very small grid cannot be adopted in the two dimensional computation of the two-step mixed finite difference method because of the stability condition.

It is concluded that the two-step mixed finite difference method is applicable for the practical simulations of transoceanic propagation of tsunami. However, when very small grid is required for the simulation, the Leap-Frog method should be adopted for the numerical procedure.

In Cho and Yoon's scheme, the summation of the numerical dispersion and the added dispersion terms coincides with the physical dispersion effect when the grid size and the time increment are selected adequately. Cho and Yoon's method is quite accurate when the grid size is chosen adequately. However, if grid size cannot be chosen adequately, the truncation error becomes large. Besides, the characteristics of the truncation error for the wave propagating to x-direction differ from that for the wave propagating to y-direction remarkably.

Thus, when the proper grid size and the time increment can be selected in the simulation, Cho and Yoon's method is very useful. However, when the proper grid is not selectable by some reason, a linear dispersive wave theory should be adopted in the simulation.

Note that the wave celerity for high-wavenumber component predicted by a numerical simulation is generally smaller than the theoretical wave celerity. In addition, the theoretical wave celerity of Boussinesq equation is smaller than that of Airy theory (small amplitude wave theory), and that of Madsen and Sorensen equation is larger than that of Airy theory. Thus, if the simulated wave celerity is compared with the wave celerity of Airy theory, Madsen and Sorensen equation may provide accurate results in comparison to Boussinesq equation.

TROPICAL CYCLONES ASSOCIATED CHANGES ALONG ORISSA COAST, EAST COAST OF INDIA

Pratap K. Mohanty[1], Uma S. Panda[1], Pravakar Mishra[2], Hideshige Takada[3] and Takashige Sugimoto[2]

1. Department of Marine Sciences, Berhampur University, Berhampur- 760 007, Orissa, India. Phone no: +91 680 242332 , Fax: 0680 243322, email: pratap_mohanty@yahoo.com
2. Ocean Research Institute, the University of Tokyo, 1-15-1, Minamidai, Nakno Ku,Tokyo 164-8639, Japan, Phone no:81-3-5351-6505, Fax: 6506, email: babuli@ori.u-tokyo.ac.jp / sugimoto@ori.u-tokyo.ac.jp
3. Department of Environmental sciences and Natural Resources, Tokyo University of Agriculture and Technology, Fuchu, Tokyo 183-8509, Japan, Phone no: 81-42-367-5825, Fax: 367-8264, email: shige@cc.tuat.ac.jp

Orissa is a maritime state along the east coast of India and has a coastline of 480km in length. The coastline is bestowed with six major estuaries besides many small rivers which debouch into the Bay of Bengal (Figure 1). During the last 109 years (1804-2000), Orissa coast experienced 128 cyclones of which six were most notable *(29 October, 1999; 23-27 May, 1989; 13-17 October, 1985; 1-5 June, 1982; 29-30 October, 1971 and 9-11 October, 1967*) and caused unprecedented loss of life and property in the state of Orissa. The frequency of tropical cyclones along the six coastal districts of Orissa (Figure 2) is very high and is responsible for shoreline changes and vegetation loss in addition to its other damaging effects. Topography of the coastline, strong littoral drift and longshore current also contribute significantly to the shoreline changes. The changes in the areas of deposition and erosion near Mahanadi delta region were estimated for the periods 1972 to 1986 and 1986 to 1993. The areas of deposition and erosion were respectively 16.52 Km^2 and 13.39 Km^2 during the former period while the depositional and erosional trend changed to 11.61 Km^2 and 6.44 Km^2 during the latter period.

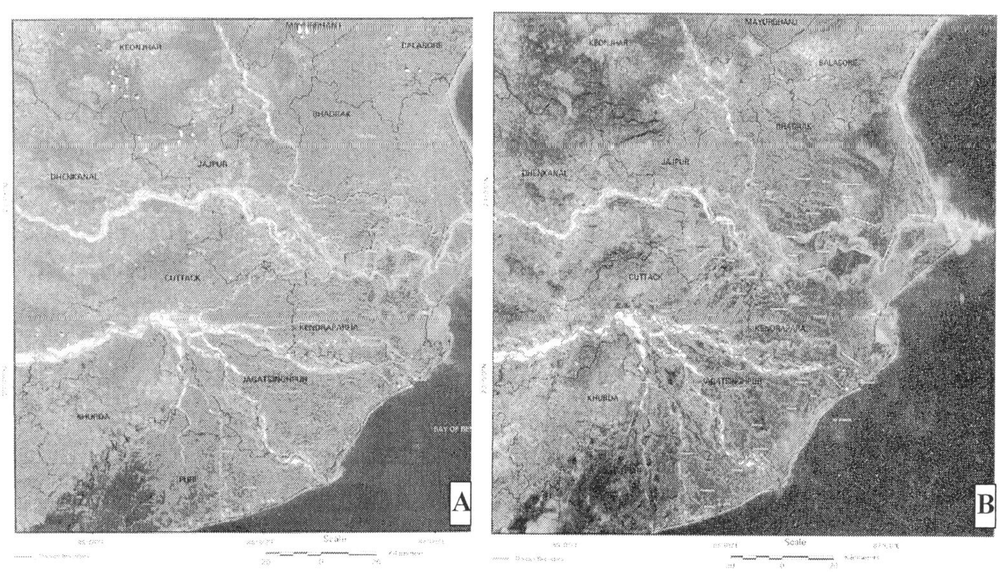

Figure 3: A) Pre Cyclone (October 11, 1999) and B)Post Cyclone (November 8, 1999) images of IRS01D WiFS showing Cyclone affected areas.

Further, it has been observed that the shoreline changes along the Orissa coasts are highly location specific and are dependent on local environmental conditions. For example; the October 29, 1999 Super Cyclone of Orissa having landfall at Paradeep with estimated pressure drop of 80 mb and maximum sustained surface wind of 140 knots was the most devastating of the century. Besides its other damaging effects, the super cyclone caused serious devastation to the natural vegetation of the coastal districts of Orissa. The quantitative estimate from the remote sensing based study using Indian Remote Sensing Satellite (IRS-ID) wide field sensor data (WiFs) in the pre and post cyclone periods (11.10.1999 and 14.11.1999) indicate increase in the areas of water body (1235.44 Km^2) and Fallow land (2602.4 Km^2) and decrease in the areas of vegetation (3394.69 Km^2) and dense vegetation (443.15 Km^2). Pre and post cyclone images of IRS-1D WiFs (Figure 3) indicate the inundated areas. Therefore, objective of the present study is to delineate the areas affected due to some of the notable storm events and to make a quantitative estimate of the areas of vegetation loss due to the cyclones.

Figure 1: Coastline of Orissa with major

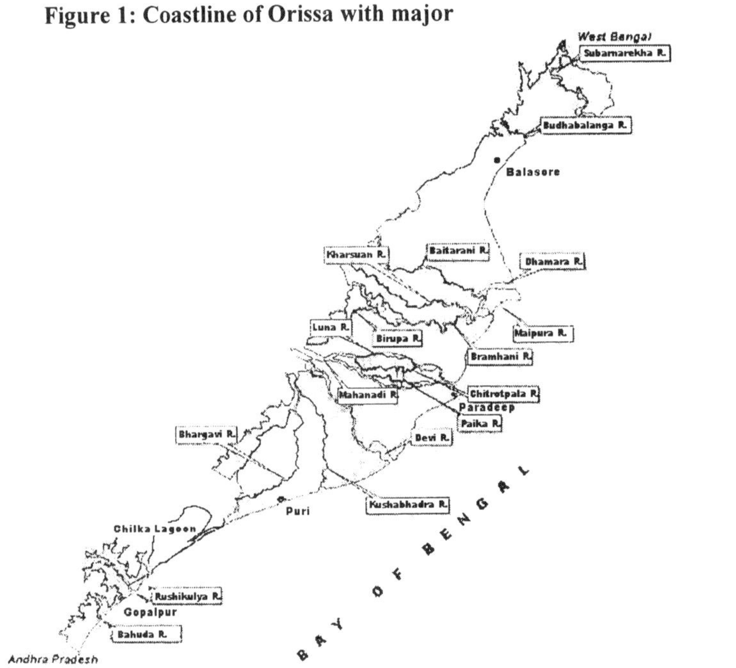

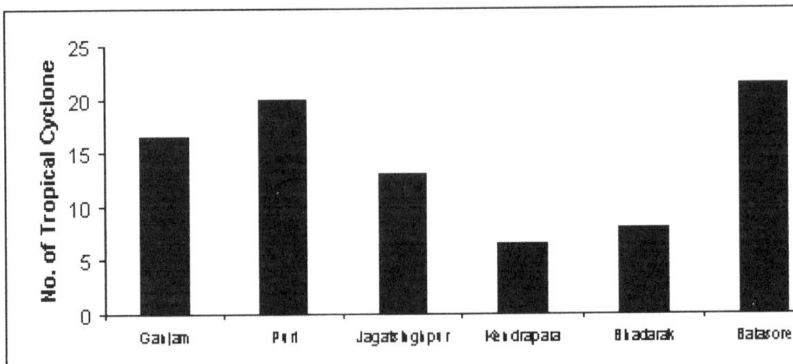

Figure 2: Frequency of Cyclones along the coastal districts of Orissa (1877-2000)

FIELD OBSERVATIONS OF BEACH EVOLUTION IN LAKE INAWASIRO, JAPAN

Naoya Okajima[1*], Hitoshi Tanaka[1] and Yutaka Fujita[2]

*Corresponding author
[1] Department of Civil Engineering, Tohoku University
06 Aoba, Sendai 980-8579, Japan.
Email: tanaka@tsunami2.civil.tohoku.ac.jp
Phone & Fax: (81)-22-217-7541
[2] Department of Civil Engineering, Nihon University
1 Nakagawara, Tokusada, Tamura, Koriyama 963-8642, Japan.

INTRODUCTION

Several studies have been conducted on shoreline change in a lake. These studies described interesting characteristics, and are effective for understanding basic processes of sand movement, because the external forces acting on lake shoreline is relatively simplified compared with those on coastal shores. In addition, these studies are useful for investigating similarity law of coastal sediment, because the scale of topographical change and external forces are smaller in a lake than those encountered along seashore, but much larger than laboratory flume. In this study, the process of longshore sediment transport around the mouth of Nagase River and the Shidahama Beach, which is located to northeast of the Lake Inawashiro, is investigated based on aerial photographs and field measurement data.

SHORELINE CHANGE AROUND THE NAGASE RIVER MOUTH

Figures 1 and 2 show the change in shorelines around the mouth of Nagase River from aerial photographs. Figure 1 shows that sediment, supplied from Nagase River be transported in the eastern direction. According to meteorological measurements, wind from WNW direction is predominant in this area. For this reason, sediment deposited at the river mouth has been transported to the east due to the high waves during winter. Furthermore, it is noted that the waves attack the shore with a large incident angle. In this regard, wave conditions are most unlike those found along coastal shores in general. Due to these wave conditions, remarkably active shoreline changes are occurring in the eastern area of the river mouth.

Figure 2 shows that sediment moved further to the eastern direction, and formed a large scale sand spit during these 50 years. In recent years, it is inferred that sediment forming the sand spit is further moving toward the Shidahama Beach.

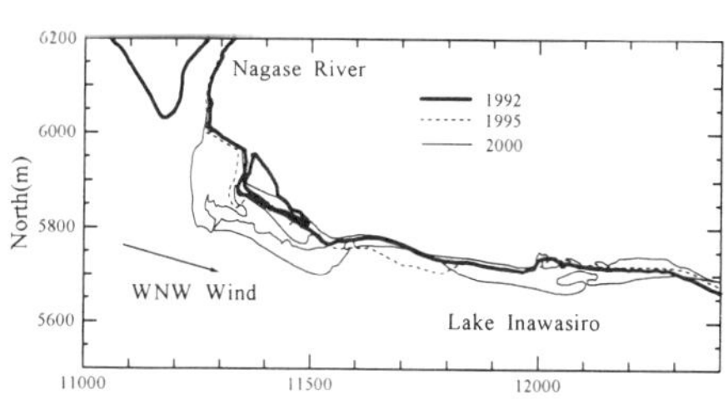

Fig.1 Shoreline change from 1992 to 2000.

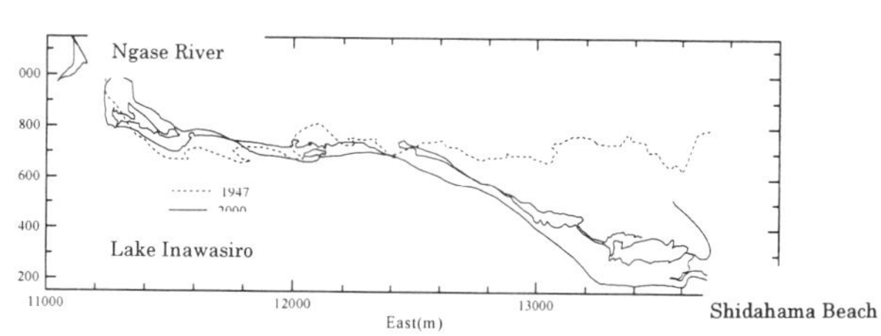

Fig.2 Shoreline change 1947 and 2000.

CHANGE IN CROSS-SHORE PROFILE ON SHIDAHAMA BEACH

Field measurement of beach profile has been carried out in Shidahama area to understand the process of longshore sediment transport. Figure 3 shows the location of each measurement section. Recently, sand spit has attached the Shidahama Beach between section 5 and section 6. Figure 4 shows longitudinal profiles along Sections 2, 4, 5 and 6. Erosion of the sand spit can be seen along Section 6, whereas accumulation is remarkable along Section 5 even within one month. This result indicates that sediment, which has formed the sand spit, further moved to the southeastern direction, and accumulated on Shidahama Beach. Along Sections 2 and 4, the change in the cross-shore profile is negligibly small, indicating equilibrium against the incident waves. Thus, it can be concluded that sediment supplied from Nagase River has reached Shidahama Beach, and is still gradually moving toward the south. At the southern end of Shidahama Beach, there is an intake of Asaka Channel, which has encountered significant sediment deposit in the past. It is supposed that severe sediment intrusion will occur the inlet in the very near future due to sediment transport to the south, described above.

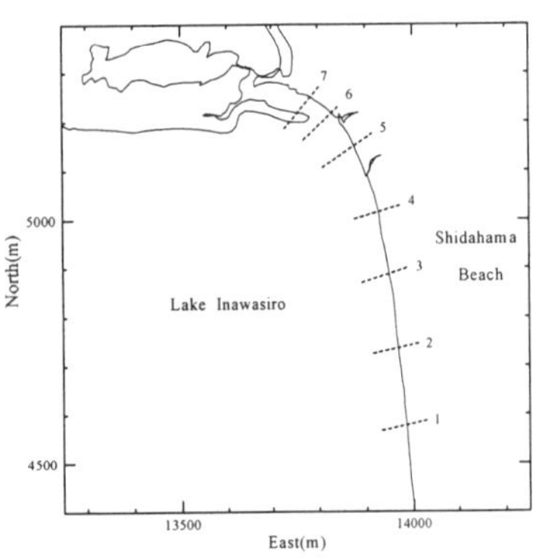

Fig.3 Field measurement sections

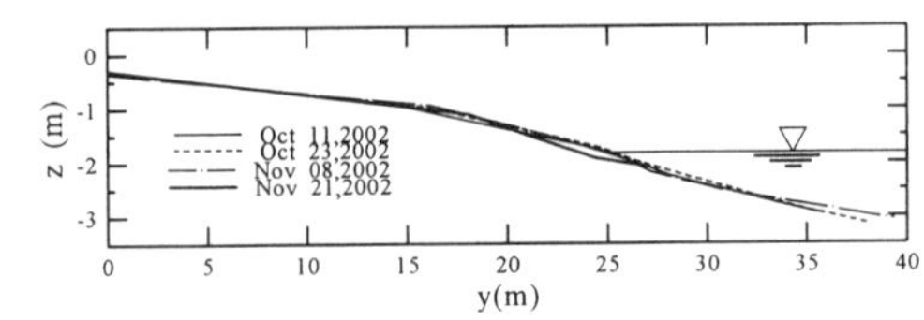

(a) Section 2

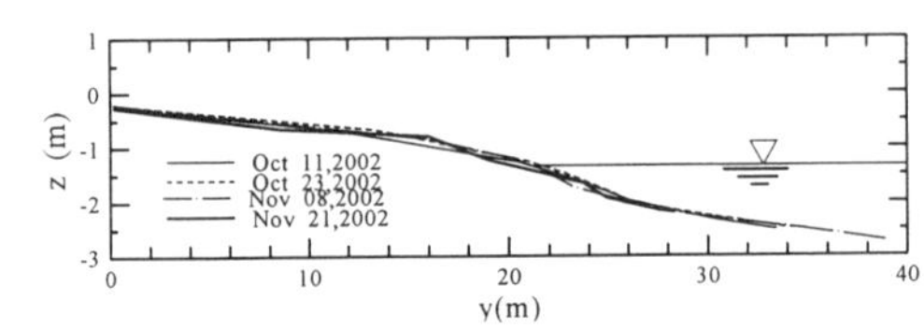

(b) Section 4

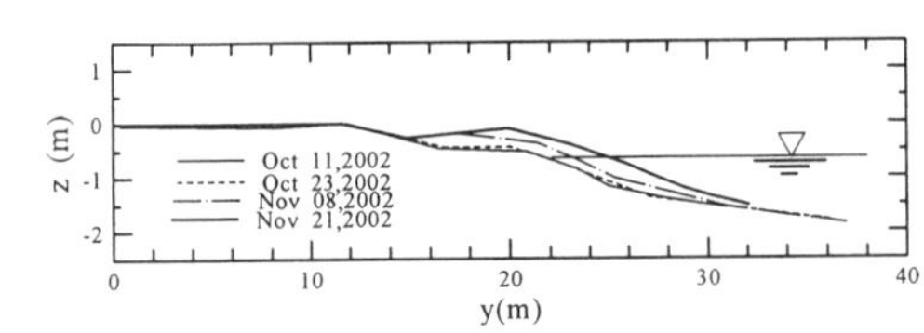

(c) Section 5

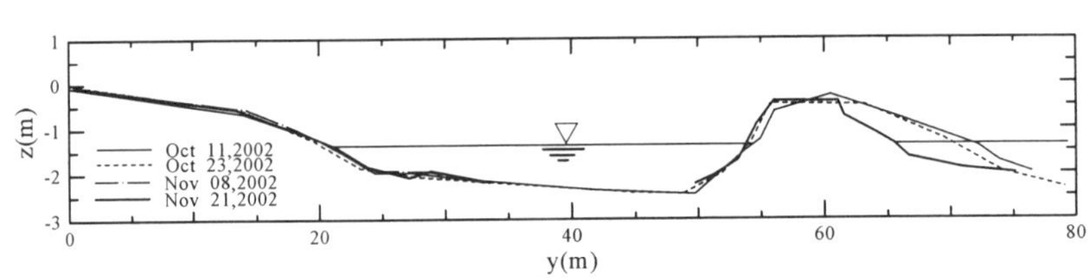

(d) Section 6

Fig.4 Cross-shore profile

COMPLEX PRINCIPAL COMPONENT ANALYSIS TO CHARACTERIZE BEACH TOPOGRAPHIC CHANGE IN SILT ISLAND, GERMANY

Hiromune Yokoki* and Magnus Larson**
*Center for Water Environment Studies, Ibaraki University, Hitachi, Ibaraki, 316-8511 Japan
Tel: +81-294-38-5219, Fax: +81-294-38-5268, Email: yokoki@civil.ibaraki.ac.jp
**Department of Water Resources Engineering, Lund University, Sweden

Objectives

To predict topographic changes is indispensable for developing a long-term coastal management plan, in particular for the coasts where sandy beaches are dominant. A number of models have been developed so far to predict beach topographic changes, based on the physical principles associated with the interaction between waves and sediment movements. However, these models cannot be applied directly to large-scale coastal areas with a length of more than several tens kilometers, for they cannot represent all the local conditions and need a huge computational capacity. On the other hand, the correlations between incident waves and long-term topographic changes are relatively easy to obtain by a statistical analysis. The correlations can be used to predict future coastal topographic changes. For the prediction, predominant patterns of the topographic changes must be analysed in details at first.

Statistical methods have been developed for analysing beach profile data since Winant et al. (1975) applied the principal component analysis (PCA). The validity of the methods of this kind is confirmed in relatively narrow coastal area, for instance, around a port (Bosma and Darlymple, 1996; Yokoki et al., 1998), whereas the applicability to much wider areas has not been verified yet. The purpose of the present study is to analyse statistically the coastal topography data obtained in Sylt Island coast, Germany, by the complex principal component analysis (CPCA), and to examine the applicability of CPCA through these analyses. To investigate the characteristics of beach topographic changes after beach nourishment in Sylt Island coast is another purpose of the study.

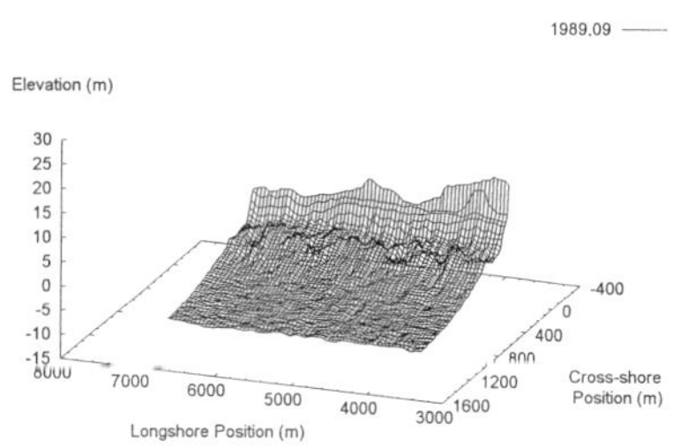

Fig.1: Kampen (1989.9)

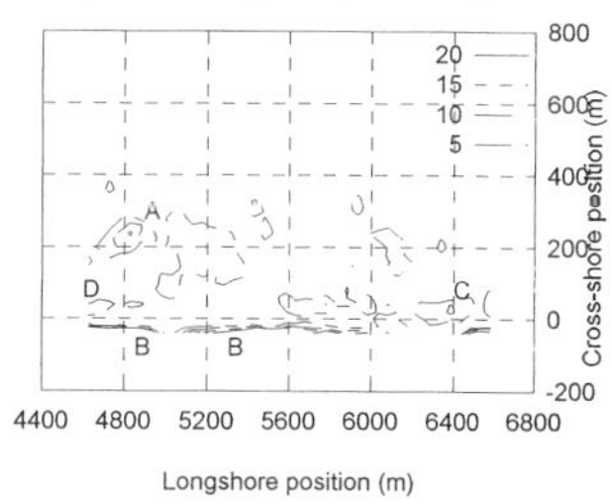

Fig.2: Variations of depths

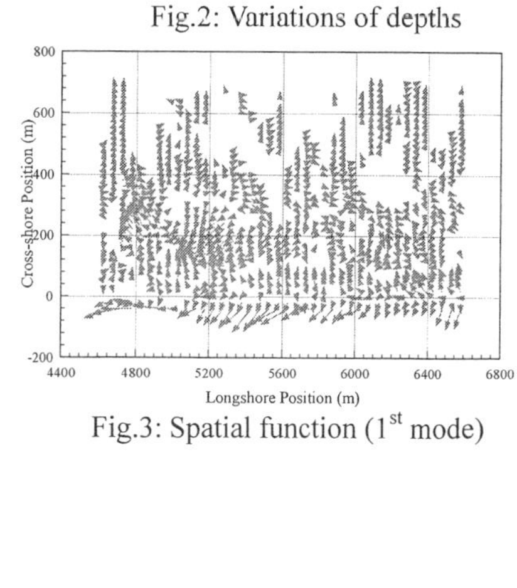

Fig.3: Spatial function (1st mode)

Theory and application data

In CPCA, the water depth, $H(x,y,t)$, is expressed as summation of products of complex temporal functions, $c_n(t)$, and complex spatial functions, $e_n(x)$, as follows (Liang and Seymour, 1991):

$$H(x,y,t) = \sum_{n=1}^{N} c_n(t) e_n(x)$$

The product of $c_n(t)$ and $e_n(x)$ is called the n-th mode of the beach topographic change. These functions are obtained as eigenvalues and eigenvectors of covariance matrix of complex water depths, which consists of a real depth, $h(x,y,t)$, and its Hilbert transform, $\hat{h}(x,y,t)$, as the imaginary part. While the ordinary PCA can detect the standing wave components only, CPCA enables to detect the progressive wave components as well in the beach profile change, since it uses the complex water depth instead of the real water depth. Each mode obtained by CPCA does not represent the simple pattern of beach topographic change, such as eroding or depositing trend. In order to deduce the real patterns of topographic change, we need to treat the magnitudes and angles of the complex temporal and spatial functions appropriately.

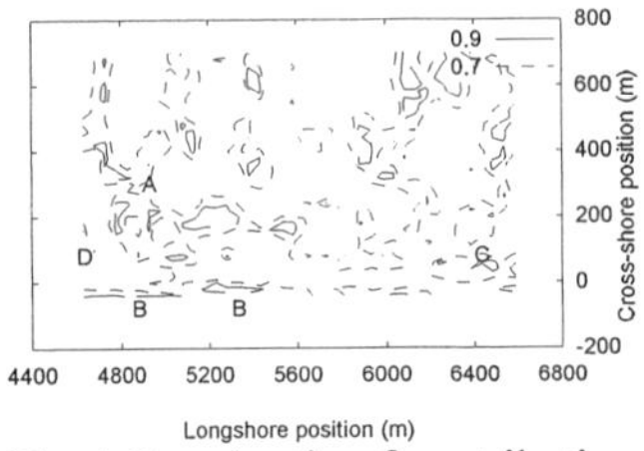

Fig.4: Local ratio of contribution (>70%)

Sylt Island is a well-known beach resort on the German North Sea coast that has suffered severe erosion at least since the middle of the last century (Dette and Gartner, 1987). The island is a sand body with a length in the north-south direction of about 35 km, bound by inlets to the north and south, and it is connected to the main land through a causeway. Since the beginning of 1980's beach nourishment has been determined a major countermeasure against erosion in the island. Kampen in the northern part of the island and Rantum in the southern part are the locations where beach fills were placed in 1985 at Kampen, and in 1987 at Rantum, respectively. In this study, we analyse the detailed topographic surveys at these two locations after beach fills were placed.

Figure 1 shows the topographic data at Kampen surveyed in September 1989. Shoreline and alongshore bar are placed at 0 and 200 m in cross-shore direction respectively in the figure. Figure 2 shows variations of depth changes through all surveyed data at Kampen. The variations are relatively large at region A, located at 4,800 m in alongshore direction, region B, the land side of the shoreline, region C and D. In these regions, water depths fluctuated sharply.

Major results and conclusions

Figure 3 shows the first mode of spatial functions in Kampen obtained by CPCA. The components of spatial function are plotted in vector format in the figure, since they are complex numbers. This figure indicates that the absolute values of spatial function are relatively large at regions A and B in Fig. 2.

Related researches so far said that the change of water depth is expressed by only one mode of spatial function where the absolute value of the mode is large. However, it has been revealed through the present study that the way of depth change is not always expressed properly by the mode of the largest absolute value. Therefore, a new parameter, local ratio of contribution of each mode to the depth change at each point was introduced in this study. The parameter enables to detect the most appropriate mode of spatial function expressing the way of depth change. Figure 4 shows the areas where the local ratios of contribution of the first mode are more than 70%. These areas correspond to the regions A, B, and C in Fig. 2. In region D, the local ration of contribution of the second mode is more than 70%.

The spatial and temporal functions obtained by CPCA indicate that the first mode at Kampen expresses the way of depth change as depositing for three years then eroding, and that the second mode expresses it as depositing for three years then eroding and depositing again largely. According to this interpretation, it is deduced that the beach fills placed around near shoreline area (regions A, B, and C) moved once offshore and then come back onshore again to region D. Furthermore, the phase difference of spatial functions shows that sediments are transported alongshore from left to right in Fig.4, with moving in cross-shore direction alternately. CPCA and the local ratio of contribution are proved to be effective to analyse the characteristics of beach topographic changes through the present study.

References

Bosma, K. F. and A. Dalrymple: Beach profile analysis around Indian River Inlet, Delaware, U.S.A., Proc. 25th ICCE, pp.2829-2842, 1996

Dette, H. H. and J. Gartner: Time history of a seawall on the Island of Sylt, Coastal Sediments '87, ASCE, pp.1006-1022, 1987

Liang, G. and R. J. Seymour: Complex principal component analysis of wave-like sand motions, Coastal Sediments '91, ASCE, pp.2175-2186, 1991

Winant, C. D., D. L. Inman and C. E. Nordstrom: Description of seasonal beach change using empirical eigenfunctions, Jour. of Geographical Res., Vol.80, No.15, pp.1979-1986, 1975

Yokoki, H., N. Mimura and K. Sato: Principal component analysis of beach topography change at Oarai Coast, Proc. Coastal Eng., JSCE, Vol.45, pp.571-575, 1998 (in Japanese)

RELATIONSHIP BETWEEN SAND BANK TOPOGRAPHY AND TIDAL RESIDUAL FLOW IN THE SETO INLAND SEA

Hiromichi Tanabe*, Satoru Takahashi* and Kazuo Murakami*
* National Institute of Advanced Industrial Science and Technology (AIST)

ABSTRACT

1. Purpose of the Study

There are many sand banks in the Seto Inland Sea. These sand banks are generally located behind islands or peninsulas. However, a large amount of sand was dredged from the sand banks in order to get an aggregate for concrete structures such as highway roads or buildings during high economic growth age. It is estimated that totally 600 million m^3 of sand was dredged from the Seto Inland Sea for last 40 years. Due to the disappearance of the sand banks from the sea bottom, several environmental impacts on fishes and/or seaweeds habitats were reported. Especially, the most serious problem influenced by the sand mining is that a number of sand eel decreased greatly in the western part of the Seto Inland Sea.

In this paper, the formation mechanisms of these sand banks and the influence of tidal current and water quality environments due to the change of topography are investigated by physical model experiments and numerical model simulations.

2. Physical Model Experiments and Numerical Simulations

Tidal current is the most important factor for transportation of sea bottom sand in the Seto Inland Sea. Detailed measurements of tidal current around area of these sand banks were carried out by a large scale of physical model of the Seto Inland Sea. The scale ratio of the physical model is 1/2000 in horizontal and 1/159 in vertical. Measurement stations were located at every 25 cm grid intervals around each sand bank area. Each measurement station, tidal currents were measured for 5 tidal cycles by using a magnetic current meter.

Figure 1 shows the directions of sand banks drawn by thick lines and the main directions of tidal currents drawn by small lines. The directions of sand banks and main tidal currents are slightly inclined and the angle difference between these directions is about 15 to 30 degree. Figure 2 shows current vectors of the tidal residual flows obtained by physical model experiments. The centers of the tidal residual circulation flows are agreed with the peak of the sand banks. Therefore, it is considered that the tidal residual circulation generated behind island and/or peninsula has some relationships to formation mechanism of these sand banks. However, the velocity of tidal residual flow is not enough to suspend and transport sand particles. The sand particles are suspended and transported by strong tidal current.

In order to know the mechanism of sand particles movement by tidal current, we carried out mathematical model experiments by numerical simulations. The model used here is the ODEM developed by Prof. Nakatsuji of Osaka University of Japan. The most important factor for sand movement in interested area is the velocity field on the sea bottom. Therefore, multi-leveled 3-D mathematical model is utilized to calculate the tidal current. At open boundaries, water surface elevations are described under the condition of sinusoidal waves of M_2 constituent. From the tidal computations, tidal residual flows are obtained by harmonic analysis, and then convergence and divergence of the tidal residual flow on the sea bottom layer. From the results, it is found that the convergence area of the tidal residual flow agrees with the area of sand bank.

3) Concluding Remarks

From the study, following conclusions are obtained. (1) Sand bank is located in the area where tidal residual circulation flow is generated. (2) The angle difference between the direction of sand bank and the main direction of tidal current is about 15 to 30 degree. (3) The area of sand bank is coincidence with the area of convergence of the tidal residual flow on the sea bottom layer. Therefore the sand particles are suspended by strong tidal current and transported by the tidal residual flow.

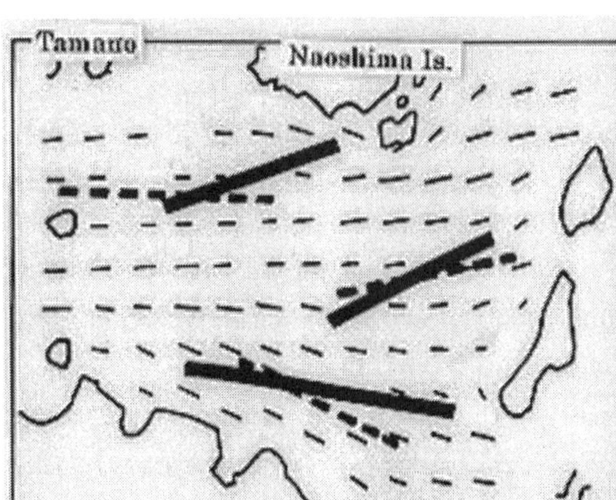

Figure 1 Directions of sand banks drawn by thick lines and directions of main tidal current drawn by dotted lines. Small lines show the current directions at each measurement point.

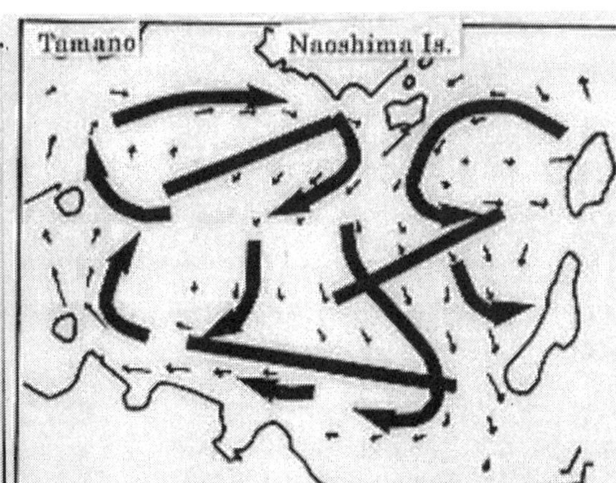

Figure 2 Tidal residual flow patterns obtained by physical model experiments.
Curved arrows : Patterns of tidal residual flow
Straight lines : Directions of sand banks.
Small arrows : Tidal residual vectors.

EROSIONAL HOT-SPOT GENERATION AT OSHIKIRI COAST IN SHIBUSHI BAY, KAGOSHIMA, JAPAN

Ryuichiro Nishi, Takaaki Uda and Takahiro Horiguchi

1) Associate Prof. Dept. of Ocean Civil Engineering, Kagoshima University, 1-21-40 Korimoto, Kagoshima-shi, Japan
Tel. (+81)-99-285-8484 Fax. (ditto), (Email: sediment@oce.kagoshima-u.ac.jp)
2) Dr. Eng., Executive Director, Public Works Research Center, 1-6-4 Taito, Taito-ku, Tokyo 110-0016 Japan
(Email: uda@pwrc.or.jp)
3) Technical chief, Coastal Division, INA Co., Ltd., 1-44-10 Sekiguchi, Bunkyoku,Tokyo,Japan,
(Email; tk-hori@ina-eng.co.jp)

PREFACE: An extension of the breakwater of Shibushi Port has been conducted at the Oshikiri coast in Shibushi Bay, Kagoshima, Japan since 1968 and its outer breakwater has been extended nearly 4.0km. This offshore breakwater created large wave shadow sheltered area behind it and non-sheltered area on the adjacent sandy beach. The typhoon damages had been caused on the non-sheltered sandy beach, thus three groins were built to reduce a deposition by an excess longshore sediment transport into a sheltered area from the non-sheltered adjacent beach. In spite of these shore protection works, one segment of the Oshikiri coast (roughly middle of Fig. 1) was especially overwashed. by the Typhoon 0111(11th Typhoon in 2001). This coastal damage seems not to be caused only by wave sheltering effect of Shibushi Port (wave diffraction effect), but formations of three dimensional sub-aerial bathymetry such as crescentic longshore bar and trough features probably enhanced the overwash damage at a certain location of the beach where a national hospital and local residential area exist. Therefore, this erosional hot-spot-like feature had to be studied in order to plan the emergency shore protection works at the site. The study shows that the erosion in the non-sheltered up drift coast was commenced mainly due to the wave sheltering effect (dH/dx; wave height gradient by wave diffraction) and the generation of 3-D geomorphological features including the crescentic longshore bars and troughs. These nearshore topographies were generated while incident significant wave heights of Typhoons exceeds nearly 6m high on this coast. In addition, it is shown that the average spacing of the troughs is around 600-700m. Finally, submerged breakwaters so called artificial reefs in Japan are recommended to reduce the further coastal damage at the site.

Fig.1 Aerial view of study area

STUDY AREA: A sandy beach in the Shibushi Bay extends nearly 16.4km and opens to the Pacific Ocean as shown in Fig.1. The Oshikiri coast locates northeast terminus where Anraku and Hishida Rivers flow into the coast, and an artificial island of 1.5km length and 1.5km width was created for crude oil storage in southwest terminus of the bay. A coastal erosion caused by the wave sheltering effect of this island was cited by Nishi et al. (2000). Coastal erosion similar to this erosion would cause the coastal erosion at the Oshikiri coast shown in Fig.2.

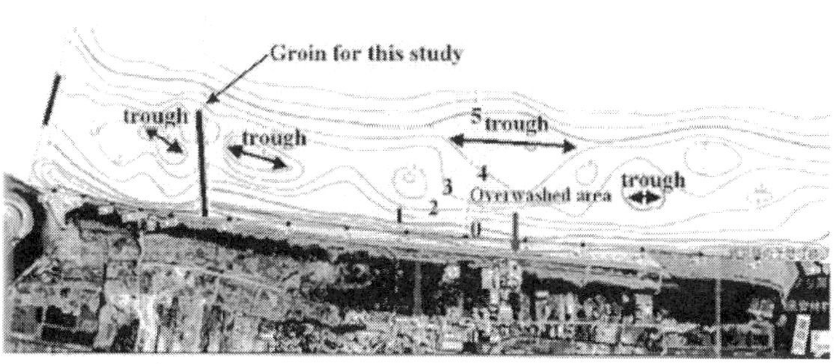

Fig. 2 Bathymetly of Oshikiri coast (zoom-up of the damaged area)

COASTAL DAMAGE: The significant beach and dune erosion and overwash around the transect 26 on the Oshikiri coast seemed to be originally caused by the sheltering effect by an offshore breakwater of the Shibushi Port which has been constructed and extending since 1968. The change in shoreline position and nearshore bathymetry have been examined by an analysis of aerial photographs and a set of sounding data. The shoreline in 1963 was selected to be the reference in Fig. 3. The change in shoreline position from 1963 to 1970 shows accretion patterns around the mouth of Hishida (x=6.6km) and Tabaru (x=10.6km) Rivers. The maximum accretion is the order of 40 to 50m at both river mouths. The shoreline change around the Anraku River (x=4.0km) shows relatively small sediment discharge compared to Hishida River. The maximum shoreline retreat in 1996 was the order of 50m, but the shoreline retreat at x=4.7km was only a half of the maximum recession even though the area was significantly damaged by overwash. Fig.4 shows a set of representative nearshore bathymetries in the recovery stage (1998) and erosional hot-spot stage (2001) as the long-term coastal processes at the study area. The subaerial topography between the transect 16 and 20 is depositional due to the sheltering effect of the extended offshore breakwater, whereas the middle of the area between the transect 20 and 30 (2km stretch of the coast) was eroded. The 4m, 5m, and 6m contours explicitly undulate with a spacing of nearly 600m. Finally, it is concluded that the erosion mechanism of this area is a superposition of crescentic longshore bar and trough formation on the sheltering effect by the offshore breakwater of the port.

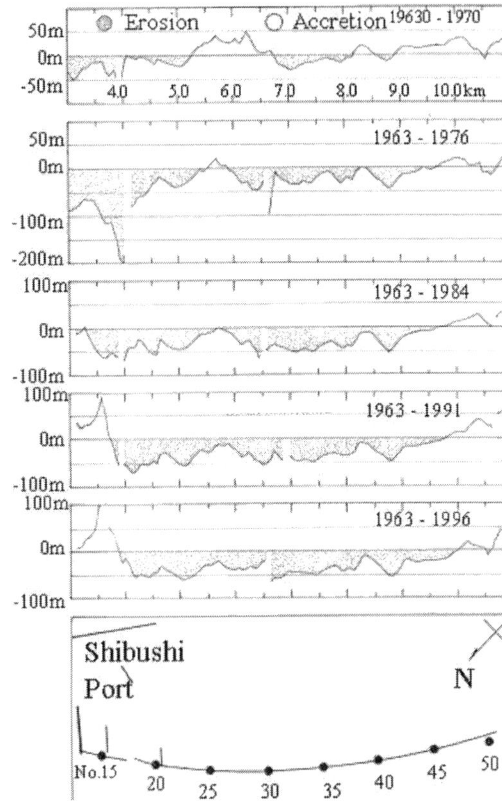

Fig. 3 Change in shoreline position.

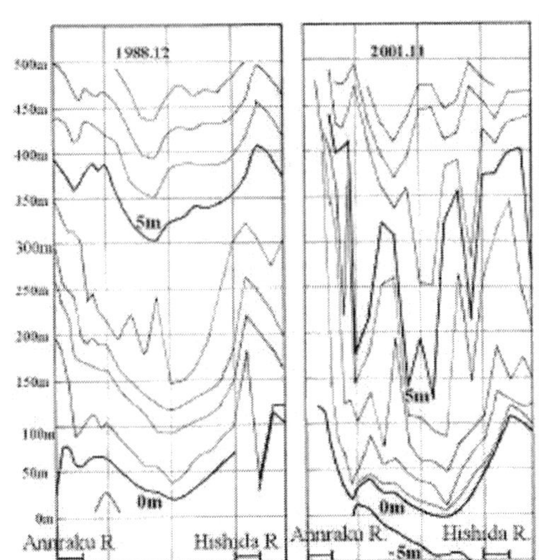

Fig.4 Change in nearshore bathymetry.

CONCLUDING REMARKS: A unique nearshore topography change on Oshikiri coast was studied based on an analysis of aerial photographs, profile data, and wave data. It is examined that the damaged area by overwash at this coast is an erosional hot-spot, which is caused by the superposition of the wave sheltering effect of the port and the generation of crescentic longshore bar and trough. The extension plan of the port is not fixed yet at the moment of this study, therefore a construction of submerged breakwater (called an artificial reef in Japan) is recommended to the prefectural government as an optimal shore protection scheme.

Wave Rotation for Coastal Protection

Shaw Mead Kerry Black

ASR Ltd
PO Box 151, Raglan, New Zealand.
Ph. +64 7 825 0380 Fax. +64 7 825 0386
enquiries@asrltd.co.nz

Often coastal protection works are carried out to mitigate the effects of erosion, but do not attempt to address the cause. For example, rock/concrete walls, breakwaters or groynes along the beach act to mitigate the effects or to realign the beach. However, end effects on the shoreline downstream and the environmental imperatives to retain natural character often preclude hard structures. Also, coastal re-alignment may not be possible in built-up areas where natural alignments would sweep though modern shoreline developments.

Wave rotation is a method of coastal protection that targets the cause of the erosion (i.e. longshore wave-driven currents). Offshore, submerged structures are oriented to rotate waves so that the alongshore current (and thus sediment transport) is reduced inshore. The realigned wave angle at the breakpoint (in harmony with the alignment of the beach) results in reduced longshore flows and sediment accretion in the lee of the rotating reef.

This paper considers two types of offshore reef that we describe as "dissipaters" and "rotators". The former acts to break the waves and protect the coast by reducing wave energy in the lee of the reef. The latter relies on wave rotation to create the chosen wave alignment. The rotator is particularly beneficial in areas of low wave climate or large tidal range when wave breaking is difficult to achieve on a submerged structure, particularly around high tide. Numerical modelling and case studies are used to demonstrate the effectiveness of wave rotation to mitigate erosion.

RECOVERY PROCESS OF SAND SPIT AT THE NATORI RIVER MOUTH

Kazuya Watanabe* and Hitoshi Tanaka

*Corresponding author
Department of Civil Engineering, Tohoku University
06 Aoba, Sendai 980-8579, Japan
E-mail: watanabe@kasen1.civil.tohoku.ac.jp
Phone & FAX: (+ 81)-22-217-7453

INTRODUCTION

Development of sand spit at a river mouth is affected by river discharge, sea waves, and tidal current. During a flood, remarkable change in river mouth morphology can be observed, followed by gradual recovery of sand spit topography. It is of practical importance to understand the behavior of sand spit at a river entrance during and after a flood, in connection with water level rise due to backwater effect caused by sand spit. In the present study, detailed surveying of sand spit recovery process has been carried out at the Natori River mouth, Japan after the flood induced by typhoon 0206.

STUDY AREA AND FIELD OBSERVATION

A map of the study area is shown Fig.1. The Natori River has catchment area of 938.9km^2, with the length of 55.0 km. The Natori River originates near the border between Yamagata and Miyagi Prefectures and pours to the Pacific Ocean. The River mouth jetties has slight curvature toward the north as seen in Fig.1, which causes erosion on the right side bottom due to spiral flow. There are Idoura Lagoon and Hiroura Lagoon on the right-hand and left-hand sides of the mouth, respectively, which indicates extreme migration of the opening in the past.

Fig.1 Study area

A big flood occurred along the Natori River due to heavy rainfall on July 10, 2002. Due to this flood, the sand spit has been completely flushed away. Immediately after the flood, the topography change has been measured every two weeks until it reach ed equilibrium stage.

RESULTS AND DISCUSSIONS

Figure 2 shows time-history of morphology at the Natori River mouth after the flood in 2002. Due to overflowing during the flood, sand spit at the river mouth was eroded to from another channel with the width of 90m, although slight recovery of the sand spit can already be seen inside the channel in Fig.2 (a). In Fig.2 (b), growth of the spit in the longshore direction is predominant. Further transportation of longshore sediment leads to reattachment of the sand spit to the jetty (Fig.2 (c)). Finally, the topography reaches equilibrium until sediment deposit occurs in between the jetties caused by overtopping waves (Fig.2(d)).

According to the study by Sawamoto et al. [1], it takes several months for the recovery of the sand spit at the Abukuma River mouth, whereas it is only two weeks at the Nanakita River mouth (Tanaka [2]). It is here noted that it takes about three weeks at the Notori River mouth for the sand spit recovery, which is intermediate between Abukuma River and Nanakita River. Such a difference of duration for recovery can quantitatively be explained using a 1-D model for river mouth migration.

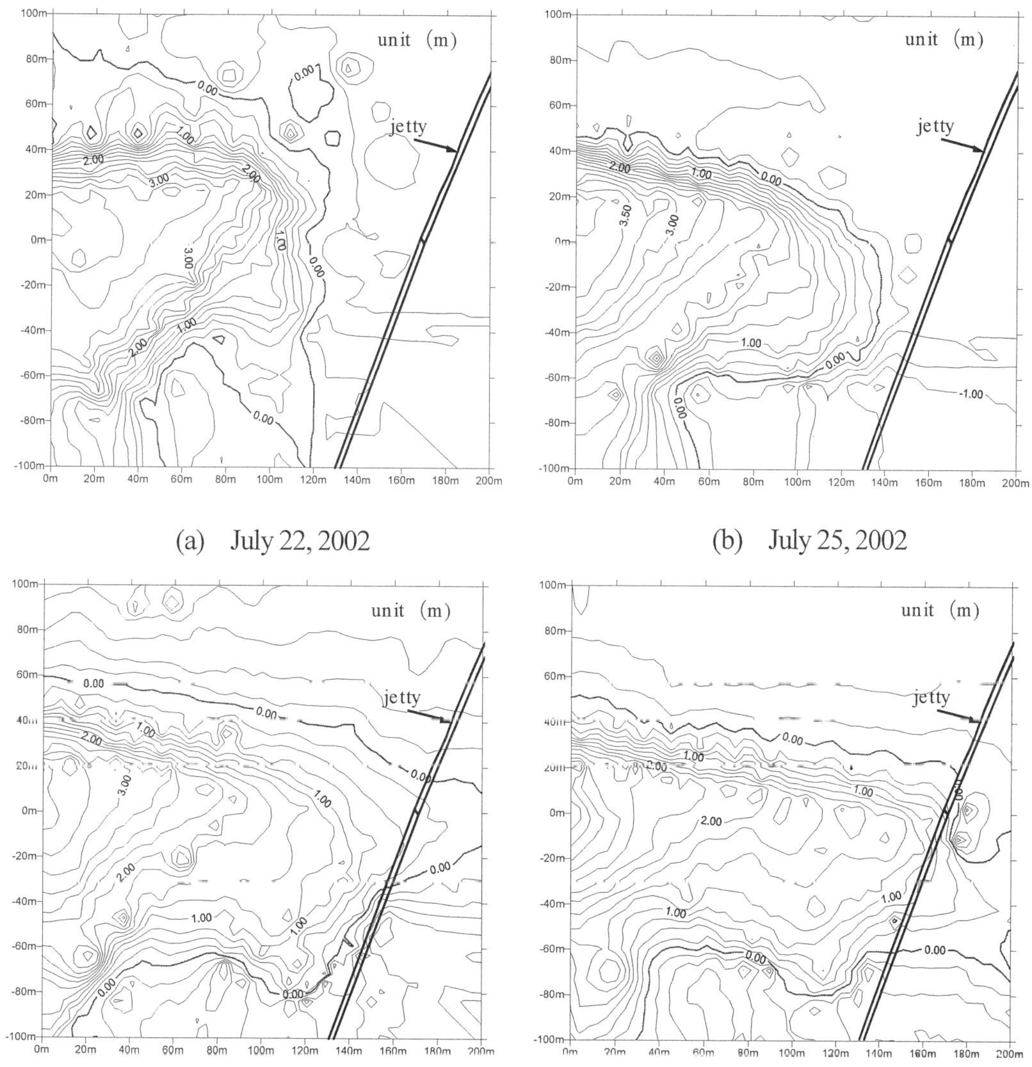

(a) July 22, 2002

(b) July 25, 2002

(c) August 8, 2002

(d) August 13, 2002

Fig.2 Contour map of Natori River mouth

REFERENCES

[1] Sawamoto, M. Shuto, N. and Taniguchi, T.: Topography change process at the Abukuma River mouth, Proceedings of JSCE, No.387/ II -8, pp.179-188, 1987.(in Japanese)

[2] Tanaka, H.: Field measurement of topography change at the Nanakita River mouth, Journal of Hydraulic, Costal and Environment Engineering, No.509/ II -30, pp.169-181, 1995.(in Japanese)

DOMINANT CAUSES OF MORPHOLOGICAL CHANGES ON INTERTIDAL FLATS

Fumihiko YAMADA[1], Nobuhisa KOBAYASHI[2], Tetsuya KAKINOKI[3]

[1] Associate Professor, Dept. of Civil & Environmental Eng., Kumamoto University, 2-39-1, Kurokami, Kumamoto, 862-0972, JAPAN, Tel +81-96-342-3546, Fax +81-96-342-3507, yamada@kumamoto-u.ac.jp

[2] Professor and Director, Center for Applied Coastal Research, University of Delaware, Newark, DE, 19716, U.S.A, Tel +1-302-831-8044, Fax +1-302-831-1228, nk@coastal.udel.edu

[3] Research Associate, Dept. of Civil Eng., Kobe City College of Technology, 8-3 Gakuen-nishi, Kobe, 651-2194, JAPAN, Tel & Fax +81-78-795-3270, kakinoki@kobe-kosen.ac.jp

1. INTRODUCTION

Tidal flats play an important roll as a habitat for the rich ecosystems. Global warming would raise the sea levels about 1 m in the next 100 years. The Intertidal areas decrease or deform with raising the sea levels. Therefore, future variation of inner bay environment has received considerable attention, because the purification abilities of tidelands depend on the ecosystems.

There is a close relationship between the ecosystems and cross-shore profiles on the tidelands. Knowledge of the morphological response on the tidal flats is essential for understanding a long-term environmental carrying capacity of inner bays. Tidal flats, however, are poorly understood compared to their sandy equivalents.

The aim of the present work has been the evaluation of the dominating causes for morphological response on the tidal flats from in-situ measurements at Shiraka river estuary. Quadratic polynomial approach has been proposed for approximation of cross-shore profiles. A relationship between topographical changes and external forces; tides, wind waves, and river discharges; has been examined.

2. FIELD OBSERVATIONS

Shiraka river estuary is located at central-eastern coast of Ariake Bay, which is famous for closed inner bay and the highest tidal range in Japan. An intertidal area on Shiraka river estuary is roughly 1.5km wide, with a mean bottom slope of 1/800, which is consistent with a mean tidal range of 3 m.

In order to maintain whole surveying conditions, ground elevations have been measured using an Electro-Optical distance meter at the fixed measuring points. We drove a 2m-wooden stake into the tideland down to 1.7m. In October 2000, two cross-shore profiles for observation have been installed both left and right bank at Shirakawa River mouth. Each profile has the 30 fixed measuring points, which are spaced every 50m. The observation has been carried out almost every one month at the low water during spring tides. Until now we have conducted the observations to 20 times over two years. (Yamada & Kobayashi, 2003)

2-1 Characteristics of Morphological Response on Tidal Flats

The main results of these observations and data analysis can be summarized as follows:

① Both measured and averaged cross-sections on tidal flats exhibit convex-upwards profiles, although the cross-sections on sandy beach usually show concave-upwards profiles.

② Examinations of tide level records measured near Shirakawa river estuary have indicated that averaged tide level changes cyclically with a period of one year. Averaged tide level rises about 40 cm from June to October. This raising rate is an order of magnitude larger than that of long-term sea level rise described before.

③ Cross-shore profiles fluctuate with changing the averaged tide level. Both accretion and erosion at offshore are larger than those at nearshore. Averaged variation range of the profiles during whole observation is ±6 cm.

The cross-shore profiles of tidal flats can be classified according to the relative contributions caused by wind waves and tidal motions to the total sediment transport (Kirby, 2000). Convex-upwards profiles have been designated as accretionary and tidally dominated flats.

Lee & Metha (1997) considers wave action as the most significant process and applies the equilibrium profile (Dean, 1991) as an approximation of tidal flat profiles. The equilibrium profile, however, is proposed originally for concave-upwards profile. The equilibrium profile assumes that a variation of the profiles equal to that of sea levels. Described above (the results of ② and ③), the averaged variation of measured profiles is at least 4 times smaller than that of tide levels. Therefore, the equilibrium profile cannot be applied for these observation results.

3. RELATIONSHIPS BETWEEN TOPOGRAPHICAL CHANGES AND EXTERNAL FORCES

Assuming that cross-shore profiles are approximated using three parameters: curvature (a), spatial gradient ($2ax+b$), and vertical deviation (c), quadratic polynomial approach $<h = ax^2 + bx + c$; h is elevation, x is offshore distance> has been proposed. Shown in Fig. 1 are comparison results between measured and approximated cross-shore profiles. This approach allows us to interpolate the measured profiles with accuracies less than $\pm$20mm

In order to clarify the dominating causes of external forces on the cross-shore profiles, the time evolutions of three parameters in quadratic polynomial was compared with those of the mean tide levels, tidal ranges, wave heights, wind velocities, and river discharges

3-1 Dominating causes for vertical deviation (*c*)

Vertical deviation (c) means an accretion and erosion of tidal flats. Figure 2 shows that time evolution of vertical deviation has a close relationship with that of mean tide levels. Vertical deviation changes cyclically with a period of one year. Tidal flats accrete with raining the tide levels, and they erode with lowering the tide levels.

3-2 Dominating causes for spatial gradient (2*ax*+*b*)

Averaged spatial gradients ($2ax+b$) have a close relationship with the changing rate of tidal energy. Tidal energy is defined as square of tidal ranges. Averaged spatial gradients changes cyclically with a period of one year. Averaged spatial gradients become steeper with increasing the changing rate of tidal energy, and they become gentler with decreasing the rate of tidal energy.

3-3 Dominating causes for curvature (*a*)

Curvature (a) have a close relationship with both tide level and tidal energy. Curvature changes cyclically with a period of half year.

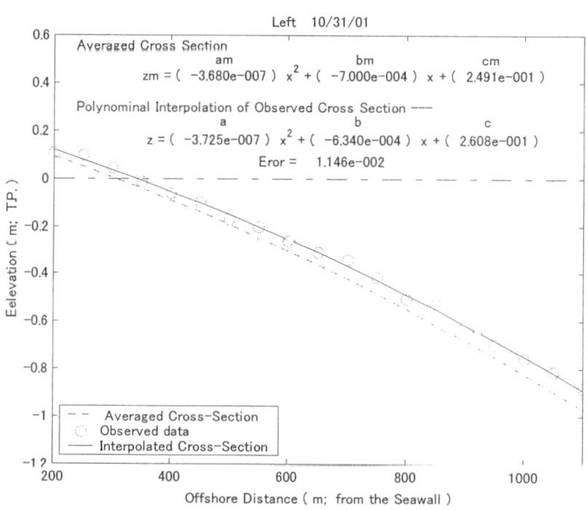

Fig. 1 Measured and interpolated cross-shore profiles

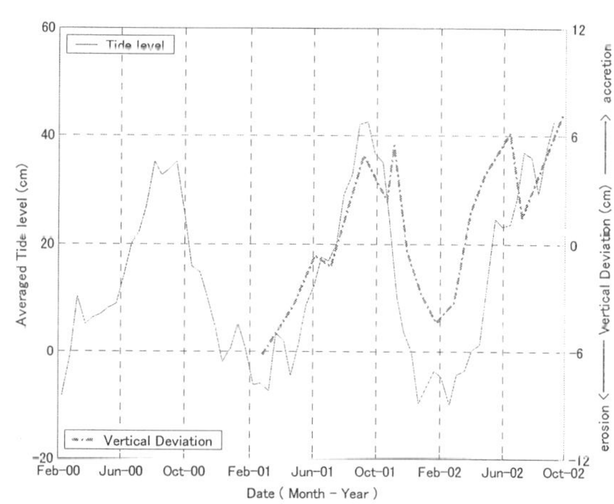

Fig. 2 Time evolutions of vertical deviation and mean tidal levels

4. CONCLUSION

In order to evaluate the dominating causes of morphological response on the tidal flats, quadratic polynomial approach has been proposed for approximation of cross-shore profiles. A relationship between three parameters of the polynomial and external forces has been examined. Consequently, tidal motions are dominant for tidal flats, which have convex-upwards profiles. This result agrees with that obtained by Kirby (2000).

A domination of wind waves and river discharges can become important at a time scale that is shorter than the present observation intervals. This warrants future observations at before and after both storm and typhoon conditions.

REFERENCES

Dean, R. G. (1991): Equilibrium beach profile: Characteristics and applications, J. Coastal Res., 7, 53-84.

Kirby, R. (2000): Practical implications of tidal flat shape, Continental Shelf Res., 20, 1061-1077.

Lee, S-C. and Mehta, A. J. (1997): Problems in characterizing dynamics of mud shore profiles, J. Hydraulic Eng., 123, 351-361.

Yamada, F & Kobayashi, N (2003): Res. Rept. of CACR, Univ. of Delaware, (in preparing)

NUMERICAL MODELING OF EROSION / DEPOSITION DUE TO TYPHOON JELAWAT IN THE NORTH PASSAGE OF YANGTZE ESTUARY

Pingxing Ding, Kelin Hu and Yazhen Kong
State Key Laboratory of Estuarine & Coastal Research, 3663 Zhongshan Rd.N.,
Shanghai 200062, P.R. China, Tel: 086-21-6223-2897, Fax: 086-21-6254-6441,
Email: pxding@sklec.ecnu.edu.cn

Introduction

On the average, there are 4-5 storms affecting the Yangtze Estuary every year. The storm-induced extreme waves and set up often make suspended sediment concentration rise. A lot of sediment is carried by currents into navigation channel and deposition occurs in the channel within and after storm, which sometimes affects ship navigating. It is very important for navigation safety to predicate accurately the storm-induced sedimentation in the channel.

No.0008 typhoon occurred from Aug.2 to Aug.12 in 2000. The typhoon path is shown as Fig.1 and the position of going ashore is quite near to the Yangtze estuary. So the typhoon is one of the most severe storms affecting the navigation channel in the Yangtze estuary. The sedimentation amount observed in the channel after the typhoon reached $5.54 *10^6$ m^3, which resulted in holding up the navigation for several days. This paper attempts to model the hydrodynamics, sediment transport, erosion and deposition caused by the typhoon.

Model

The predication model of storm-induced erosion and deposition developed by the paper's authors (Ding et al 2003) are utilized to compute sedimentation of navigation channel in the Yangtze estuary caused by No.0008 typhoon. The model consists of following submodels: (1) nearshore typhoon model, (2) wave model, (3) current model, (4) salinity model, and (5) sediment transport model. In order to fit the complicated coastal shape and improve the resolution in the interesting area, the generalized curvilinear coordinates proposed by Shi et al (1999) are utilized. There are three different scale computational regions (Figs. 2-4) in our predication model.

Results

The comparison of the calculated results and the observed data is shown as Fig.5. There are 4 curves in this figure. They indicate the observed distribution along the navigation channel, the contributions from the suspended sediment and bed load transport, and from the sum of suspended and load transports, respectively. It can be found from the figure that: (1) there are two peaks of sedimentation. The first peak is located near the upstream of the channel, and the second peak is located near the downstream of the channel, (2) the composition in the accretions is different. The first peak is mainly due to the bed load transport, whereas the second peak is mainly due to the suspended sediment transport, which had been proved by the observed data. The median grain diameter of sediment on the first accretion is 0.0238 mm, but the median grain diameter on the second accretion is 0.0047 mm, and (3) the calculated results are in good agreement with the observed ones.

References

Pingxing Ding, Kelin Hu and Yazhen Kong, Numerical simulation of total sediment transport in the Yangtze estuary, Acta Oceanologica Sinica, in press.

Fengyan Shi, Pingxing Ding and Yazhen Kong, Numerical tidal current modeling using fine boundary-fitted grids for tidal flats, China Ocean Eng.,1999, 13(2): 115-123.

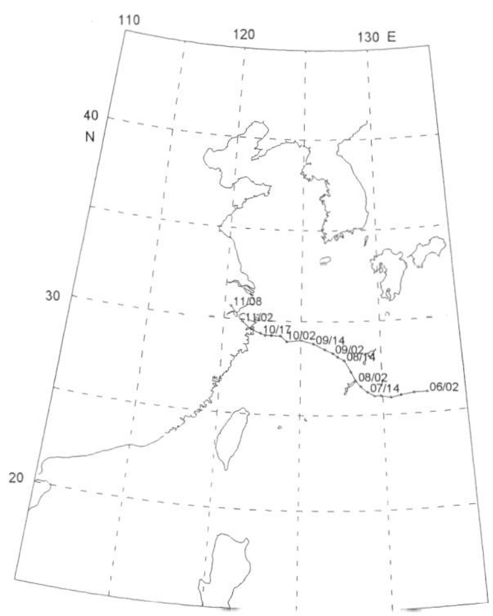

Fig. 1 The path of No. 0008 Typhoon

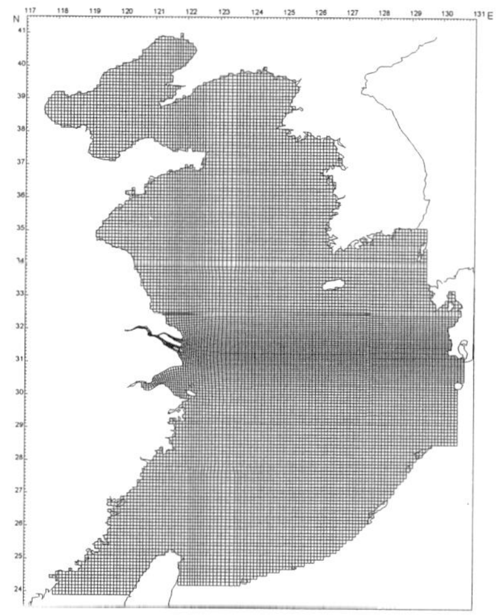

Fig. 2 Region of East China Sea and its grids

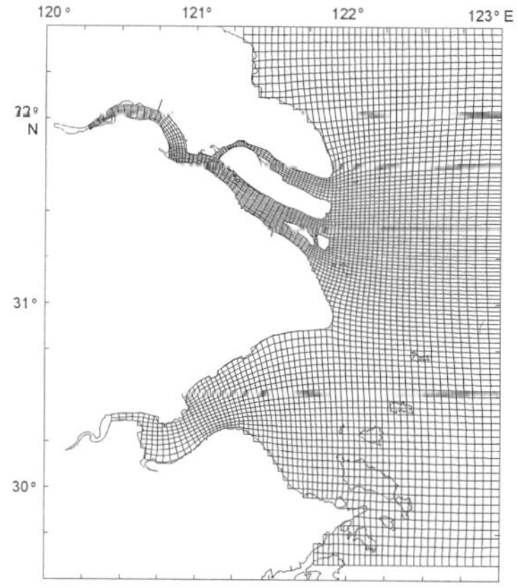

Fig. 3 Region of Yangtze Estuary and Hangzhou Bay and its grids

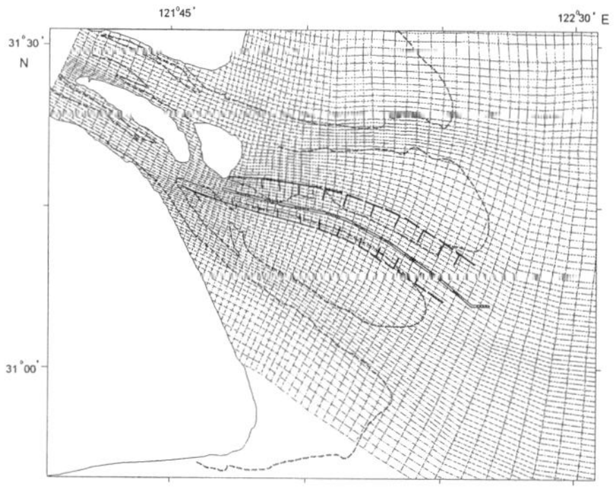

Fig. 4 Region of Yangtze Estuary and its grids

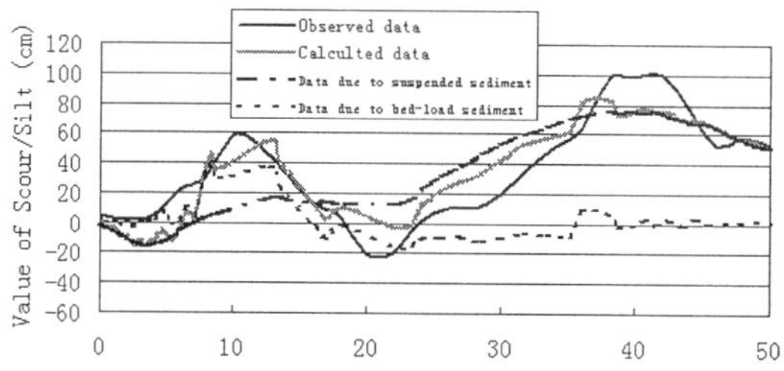

Fig. 5 Comparison of the calculated results and the observed data

LOCAL SCOUR AROUND A VERTICAL PILE WITH A CAISSON FOUNDATION

Kervin Yeow

Department of Civil & Resource Engineering, The University of Western Australia

Liang Cheng[1]

Senior Lecturer of Department of Civil & Resource Engineering, The University of Western Australia

35 Stirling Hwy, Crawley, Western Australia 6009, Australia

Tel: +61.8.9380.3076

Fax: +61.8.9380.1018

Email: cheng@civil.uwa.edu.au

ABSTRACT

The current emphasis on triple bottom line has sparked the growth of offshore wind farms as an alternative source of electricity. Caisson foundations have been typically adopted to support these offshore structures. This new foundation scheme has proved to be extremely effective in terms of cost and method of installation. However, the integrity of the foundation is found to be severely affected by the process of local scour around the structure. The current practice of designing the protection around such structures is likely conservative due to the limited researches done in this area.

The process of scour around structures occurs when the sediment transport capacity is larger in the vicinity of the structure as compared to the area away from it. The process can be classified into live-bed scour and clear water scour. The former takes place when the shear stress is greater than the critical shear stress of the bed ($\sigma>\sigma_{cr}$) while the latter occurs when the condition is reversed ($\sigma<\sigma_{cr}$). Only live-bed scour would be examined in this study since its effect is of higher severity to the stability of the structures. The flow mechanism involved in a pile sitting on a caisson foundation is assumed to be similar to that of a vertical pile (Fig. 2). Hence, analysis will be carried in the same light as that of a vertical pile.

Scour around vertical piles in both steady currents and waves have been extensively investigated in the last decade. The flow mechanisms, namely the horseshoe vortex formed in front of the pile and the lee-wake vortex formed at the lee side of it, have been relatively well understood (Sumer & Fredsoe, 1997). The horseshoe vortex is caused by the rotation in the incoming flow while the lee-side vortex is caused by the rotation in the boundary layer over the surface of the pile (Fig. 2). These two mechanisms are governed primarily by the Keulegan-Carpenter number (KC) ($KC=U_mT/D_c$, U_m: maximum value of oscillatory flow velocity; D: diameter of pile; T: wave period). The findings from Sumer et al. (1997) provide the indication that no horseshoe vortex exists for KC< 6. The size and life span of the horseshoe vortex were also found to increase with KC. These flow mechanisms are capable of removing significant amount of sediments from the vicinity of the pile. In another earlier study conducted by Sumer et al. (1992), the rate and extend at which a scour reaches equilibrium in waves were investigated. The following formula was suggested from the study:

$$\frac{S}{D}=1.3\{1-\exp[-0.03(KC-6)]\} \qquad KC\geq 6 \qquad (1)$$

In spite of the success of previous studies in examining the mechanism and effect of scouring, there are still certain aspects that have not been well understood. This paper aims to investigate the factors that have been

[1] Corresponding author

disregarded by various studies to date. The relationship between the exposure of the caisson and the scour depth presents a crucial consideration in the effect of scouring around a caisson type structure. The effects of pile to caisson diameter ratio on scour will also be investigated in this study. The effects of lee-wake and horseshoe vortex are demonstrated to be the two key mechanisms in the scour process. The parameters governing the development of equilibrium scour under a constant experimental condition are the exposure of the caisson (L_c) and the ratio between the diameter of pile to the diameter of caisson (D_p/D_c) (Fig. 1).

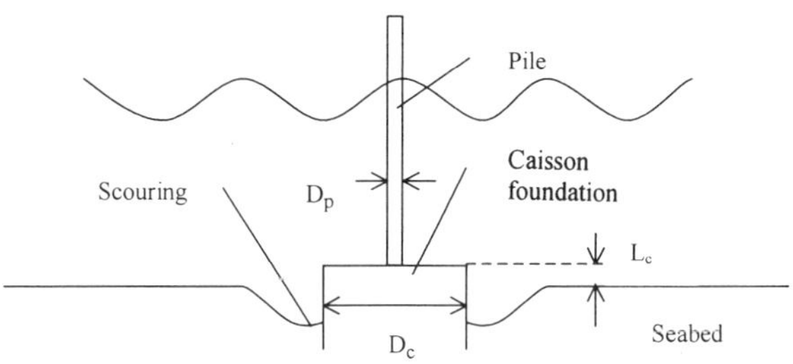

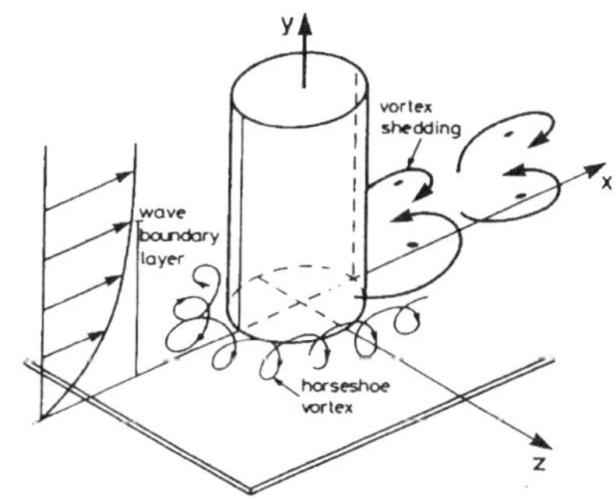

Figure 2: Near-Bed flow around pile

Figure 1: Cross section of a pile sitting on caisson foundation

REFERENCES

Sumer, B.M, Christiansen, N. and Fredsoe, J. 1992. Scour around vertical pile in wave. ASCE, J. Waterway, Port, Coastal and Ocean Engineering, Vol. 118, No. 1, pp. 15-31.

Sumer, B.M, Christiansen, N. and Fredsoe, J. 1997. The horseshoe vortex and vortex shedding around a vertical wall-mounted cylinder exposed to waves. J. Fluid Mech.,Vol. 332, pp. 41-70.

THREE DIMENSIONAL SCOUR BELOW OFFSHORE PIPELINES

Meagan Frame
School of Civil & Resource Engineering, The University of Western Australia

Liang Cheng[1]
Senior Lecturer, School of Civil & Resource Engineering, The University of Western Australia
35 Stirling Hwy, Crawley, Western Australia 6009, Australia
Tel: +61.8.9380.3076
Fax: +61.8.9380.1018
Email: cheng@civil.uwa.edu.au

ABSTRACT

The rapid development of the offshore oil and gas industry in past decades has required the extensive use of offshore pipelines to transport these petroleum resources. Erosion of seabed materials beneath a pipeline due to the scouring phenomenon will eventually result in the growth of free spans, leaving sections of a pipeline unsupported. Three-dimensional scouring has been identified as one of the major causes of pipeline failure.

Introducing a structure, such as an offshore pipeline, into a marine environment will immediately disturb the flow path surrounding the structure. For the flow to pass the structure it must alter its path, which leads to the formation of vortices. These vortices induce an increase in local sediment transport and consequently the seabed may be eroded beneath and around the structure, the *scour process*. For a pipeline this scouring action occurs in a three-dimensional manner. A scour hole may break out at some point beneath the pipeline and propagate along the length of the pipe resulting in the development of a free span.

Early investigations considered scour around a pipeline as a two dimensional process, looking only at a cross section of the pipeline under current, wave and combined wave and current conditions. Under wave conditions Sumer & Fredsoe (1990) proposed the Keulegan-Carpenter (KC) number governs the equilibrium scour depth. The KC number relates to the wave conditions and is defined as:

$$KC = \frac{U_m T_w}{D} \tag{1}$$

Where U_m= maximum flow velocity at seabed, T_w= wave period and D= pipeline outer diameter.

Recent research efforts have been directed towards understanding the factors influencing the development of free spans and the behaviour of the pipeline once it is suspended. Sumer & Fredsoe (1991) developed a semi empirical model to estimate the time development of free spans in currents, considering a range of parameters. This model identified the significance of pipeline embedment depth (e) on the rate of scour propagation. The longitudinal free span development is described by the volumetric rate of sediment transport:

$$c = \frac{1}{eD}\frac{d(Vol)}{dt} \tag{2}$$

Furthermore Sumer & Fredsoe (2002) proposed that a third spiral type vortex, in addition to lee wake and tunnel vortices, was responsible for the longitudinal scour development. Recent three-dimensional numerical modelling (Chen & Cheng 2001) supports the assumption that a third spiral vortex is present and contributes to the three-dimensional scour process.

Longitudinal free span development under wave actions has been investigated by Grass et al. (1994). Based on the volumetric rate of sediment transport and directly related to the vertical rate of scour,

[1] Corresponding author

Grass et al. (1994) have developed a simple theoretical model for the rate of free span development under wave. However this research is only in the beginning stages and has only been applied to small scale modelling.

This research project attempts to characterize the rate of three-dimensional scour below offshore pipelines under wave action. Scale model tests were conducted with parameters of wave height, wave period and pipeline embedment depth being modified to identify the influence these parameters have on the rate of scour. Investigations have found that the rate at which the free spans develop decreases as the depth of embedment of the pipe increases. Increasing the wave period and/or wave height is also found to increase the rate of free span development. The lateral free span development progresses at a linear rate along the pipe. Research has further identified that a new mechanism may contribute to the three-dimensional scouring process.

REFERENCES

Chen, B. & Cheng L. 2001, 'Three-Dimensional Modelling of Flow Around a Free Spanned Pipeline', Asian and Pacific Coastal Engineering, China, pp721-727.

Grass, A.J., Simons, R.R., Paskin, S., and Baker J.H.A., 1994, 'Scour, Spanning and Self-Burial of Seabed Pipelines' SOSC94.

Sumer, B.M., and Fredsoe, J., 1994, 'Self Burial of pipelines at Span Shoulders' International Journal for Offshore and Polar Engineering, Vol. 4, No.1.

Sumer, B.M & Fredsoe, J. and Hansen & Staub. 1991, 'Time Development of Scour Induced Free Spans of Pipelines', Vol.V, Pipeline Technology, OMAE, pp.25 32.

Sumer, B.M, Fredsoe, J. 1990, 'Scour Below Pipelines in Waves', Journal of Waterway, Port, Coastal and Ocean Engineering, Vol.116, No.3, pp.307-323

Sumer, B.M. and Fredsoe, J. 2002, 'The Mechanics of Scour in the Marine Environment', World Scientific Publishing Co. Pte. Ltd., Singapore.

LAGRANGIAN GRIDLESS MODEL OF TOE SCOURING OF SEAWALL DUE TO TSUNAMI RETURN FLOW

Minoru HAYASHI[1], Hitoshi GOTOH[2], Tetsuo SAKAI[3] and Hiroyuki IKARI[4]

[1] *Engineer, Dept. of Construction, Nishimuro Promotions Bureau,*
Wakayama Prefectural Government, Japan
[2] *Associate Professor, Dept. of Civil Engrg., Kyoto Univ.,*
Yoshida Honmachi, Sakyo-ku, Kyoto, 606-8501, Japan
tel:+81-75-753-5098, fax: +81-75-761-0646, e-mail: gotoh@coast.kuciv.kyoto-u.ac.jp (correspondence)
[3] *Professor, Dept. of Civil Engrg., Kyoto Univ., Japan*
[4] *Graduate Student, School of Civil Engrg., Kyoto Univ., Japan*

Objectives

A foreshore erosion due to a return flow caused by significantly overtopping waves, such as tsunami flood, is difficult to be numerically simulated, because of its complicated characteristics. There are two key factors of this phenomena: (i) the description of water surface behavior including dropping jet from the top of a seawall and (ii) the description of local scouring driven by dropping jet. For the description of water surface behavior, the MPS (Moving-Particle-Semi-Implicit) method, which can track the fluid motion with very high resolution without using computational grids even under the existence of the fragmentation of fluid, is appropriate choice. To describe the local scouring, new model of scouring on the concept of particle tracking is introduced, to extend standard MPS method, which only can treat fixed wall boundaries, to fit the description of movable bed, or deformable wall.

Simulation Model

The governing equations, or continuity equations and momentum equations are discretized by the MPS method, which is categorized as the gridless Lagrangian approach of flow simulation. In the MPS method, all of the terms of the momentum equations (the convection term, the pressure-gradient term, the viscosity term, the gravity term and the solid-liquid interaction term) are described by the interaction of particles; in other words, the governing equations are discretized without using the computational grids. To calculate the gradient and the Laplacian, the interaction zone, in which the magnitude of interaction between different particles are evaluated by the weight function, is supposed to be located around each individual particle. Further, the convection term is calculated by tracking motion of each individual particles, hence this method is free from the numerical diffusion, which brings the difficulties in the ordinary Eulerian approaches. As for the model of scouring, the concept of the pick-up function is employed. The local flow velocity is assessed in the vicinity of the fixed wall, which is constituted by the rows of particles. When the local flow velocity excesses the threshold of pick-up, the wall surface particle is released from the fixed wall. After the pick-up, the wall particle is transformed into the sediment particle, and is transported by the flow.

Results

The calculated domain is the vertically-two-dimensional water basin, at the center of which the seawall in the height of 0.6 m is located. The righthand side of the seawall is a movable bed, which has sufficient thickness in comparison to the fully-developed depth of scour hole. The setup of the calculation is determined to fit the indoor experiment of the foreshore erosion conducted by Noguchi et al.(1997). Totally 18,500 numbers of particles, the diameter of which is 1.0 cm, are tracked. Fig. 1 shows the time series of the foreshore scouring. In the initial stage (t=1.1 s), the water depth over a movable bed is sufficiently large to cut the direct effect of the dropping jet collision onto the movable bed surface. In early stages, with the decreasing water depth, water on the movable bed is being accelerated. In this moment, some of surface particles are picked up due to the increase of the bottom shear stress. With further decreasing water depth, the movable bed is subjected to the attack of the dropping jet, which promotes the toe scouring of seawall. In Fig. 2, the time series of the maximum depth of score hole is shown in comparison to the experiment conducted by Noguchi et al.(1997). The fundamental tendencies of the scour hole development, which is predicted by the present model, agree with the experiment.

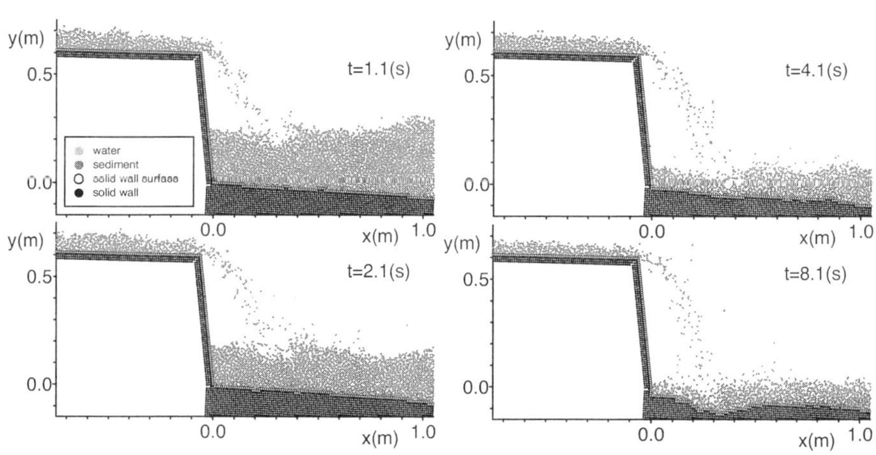

Fig. 1 Typical snapshots of solution

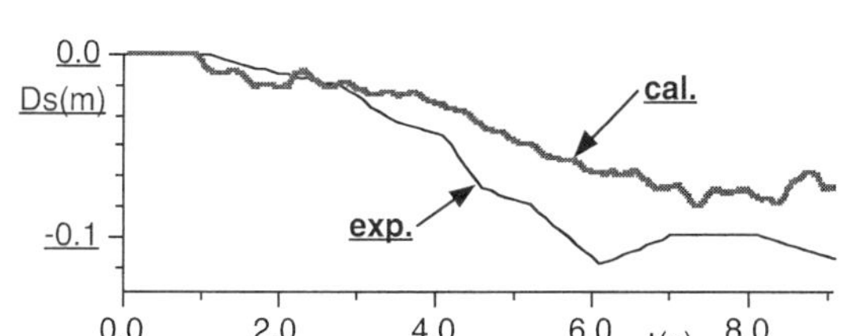

Fig. 1 Time series of the maximum depth of score hole

LOCAL SCOUR AROUND LARGE CYLINDERS UNDER WAVE ACTION

Chen Quoping   Zuo Qihua

(Nanjing Hydraulic Research Institute, Nanjing, 210029 China)

ABSTRACT

This paper presents the results of a series of experimental study on the local scour around a large circular cylinder under wave action. The characteristics of scour geometry in different conditions are described. Like the scour in front of a breakwater, there are two basic types of scouring pattern in front of a large circular cylinder. The experimental data together with the mechanism analysis indicate that the scouring patterns are dependent on the sand grain sizes as well as the wave conditions. The scour depth is found to depend mainly on the wave number, the relative cylinder diameter (the cylinder diameter to the wave length ratio), sediment diameter and the maximum velocity of the wave relations are proposed to expect the ultimate maximum scour depth and the developments of scour in front of the large circular cylinder. The applicability of the proposed formula is confirmed by experimental data obtained in a laboratory basin with a wave generator.

Keywords: scour, large cylinder, wave action

A CRITERION FOR THE INITIATION OF SEDIMENT MOVEMENT IN OSCILLATORY WAVES

Yiren Zhou and Sun Linyun
Senior Engineer, Nanjing Hydraulic Research Institute, Nanjing, P.R. China

ABSTRACT

In assessing the impact of large structures in offshore regions, the first problem is to predict the threshold of sediment. By reanalyzing some of the previously published results and making use of Jonsson's(1966)[1] wave friction factor concept which enabled one to determine the magnitude of the maximum bottom shear stress associated with an oscillatory fluid motion, Madsen and Grant(1976)[2] suggested that the shields criterion obtained from unidirectional steady flow was applicable as a general criterion for the onset of sediment movement also in oscillatory unsteady flow. In their Fig, however, the general trend of the data obtained from oscillatory flow conditions indicates that the critical value of the shields parameter is all above the shields curve determined from unidirectional steady flow experiments in the boundary layer of smooth turbulent-rough turbulent transition regime. It is obvious that this difference is systematic and can not be attributed to individual differences in determining just when initial sediment movement occurs. The purpose of the present paper is to contribute to the improved sediment criterion under wave action by analyzing the different boundary layer conditions between steady and unsteady flow. In this paper, by analyzing the mechanism of initiation of sediment in oscillatory flow and blending criteria in different boundary layers, a new and seasonal critical curve which can predict the sediment threshold under wave action is given, A vast amount of experimental data is found to show good agreement with the curve suggested by the author. From the criterion curve, a straightforward formula, which has better physical explanation, can be obtained. This analysis method will provide a new way for the study of inception of sediment transport in offshore region.

REFERENCES:

1. Jonsson I G. Wave Boundary Layer And Friction Factor. Coastal Engrg, 1966
2. Madsen O S, and Grant W D. Quantitative Description Of Sediment Transport By Waves. Coastal Engrg, 1976

A 3D Morphodynamic Model with Shoreline Change Based on Quasi-3D Nearshore Current Model

Masamitsu Kuroiwa[1], J. William Kamphuis[2], Takayuki Kuchiishi and Yuhei Matsubara

1)Research Associate, Dept. of Civil Engineering, Tottori University,4-101,Koyama,Tottori,680-8552,Japan, Tel:+81-857-31-5300,Fax:+81-857-28-7899,e-mail:kuroiwa@cv.tottori-u.ac.jp
2)Professor Emeritus, Dept. of Civil Engineering, Queen's University, Kingston, ON, Canada, K7L 3N6 Tel:+1-613-533-2148,Fax:+1-613-533-2128,e-mail:kamphuis@civil.queensu.ca

1. INTRODUCTION

When we construct coastal structures such as the breakwaters, jetties and harbours, the prediction of medium or long term beach evolution around the coastal structures is required. Some mophodynamic models for predicting the medium-term beach evolution(coastal area models) have been proposed such as, Watanabe *et al(*1986),de Vriend *et al*(1993). A coastal area model coupled with the shoreline model have been proposed by Shimizu *et al.*(1996). In the coastal area model, the nearshore current field has been determined using a depth-averaged (2DH) model. Recently, some coastal area models using a quasi-3D nearshore current model have been also proposed (e.g. Kuroiwa *et al.* 2000). Although these coastal area models have been applied to some practical problems, these models are still uncompleted because shoreline changes have not been sufficiently incorporated. Therefore, in order to predict 3D beach evolution with better accuracy, a 3D morphodaynamic model, that considers the effect of nearshore current profile and estimates the shoreline changes, is needed.

The final goal of this study is to develop a medium-term coastal area model using a quasi-3D current model that can predict the shoreline changes. As the first phase in the development of a new morphodynamic model, it is applied to the prediction of the bottom topography and shoreline changes due to nearshore currents and undertow around coastal structures.

2. NUMERICAL MODEL

The morphodynamic model presented in this study consists of four modules; 1)a wave module, 2)a nearshore current module, 3)a sediment transport module, and 4) a beach evolution module. **1)Wave module :** Wave fields are determined using a multi-directional random wave model proposed by Mase(2001), which is based on the energy balance equation with an energy dissipation term due wave breaking and a wave diffraction term. The energy dissipation term is formulated by using Goda's breaking criterion and the Rayleigh distribution's assumption for wave heights. However, in the case of arbitrary beaches such as barred beaches, the accuracy of prediction is not sufficient. Therefore, the dissipation term is modified based on Dally *et al.*(1984). In this computation, the JONSWAP spectrum and the Mitsuyasu angular spreading function are employed. **2)Nearshore current module:** Nearshore current field is determined based on the Q-3D model proposed by Kuroiwa *et al.*(1998). The previous model did not sufficiently account for the effect of the random waves. Therefore, in this study the characteristic values for calculating the nearshore currents; radiation stress, eddy viscosity, shear stress due to wave breaking at the water surface, and bottom shear stress, are determined using the root-mean square wave height and peak period of the JONSWAP spectrum obtained from the results of the wave module. Furthermore, the shear stress at mean water surface due to the breaking wave (the boundary condition at the mean water surface) is determined by using a function of Dally's the energy dissipation rate, taking into account the effect of the wave decay, recovery and secondary breaking on the arbitrary beaches such as barred beaches. **3)Sediment transport module:** In order to determine the 3D beach evolution due to nearshore currents including undertow effect, a simple time-averaged sediment concentration model proposed by Briand and Kamphuis(1993) is adapted. The total local sediment transport rates are determined by multiplying local vertical profiles of current velocities and sediment concentrations, and integrating from sea bottom(z'=0)to mean water surface ($\bar{\varsigma}$). The sediment transport rates in the cross-shore(x) and longshore(y) directions are expressed as

$$q_x = \int_0^{\bar{\varsigma}} UC(z')dz' \text{ ----- (1)}, \qquad q_y = \int_0^{\bar{\varsigma}} VC(z')dz' \text{ -----(2)}$$

where z' is the level from sea bottom, q_x and q_y are the local sediment transport rate in cross-shore and longshore direction. U and V are the local nearshore current velocities, which are determined from Q-3D model. $C(z')$ is determined by using the model proposed by Briand *et al.*(1993). The concentration can be expressed as

$$C(z') = C_A \exp\left[Ke\frac{(z'-z_A)}{z_A}\right] \text{---(3), in which } Ke = -\ln\left(\frac{C_A}{C_B}\right) \text{---(4)}, \quad C_A = \frac{K_{A1}(\nu_t/h) + K_{A2}(\tau/\rho)^{1/2}}{w_f} \text{---(5)}$$

where C_A is the concentration at $z' = z_A$ which is taken as the bottom boundary layer thickness. The C_A is given as a function of eddy viscosity coefficient ν_t, bottom shear stress τ_b and fall velocity of sand w_f. C_B is the concentration at which the sediment starts moving. Ke is a dimensionless parameter that determines the curvature of the exponential profile from sea bottom to the top of the boundary layer. The unknown coefficients K_{A1}, K_{A2} and C_B in Eqs.(4) and (5) are evaluated by comparing with the results of laboratory tests and field investigations when the beach evolution are computed. In order to predict the shoreline changes, the sediment transport rates in the run-up region are considered. The transport rates are determined using a method proposed by Larson *et al.*(1995),which is a simple method of linear decay from reference point at shoreline to limit run-up heights. The run-up heights are estimated by using Hunt formula. **4)Beach evlolution module:** The changes in bottom elevation are calculated using the continuity equation of sediment transport rate proposed by Watanabe et al. (1986). In our computation, the profile changes in the run-up region are also determined from the continuity equation. The shoreline is treated as a moving boundary. The new bottom topography is fed back into the hydrodynamic and sediment transport computations.

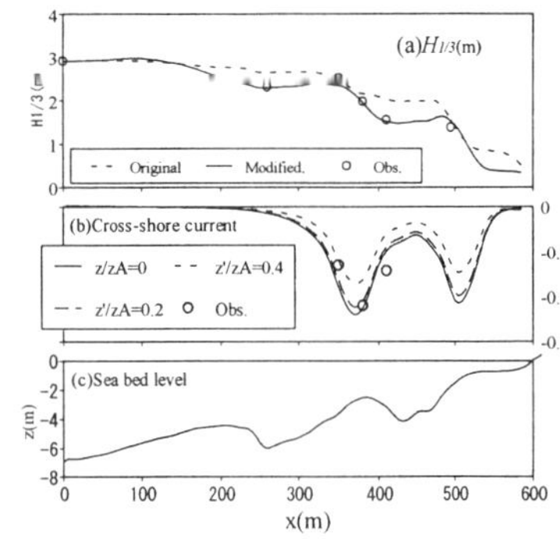

Fig.1 An example of computed results($H_{1/3}$=2.91m,$T_{1/3}$=11.81s, α=19 at 6m depth)

3. COMPUTED RESULTS AND VERIFICATIONS

At first, the modified wave and nearshore current model were calibrated and verified using the results of field observation performed over a barred beach from Jan.29 to Feb.3, 1997, conducted by Kuriyama.(1998). Significant wave heights, cross-shore (undertow) and longshore current velocities on the sand bar were measured. Beach profiles were also measured from Jan. 31 to Feb.3. The sand bar shifted in the offshore direction during Jan.31 to Feb.1. Fig.1 shows an example of comparison of significant wave heights and cross-shore currents between the measured and predicted result. It is found that both modified wave and nearshore current model give good agreement with the measured data. Furthermore, the present morphodynamic model was applied to the sand bar movement on Jan. 31. Fig. 2 shows the predicted sand bar movement which is represented by bold solid line. The predicted sand bar profile was reasonably consistent with the measured profile. As a first phase of this study, it was confirmed that the present model can predict the sand bar movement to offshore direction due to undertow.

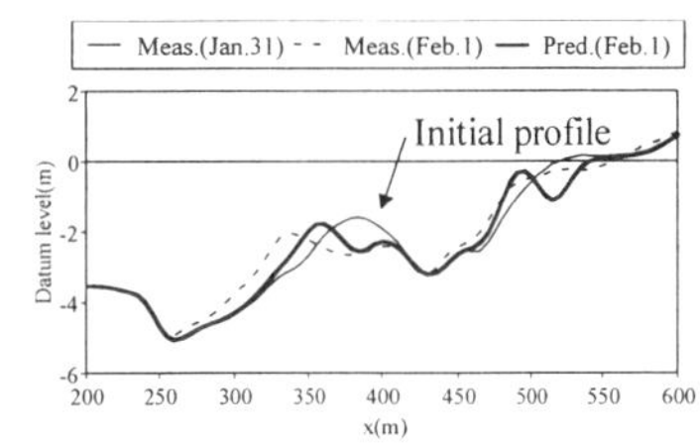

Fig.2 Computed beach profile ($H_{1/3}$=2.32m,$T_{1/3}$=10.47s, α =18.8 at 6m depth)

Secondly, the present model was applied to the prediction of beach evolution around a detached breakwater with the length of 200m. Fig.3 shows preliminary results of sediment transport rate vectors and water depth contours. In this computation, the significant wave height and period were 1.2m and 7s, respectively. The initial bottom profile was a uniform shape beach slope of 1/40. In Fig.3 (b), solid lines and broken lines are at 3 and 6 months later, respectively. It is clear that the shoreline and bottom level changes can be computed. It was also confirmed that a tombolo was formed by using the present model.

From these results mentioned above, it can be concluded that the present model can predict beach evolution with shoreline changes due to undertow and nearshore currents. Furthermore in this study, the present model will be verified using the laboratory and field data associated with beach evolution around coastal structures such as detached breakwaters and jetties. The applicability of the model will be also discussed. The results will be presented in the final paper and at the conference.

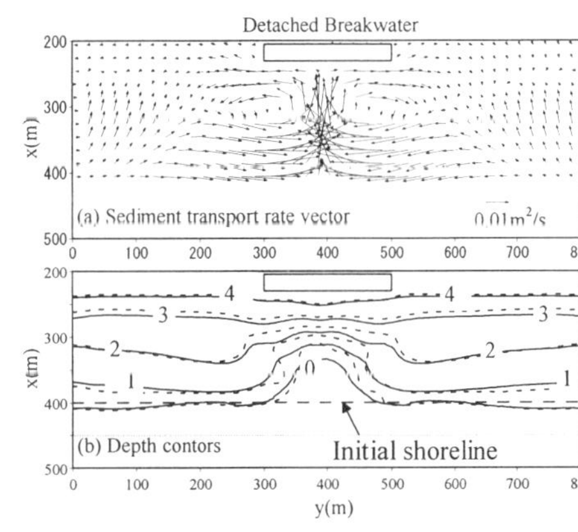

Fig.3 Examples of computation of sediment transport rate (a) and water depth contours (b)

ACKNOWLEGEMENT

We would like to thank to Associate Professor, Dr.Hajime Mase, Kyoto University,Japan,for advising about the computation of multi-directional random wave using the energy balance equation proposed by himself.

REFERENCES

Briand *et al.*(1993):Sediment transport in the surf zone:A Quasi-3D Numerical Model, Coastal Eng.,Vo.20. .**Dally *et al.* (1984)**: A model for breaker decay on beaches, Proc. 19th ICCE. ,**de Vriend *et al.*(1993):** Medium-term 2DH coastal are modeling, Coastal Eng.,Vol.21, **Kuroiwa *et al.* (1998)**: Applicability of a quasi-three dimensional numerical model to nearshore currents, 26th ICCE, **Kuroiwa *et al.* (2000):** Numerical prediction of bottom topographical change using Q-3D nearshore current model, 27th ICCE. **Kuriyama,Y (1998):** Field measurements of undertow on longshore bars, 27th,ICCE, **Larson *et al.*(1995)**: Prediction of cross-shore sediment transport at different spatial and temporal scales, Marine Geology,Vol.126, **Mase,H.(2001)**:Multi-Directional random wave transformation model based on energy balance equation, Coastal Eng. Journal, Vol.43, **Shimizu *et al.*(1986)**:Improved 3D beach evolution model coupled with shoreline model,26th ICCE, **Watanabe *et al.* (1986):** A numerical prediction model of three-dimensional beach deformation around a coastal structure, Coastal Eng. in Japan, Vol. 29.

Long-Term Shoreline Changes using Aerial Photos on the Namhangjin Coast

Seungjin Jung[1], Kyuhan Kim[2] and Chongkun Pyun[3]

Erosion of the shore and deposition usually occur in the beach area. These two phenomena, in turn, arise in various forms such as the closure of river mouth, development of estuary and coastal sandbanks, and shoreline changes. Especially, withdrawing of the shoreline caused by the erosion now threatens to destruct the environment around the shore, and establishments and residences bordering the coast. Aware of these problems, a considerable number of researches have been conducted to quantitatively predict the future of the shoreline change, which is affected by natural surroundings and construction of establishments. These efforts have been made in order to preserve the environment of the shore, and to enhance its function as a tourism attraction and shore protection.

Previous researchers, however, had to face numerous difficulties. For example, to conduct the hydraulic model test, a huge amount of money is required. On the other hand, researchers who attempted the numerical model had to deal with problems arising out of lack of the literature and data on the changes that a certain subject shoreline has gone through. Consequently, they were, in most cases, unable to represent or predict future shapes of a certain subject area. On the other hand, aerial photo has provided a new, valuable source to the researchers in examining the shoreline change. As a result, researchers could not only overcome the insufficiency of the past observation literature, but also closely monitor the changes a certain shore undergoes.

Therefore, this researcher analyzed the aerial photographs of a subject area and derivative data therefrom by means of the digital orthophoto, thereby securing accuracy of evaluation and reliability of the data used. Moreover, analysis of aerial photo served as the basis for the further analysis of the long-term shoreline changes, and as the basis for representation and prediction of the subject area. Adopting these approaches, this researcher endeavored to exam various problems, which might possibly be caused by the changes of a particular subject shoreline in the future.

[1]Doctoral Student, Dept. of Civil and Envir. Eng., Myongji Univ., Yongin, Kyonggi 449-728, Korea
(Tel : 033-670-3481, Fax : 033-672-6156, E-mail : jungsj@mju.ac.kr)
[2]Prof., Division of Civil and Envir. system Eng., Kwandong Univ., Yangyang, Kangwon 215-800, Korea
(Tel : 033-670-3317, Fax : 031-672-2055, E-mail : kkhkim@kdccs.kwandong.ac.kr)
[3]Prof., Dept. of Civil and Envir. Eng., Myongji Univ., Yongin, Kyonggi 449-728, Korea
(Tel : 031-330-6409, Fax : 031-336-9705, E-mail : ckpyun@mju.ac.kr)

SHORELINE CHANGES OF A POCKET BEACH TRIGGERED BY CONSTRUCTION OF PORT BREAKWATERS AND FUTURE MEASURES – THE EXAMPLE OF SHIRATSURU BEACH

Takaaki UDA[1], Ryu-ichiro NISHI[2], Akio KIKUCHI[3], Toshiro SAN-NAMI[4] and Takayuki KUMADA[5]

[1]Dr. Eng., Executive Director, Public Works Research Center, 1-6-4 Taito, Taito, Tokyo 110-0016, Japan (+81-3-3835-3609, +81-3-3832-7397, email uda@pwrc.or.jp)

[2]Dr. Eng., Associate Prof., Dept. of Ocean Civil Engineering, Kagoshima University 1-21-40 Korimoto, Kagoshima, Japan

[3]Coastal Techno Company Ltd., 4-329-3 Ishimine, Shuri, Naha 903-0804, Japan

[4]Research Engineer, Coastal Engineering Laboratory Co. Ltd, 1-22-208 Wakaba, Shinjuku-ku, Tokyo 160-0011, Japan

[5]M. Eng., Ph.D. Student, Department of Oceanic Architecture & Engineering, College of Science & Technology, Nihon University, 7-24-1 Narashinodai, Funabashi, Chiba 274-8501, Japan

Ⅰ. INTRODUCTION

In Japan there are many pocket beaches of a variety of scales, and many fishing villages with a small port locate there. Historically these fishing ports have been located on the lee side of the headland, because of the wave-sheltering effect of the headland itself. In recent years, in order to modernize the fishing port, the facilities of the fishing port have been expanded, with the frequent construction of the new port in the outer region of the old fishing port. Although the enlargement of the fishing port was attained as a plan by this activity, many cases have been reported, in which the wave-sheltering effect affected the overall wave field on the pocket beach by protrusion of the breakwater from the lee side of the headland and induced severe beach erosion in the surrounding area. This study aims at investigating this problem, taking the Shiratsuru beach in Kumamoto Prefecture in Kushu in Japan as the example, and future measures are considered.

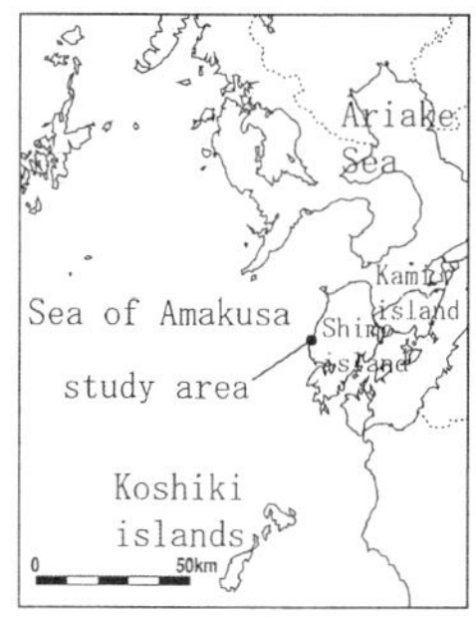

Fig. 1. Location of Shiratsuru beach in Amakusa in Kumamoto Prefecture.

Ⅱ. LOCATION AND ENVIRONMENT OF SHIRATSURU BEACH

The Shiratsuru beach is located at the central part of Amakusa District in Kumamoto Prefecture in Kushu in Japan as shown in **Fig.1** and it is a sandy beach of a stretch of around 900m length, facing the Sea of Amakusa. The Takahama River flows into the Shiratsuru beach and pours into the sea at the southern end of the pocket beach. The Shiratsuru beach is a typical pocket beach surrounded by Kami-ose and Shimo-ose Points at the northern and southern ends, respectively. Behind the shoreline, towns concentrated in the northern half of the Shiratsuru beach, whereas the environment of the natural sand dune was left behind as the coastal forest area along the right bank of the Takahama River in the southern half. The Takahama fishing port, being an important port for evacuation in this district, is located on the southern rocky coastline. In recent years, the wave conditions on the pocket beach have greatly changed due to the extension of the breakwater of the Takahama fishing port at the southern end of the pocket beach. In the northern part, the shoreline retreated in contrast to the shoreline advance in the southern part of the pocket beach.

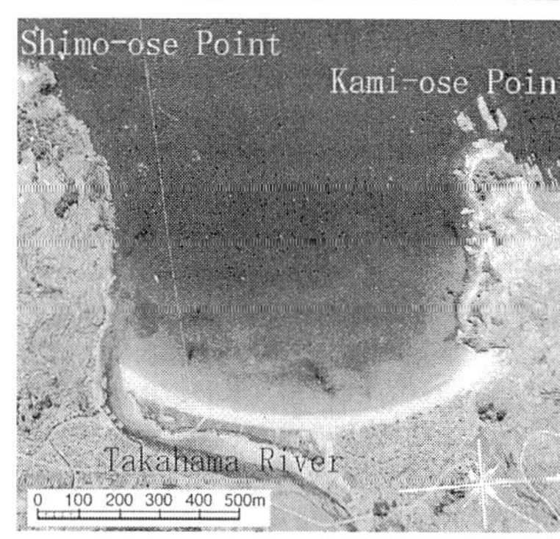

Fig.2(a) Aerial photograph of Shiratsuru beach in May, 1947.

Fig.2(b) Aerial photograph of Shiratsuru beach in October, 1999.

Ⅲ. COMPARISON OF AERIAL PHOTOGRAPHS

On the Shiratsuru beach, eight sets of aerial photographs have been taken during 52 years between 1947 and 1999. Here, the long-term changes of the coastline are investigated through the comparison of aerial photographs taken in 1947, 1965, 1994 and 1999. For example, **Fig.2(a)** shows the aerial photograph of the Shiratsuru beach taken in May, 1947 before the large landform change. At this time, large-scale structures affecting wave conditions of this pocket beach did not exist. **Figure 2(b)** shows the aerial photograph in October, 1999. The breakwaters of the new Takahama fishing port were built, and the surrounding shoreline largely changed after the construction of the breakwater of the fishing port. In the northern part of the pocket beach, the shoreline retreated and it advanced in the southern part. The configuration of the natural pocket was artificially changed due to the construction of the breakwater and the shoreline changed responding the newly formed wave field.

The shoreline positions in 1947 and 1999 were read from the aerial photographs as shown in **Fig.3**. The shoreline greatly advanced mainly in the wave shadow zone of the Takahama fishing port and the amount of the maximum shoreline advance attained 95m. On the contrary, the shoreline recession in the northern part was comparatively small compared with the shoreline advance in the southern part. The stable shoreline was predicted by using Hsu and Evans model improved by Serizawa et al. (1996) in order to satisfy sand budget. The predicted shoreline in 1947 agreeed very well with the measured one. On the contrary, predicted shoreline change in 1999 is larger than measured one, implying further shoreline changes in the future.

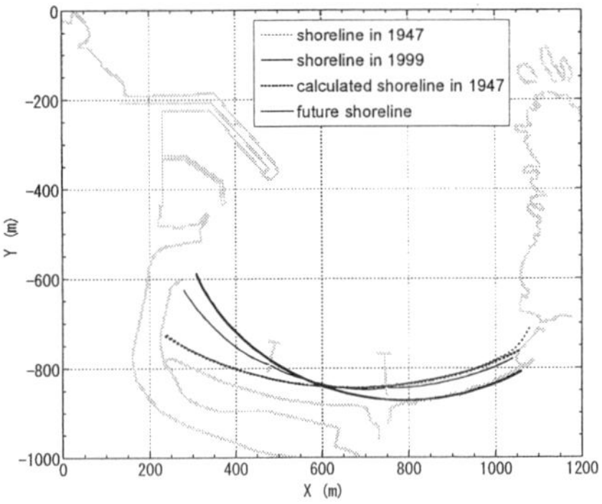

Fig.3. Comparison of measured and predicted shoreline positions.

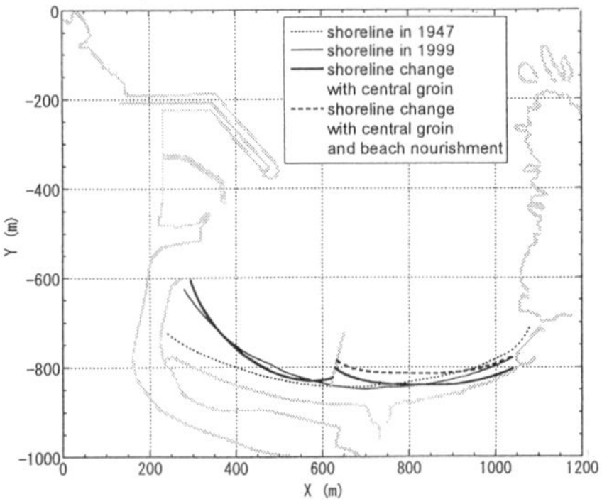

Fig.4. Prediction of shoreline changes in case of the construction of a groin at the center of the pocket beach.

Ⅵ. MEASURES

A sequence of phenomena occurred on the Shiratsuru beach was summarized first. Then, under those circumstances, various measures have been considered. Important notes in considering the future practical measures without large changes of existing coastal facilities and town area located behind the shoreline are summarized as follows:

(1) Rise of crown height of seawall to prevent wave overtopping

The shoreline recession was caused inevitably associated with the construction of the breakwater. Accordingly, urge people to be patient by explaining the reason of the beach changes, although considerable damages are caused to the coastal view and coastal utilization. This is the most economic way.

(2) Set-back of the seawall in accordance with the shoreline recession

This method is useful in leaving the natural sandy beach and restraining the total construction cost, but houses in the town must be moved.

(3) Separation of the beach by constructing a long groin at the central part of the pocket beach

The shoreline form in this case predicted by the modified method of Hsu and Evans (Serizawa et al., 1996) is shown in **Fig.4**. A long stretch of natural coastline of the Shiratsuru beach is separated into two portions, and this will lose popularity for many bathers and surfers of this beach. However, in order to recover and maintain sandy beach in the northern part of the pocket beach, this is the only way. If the foreshore is expected to be further widened in the northern part of the groin, beach nourishment is also required as shown in **Fig.4**. The weak point of this method is the disappearance of the natural beauty of the Shiratsuru beach by the construction of the large structure. Considering that the past measure being adopted for a long time on this beach was the construction of the groin, this must be the most orthodox method, if the groin of optimum scale is built at the appropriate location.

(4) Sand recycling

A short groin is newly built at the central part of the beach to control the sand transport toward the fishing port, and sand transported over this short groin and accumulated near the fishing port is dredged and returned to the northern beach again artificially. However, sand deposition at the Takahama fishing port and the Takahama River mouth is extreme at present, and the selection of this way is considered to be difficult in the long-term.

Ⅴ. CONCLUDING REMARKS

The measure against beach erosion occurring on the Shiratsuru beach is insufficient in the long-term, because it does not intend to stabilize the shoreline. On the other hand, the construction of the new groin at the central part of the beach denies the desire that the natural sandy shore of the Shiratsuru beach is preserved as far as possible. The postponing of the selection of the fundamental measure results in the failure of the project in spite of the large expenditure. It is required for people and related engineers to accurately recognize the present problem and discussions open to the public should be made among stakeholders. Based on the accurate recognition of the beach changes on a pocket beach associated with the construction of the artificial structures, inexpensive measures should be considered including any possibilities and the decision must be made under opening of the accurate information to the public.

A NUMERICAL MODEL OF THREE-DIMENSIONAL BEACH DEFORMATION FOR GRADED SEDIMENTS

Takahide HONDA*, Hiroshi KOBAYASHI, Shinji SATO*, Akira WATANABE*, Masahiko ISOBE* and Toshimasa ISHII

Laboratory experiments were performed on beach evolution and sediment sorting around coastal structures on a initially plane 1/20 sloping beach of uniformly graded sand. One experiment was for a detached breakwater, and the other for groins. Spatial and temporal variations of beach topographies and grain size distribution of surface sediments were described. A numerical model was developed on the basis of the enhanced Boussinesq equation and a sand transport model for mixed-grain size sands under atilt waves. The validity of the model was verified with experimental data.

Wave basin was 6.5m long and 11m wide. 2 cases off-shore structures, a detached breakwater and groins, were set. Regular waves were used. Mixed-grain size sands have the medium grain size d50=0.36mm. The experiment was performed under following conditions: incident wave height H_0=4.4cm, wave period T=1.0s, off-shore depth h_0=23.0cm and incident wave angle $\alpha=0^\circ$ for a detached breakwater, 30° for groins. At first, the wave level and the wave velocity at the lower layer (along-shore and cross-shore direction) were measured at 200 pt. (15cm mesh) without sands. Then mixed sands were spread over the basin. The beach topographies were measured and mixed sands at the surface and vertical cores were sampled every 1 hour. After 5 hour, we confirmed the beach evolution to have reached equilibrium. For all sampling sands, the grain size distribution was measured each other. Fig3-4 were the example for the experiment result.

The improved numerical model introduced the effect due to the atilt waves and extended the mixed sand model to three-dimension. And investigating for the swash model, we choose the porous swash model. First, we calculated the two-dimensional long-scale experiments using the improved model. This model showed the almost full stage for experiments, where former model could only show the initial stage. Next we applied the improved model to three-dimensional laboratory experiment. The results proved that coarse sands were accumulated near the on-shore.

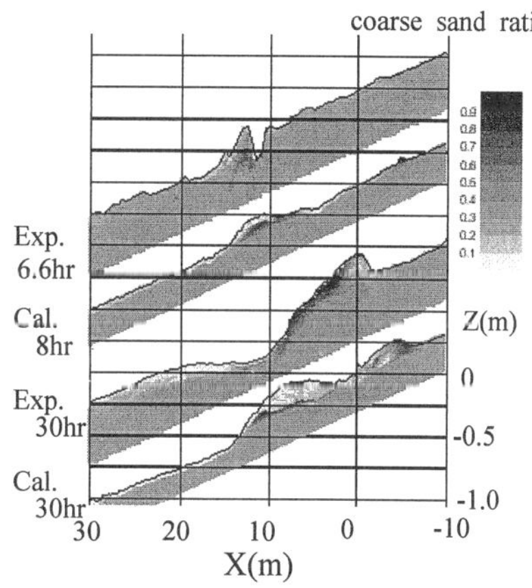

Fig.1 Calculation for the two-dimensional experiment

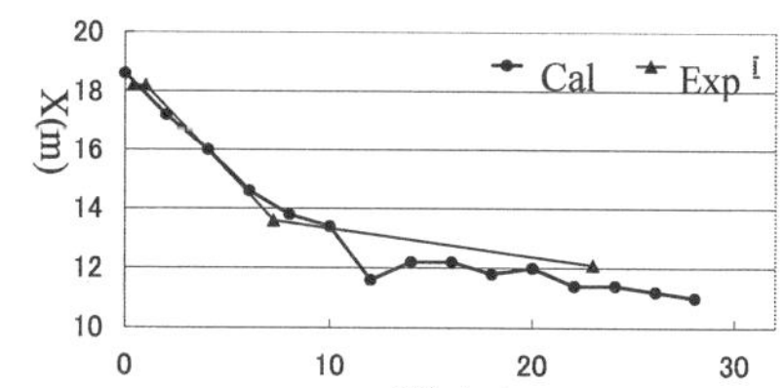

Fig.2 wave breaking point transfer

*The University of Tokyo (7-3-1 Hongo Bunkyoku Tokyo, 03-5841-6114, FAX 03-5841-8503)

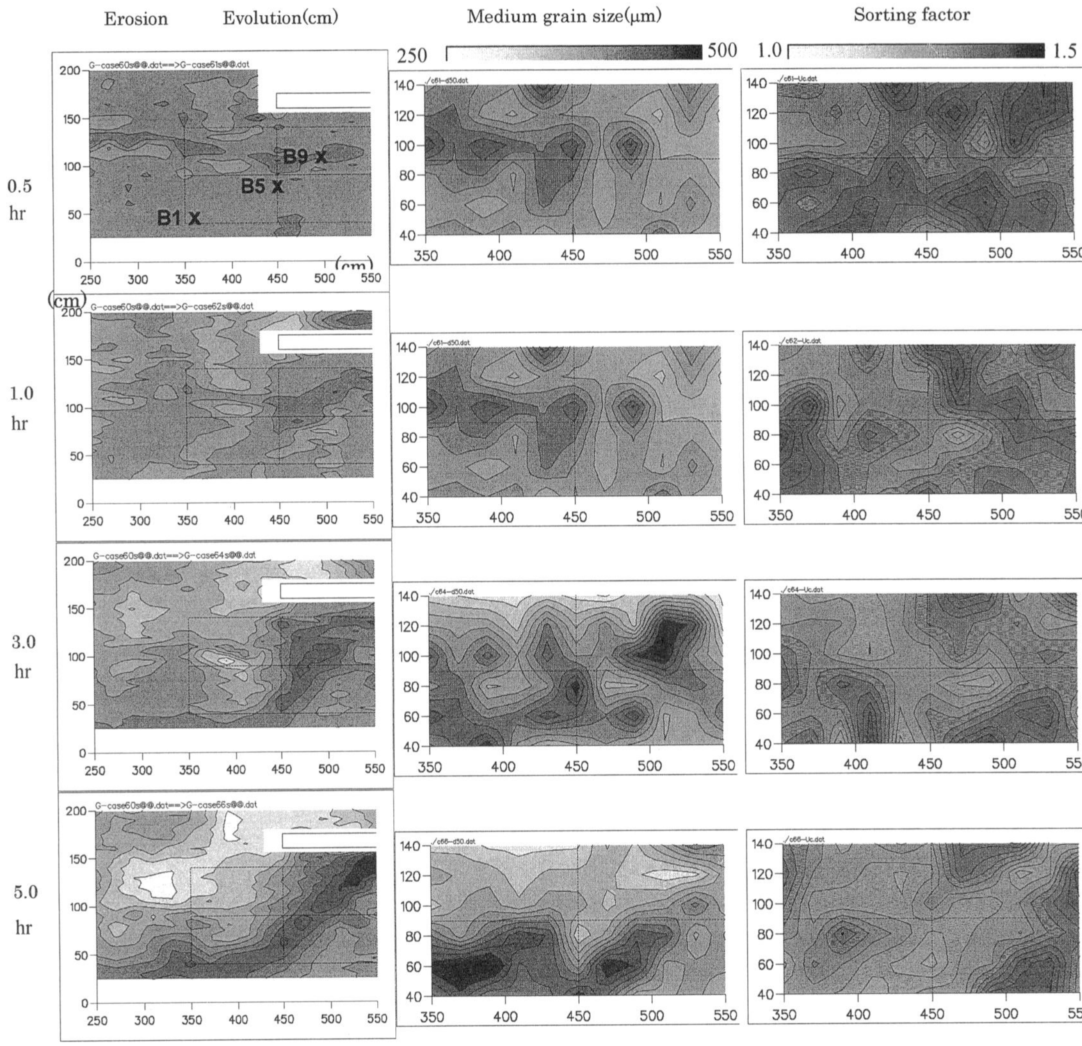

Fig.3 Beach evolution around the detached breakwater

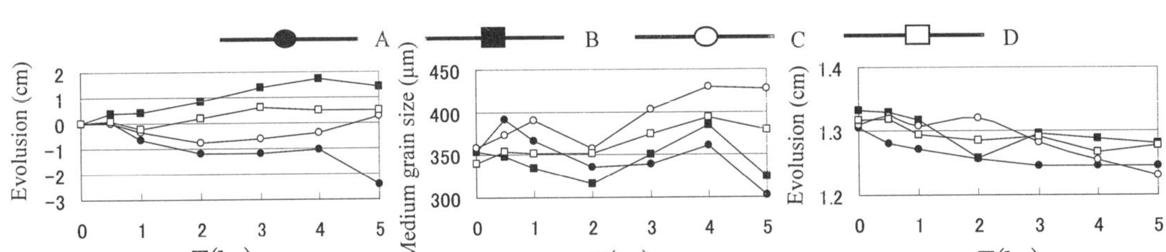

Fig.4 Time change of beach evolution and sediment sorting

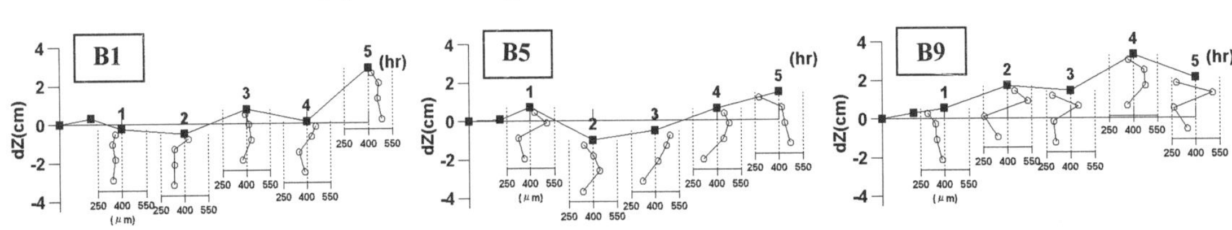

Fig.5 Time change of vertical sediment sorting

Field Measurement on Sorting of Bed Material and Topography Change on Mixed Grain Size Beach

Susumu Araki Research Associate, Dept. of Civil Engineering, Osaka University, Japan
Ichiro Deguchi Professor, Dept. of Civil Engineering, Osaka University, Japan
Tomohiro Ikeda Graduate Student, Dept. of Civil Engineering, Osaka University, Japan
Tsutomu Mitsuishi Undergraduate Student, Dept. of Civil Engineering, Osaka University, Japan
(Address) Department of Civil Engineering, Osaka University
2-1 Yamada-oka, Suita, Osaka, 565-0871 Japan
(Tel) +81–6–6879–7614 (Fax) +81–6–6879–7616
(E–mail) araki@civil.eng.osaka-u.ac.jp

1. BACKGROUND AND AIMS

Bed material in the field often has widely spreading grain size distribution. In general, topography change is predicted by evaluating sediment transport rates using representative grain size, for example, the mean grain size. In a beach where grain size distribution is narrow, such a procedure can apply for evaluateing sediment transport rates because almost all of the particles are transported with the same mode (bed load, suspended load or sheet flow). On the other hand, in a beach where grain size distribution is wide, direction of bed material transport depends on its grain size because the mode of transport of fine bed material is not always the same as that of coarse bed material. In addition, the interaction in sediment transport between fine and coarse bed materials have to be taken into account.

Some procedures have been proposed to take account of the effect of the grain size distribution on the sediment transport (Sakai *et al.*, 2000, Sato *et al.*, 2000, Dibajnia *et al.*, 2001, Tanaka *et al.*, 2001) on the basis of laboratory experiments. On the other hand, transport of bed material with widely spreading grain size distribution in the field has not been investigated yet due to the complex sediment transport.

In the present study, sorting process of bed material and resulting topography change on a sloping beach where bed material is widely spreading grain size distribution were investigated through field measurements for one year.

2. OUTLINE OF FIELD MEASUREMENT

Field measurements were carried out on Narugashima Beach located in the southeastern part of Awaji Island, facing the Kitan Strait. Bed material in the objective beach is mainly composed of sand of mean grain size d_{50}=0.5mm and small pebbles of mean grain size d_{50}=10–20mm. We measured bottom profiles near the shoreline on three measuring lines set at the longshore interval of approximately 20m in September, November, December 2001, March, May and August 2002. Bed material on each measuring line was also sampled at 2.5–5m interval in the cross-shore direction and 5–20cm interval in the downward from the surface. Incident waves were measured by pressure-type wave gauge in December 2001 and March 2002.

3. MAIN RESULTS

Figure 1 shows a measured bottom profile and a cross-shore distribution of mean grain size d_{50} at the surface in September, November, December 2001 and March 2002. From the change in bottom profile, the beach was eroded during this period and the berm disappeared by November. At the beginning of erosion (from September to November), the mean grain size d_{50} at the surface around the shoreline (x=20–25m) became small because fine sand behind the berm in September moved offshore by erosive wave. However, from November to March, the mean grain size d_{50} at the surface around the shoreline became large because the fine sand moved farther offshore. This transport of the fine sand is also verified from the time history of the vertical distribution of bed materials sampled around the shoreline.

Figure 2 shows a measured bottom profile and a cross-shore distribution of mean grain size d_{50} at the surface in March, May and August 2002. In the figure, the bottom profile and the distribution of d_{50} measured in September 2001 is also plotted for comparison. Although bed material accreted from March to August, the erosion taken place during September 2001 to March 2002 is not completely recovered, especially around the berm. During this period, the mean grain size d_{50} at the surface of x=17.5m became large and the d_{50} at the surface of x>20m became small. This means that coarse bed material and small pebble moved onshore by accreting wave. The cross-shore distribution of d_{50} at the surface in August 2002 is coarser than that in September 2001 in a large part of sampling points.

REFERENCES

Dibajnia, M., Y. Hashimura and W. Kioka (2001). Proposal for new sheet flow transport model for mixed grain size sand, *Proc. of Coastal Engineering*, JSCE, Vol. 48, pp. 501-505. (in Japanese)

Sakai, T., H. Gotoh, E. Harada, T. Takahashi and Y. Hamada (2000). Unsteady armouring in vertical grading of mixed size grains under oscillatory sheet flow, *Proc. of Coastal Engineering*, JSCE, Vol. 47, pp. 496-500. (in Japanese)

Sato, S., M. Tanaka, N. Hikawa, A. Watanabe and M. Isobe (2000). Development of a sheet flow transport model based on the movement mechanism of mixed grain size sand, *Proc. of Coastal Engineering*, JSCE, Vol. 47, pp. 486-490. (in Japanese)

Tanaka, M., H. Kobayashi, S. Sato, A. Watanabe and M. Isobe (2001). Effect of sands mixing ratio on deformation of beach profile, *Proc. of Coastal Engineering*, JSCE, Vol. 47, pp. 506-510. (in Japanese)

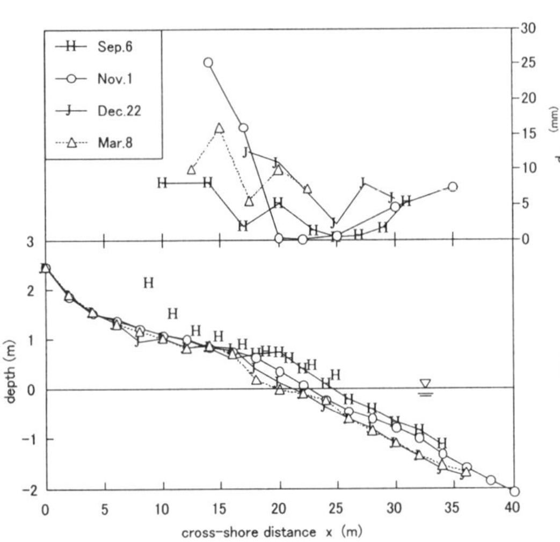

Fig. 1. Measured profile and cross-shore distribution of d_{50} (Sep.6, 2001–Mar.8, 2002)

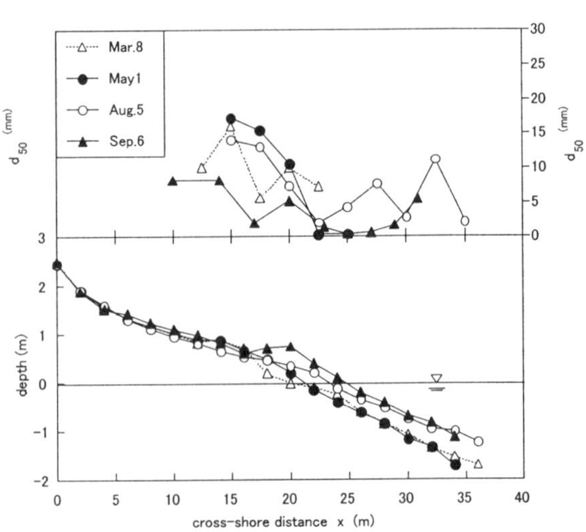

Fig. 2. Measured profile and cross-shore distribution of d_{50} (Mar.8, 2002–Aug.5, 2002)

TRANSPORT MECHANISM OF NONUNIFORM SANDS AND ITS ONSET OF SHEETFLOW UNDER ASYMMETRIC OSCILLATORY FLOWS

Ahmed S. M. AHMED[1] and Shinji SATO[2]

1. Introduction

Various criteria of onset of sheetflow were developed but for uniform sediments under sinusoidal oscillatory flows. A criterion of onset of sheetflow for nonuniform sands under nonlinear oscillatory flows is not yet comprehensively investigated, despite the criterion proposed by Dibajnia and Watanabe (1996) that required further verifications. On the other hand, movement mechanism of individual sediment particles in sediment mixture is worthwhile to be investigated.

2. Objectives

The main objective of this paper is to understand the transport mechanism of individual sediment particles in the sediment mixture leading to the bed formation. Enhanced PIV technique developed by the present authors, Ahmed and Sato (2001), and verified by Ahmed and Sato (2002a), was utilized. A comparison between seven onset of sheetflow criteria was achieved after conducting a sensitivity analysis of determining the parameters representing the mean diameter of nonuniform sands and asymmetric oscillations.

3. Experiments

Experiments with uniform sands, conducted by Ahmed and Sato (2002b), and with nonuniform sands composed of three-grain size sands, conducted by Ahmed and Sato (2002c), were utilized in which rippled bed and sheetflow regimes were observed. Experiments with uniform fine sand, D = 0.21, uniform coarse sand, D = 0.74, and with mixed sand composed of the former sizes were conducted to measure the sand particle velocity.

4. Results

Figure 1 shows the temporal variation of sand particle velocity at the still bed for uniform fine sand and uniform coarse sand. It is clearly observed the velocity of coarse sand is slower than that of fine sand. Figure 2 shows the temporal variation of sand particles velocity of uniform fine sand and fine sand in sand mixture at left panel. That clearly demonstrates the reduction of the fine sand velocity as it is mixed with coarse sand. On the other hand, the velocity of the coarse sand in the sand mixture increased compared to the velocity of uniform coarse sand, as shown in right panel of Fig. 2. Both fine sand and coarse sand movement resulted in local net transport to onshore or offshore. Based on the probability of presence of the coarse sand, the onshore side was coarsened leading to bed undulation and finally rippled bed was formed at the onshore side even while flat bed was formed as the fine sand transported to offshore. This ensures the significant parameters to be utilized to propose a criterion of onset of sheetflow for nonuniform sands under asymmetric oscillation.

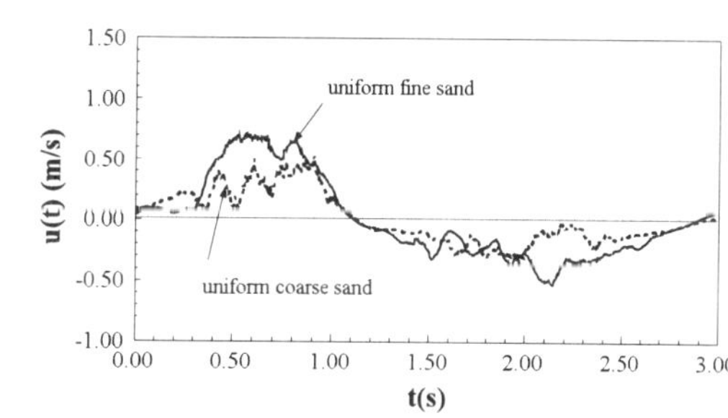

Figure 1: Temporal variation of particles velocity of uniform fine sand and coarse sand at still bed

[1] Researcher, Hydraulics Research Institute, National Water Research Center, Delta Barrage 13621, Egypt
Fax No. +(202)-2189539
Email: ahmed_say@hotmail.com or ahmed@coastal.t.u-tokyo.ac.jp (until September 2002) all correspondence are addressed to the first author

[2] Professor, ditto
Email: sato@coastal.t.u-tokyo.ac.jp

In addition, Dibajnia and Watanabe (1996) criterion was examined which failed to classify the present experimental data. As the criterion of Sato and Horikawa (1988) implies the Shields parameter, Ψ, and sediment flow acceleration (d_o/D) in which d_o is the water particle orbital diameter and D is the sediment diameter that are of importance parameters. Sato and Horikawa (1988) criterion showed successful performance to classify the present experimental data as well as previous experimental data, as shown in Fig. 3.

5. Conclusions

PIV technique demonstrated the reduction of fine sand velocity and the increased velocity of coarse sand, as they exist in a sand mixture. Sato and Horikawa (1988) onset of sheetflow criterion could be extended to present the onset of sheetflow for nonuniform sand under nonlinear oscillation.

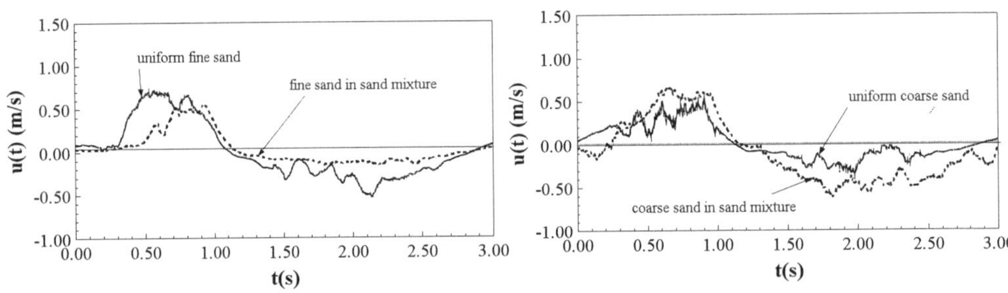

Figure 2: Temporal variation of particle velocities of uniform fine sand and fine sand in sand mixture (left) and of uniform coarse sand and coarse sand in sand mixture (right)

6. References

Ahmed, A. S. M. and Sato, S., (2001). Investigation of bottom boundary layer dynamics of movable bed by using enhanced PIV technique, *Coastal Engineering Journal*, Vol. 43, No. 4, pp. 239-258.

Ahmed, A. S. M. and Sato, S., (2002a). Enhanced PIV technique to investigate bottom boundary layer dynamics, *An International Symposium for PIV and Modeling Water Wave Phenomena*, Peterhouse, Cambridge, UK.

Ahmed, A. S. M. and Sato, S., (2002b). A sheetflow transport model for nonlinear oscillatory flows, Part I: Uniform grain size sediment, *Coastal Engineering Journal*, *(Submitted)*.

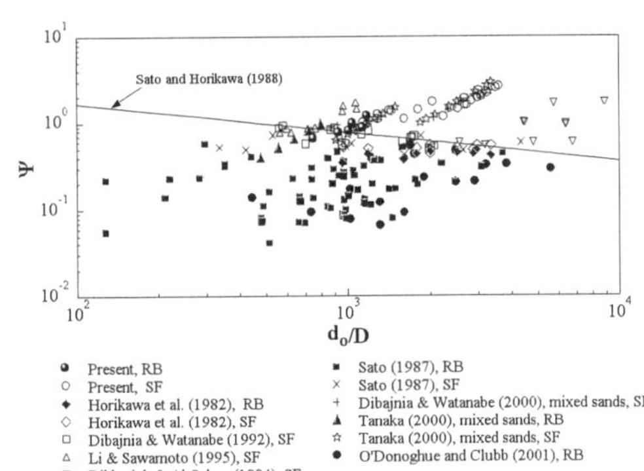

Figure 3: Verification of onset of sheetflow criterion of Sato and Horikawa (1988)

Ahmed, A. S. M. and Sato, S., (2002c). A sheetflow transport model for nonlinear oscillatory flows, Part II: Mixed grain size sediment, *Coastal Engineering Journal*, *(Submitted)*.

Dibajnia, M. and Watanabe, A. (1996). A transport rate formula for mixed-size sediments, *Proc. of 25th ICCE*, ASCE, pp. 3791-3804.

Sato, S. and Horikawa, K. (1988). Sediment ripple geometry and sediment transport mechanism due to irregular oscillatory flows, *Proc. 21st ICCE, ASCE*, pp. 1748-1762

Numerical Model on Filtration Flows across a Beach Face and Sediment Transports in the Swash Zone

Md. Azharul Hoque
Graduate Student, Kagoshima Univ.

and

Toshiyuki Asano
(Corresponding Author) Prof., Dept. of Ocean Civil Engrg.
Kagoshima Univ., 1-21-40, Korimoto, Kagoshima, 890-0065, JAPAN
Tel. & Fax : +81-99-285-8482
e-mail : asano@oce.kagoshima-u.ac.jp

1.Aims

The importance of sediment transport in the swash zone has been widely recognized, however, the present understanding is not satisfactory. The main reason for this comes from the complexity of the sediment transport mechanism. One of the important and unsolved factors is the in-/ex-filtration effect on the hydrodynamics and the sediment transport.

In this study, a numerical model is proposed by coupling an analysis on ground water flows in a sandy ground with an analysis on swash wave motion over a uniform slope by inserting an imaginary thin layer of high conductivity at the interface. The hydrodynamic properties of in-/ex-filtration flows across the beach face have been discussed.

Next, using the obtained velocity fields over the permeable beach slope, the sediment transport mechanism has been investigated. For the stabilizing/destabilizing effects that the filtration flows act on the sediment transport, two opposing processes, modification of the effective weight of the sediment particle and to modification of the boundary layer thickness have been considered. These effects of the filtration flows on the sediment transport rate have been quantified based on a bed load transport model with a modified Shields parameter.

2. Numerical model on wave-induced in-/ex-filtration velocity

The motions of groundwater flows influenced by the boundary condition at the beach face such as tidal fluctuations and waves driven sea level oscillations, are sluggish. In contrast, the wave induced up-rush and downwash motions on the beach face are quick with a time period of a few seconds. Here, a problem in the numerical analysis will arise of how to connect the slow motion under the beach surface and the quick motion over the beach surface. In this study, a numerical procedure, so called an artificial flow-deflecting layer proposed by Sugio and Desai (1987) was adopted. This layer is a strip of a finite thickness with high conductivity about ten times greater than the adjacent flow domain.

A schematic diagram of the problem is shown in Fig. 1.The thin layer ACFD is the flow deflecting layer.

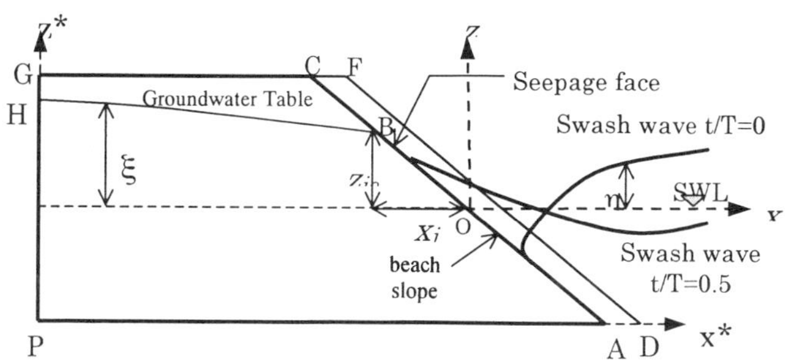

Fig. 1

Figure 2 illustrates the calculated vertical filtration velocity at various on-offshore locations in the swash zone. The point at x=200 is equal to the maximum rundown point. As a result, the point is situated on the in-outflow face and subjected to swash wave motion for entire duration of the wave period. It can be noticed that the points, x=100 cm and x=-100 cm are subjected to the swash wave motion for the partial duration of the wave period. The filtration velocity components at onshore ward locations are found to be higher than that at offshore ward locations.

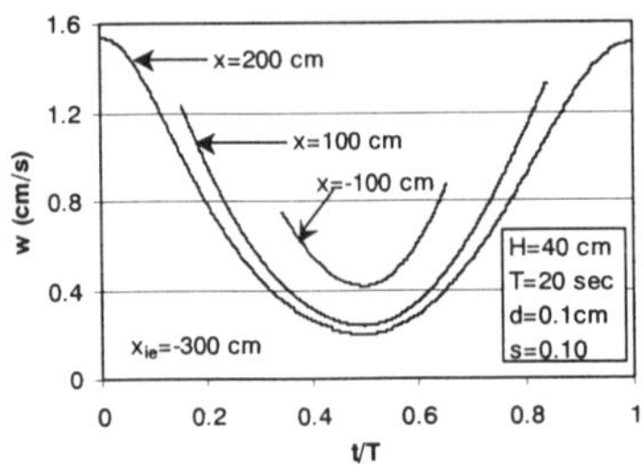

Figure 2

3 Effects of filtration velocity on sediment transport

Based on the obtained velocity data, the instantaneous and phase averaged sediment transport rates under the filtration effects are discussed. To take into account of the in-/ex-filtration effects on the sediment transport, two opposing processes on the stability of the sediment surface layer are considered; that is, effective weight modification and boundary layer modification. These effects of the filtration flows have been quantified based on a bed load transport model with a modified Shields parameter proposed by Nielsen(2001). By systematic calculations varying the individual input parameters, the effects of the filtration flows on altering the sediment transport rate in the swash zone have been investigated.

Spatial distributions of time averaged transport rate for different sediment sizes are shown in Figure 3. The notation of Q_w indicates the time averaged sediment transport with the filtration effects, and Q_0 without the filtration effects. The changes are less significant with approaching to the onshore-ward, because the existence time of the wave front is very small compared to the whole wave period. The substantial changes are observed at offshore-ward locations where maximum sediment transport occurs. Without filtration effects, sediment transport for all the particle sizes direct to the onshore. Fig. 3(a) shows the time averaged sediment transport for the finer particle case, where offshore-ward sediment transport dominates. In contrast, Fig. 3(b) illustrates the transport direction turns to the onshore-ward by considering the filtration velocity, although the magnitude is not large. The variations are mainly attributed to the boundary modification effects.

The difference between Q_w and Q_0 ($Q_{filtr}=Q_w-Q_0$) will isolate the contribution of the filtration velocities on the dimensionless time averaged sediment transport. The grain size dependency of the transport Q_{filtr} is investigated. It is found that the direction of Q_{filtr} becomes more onshore with increasing grain size. The critical grain size at which the balance changes; 'crossover' point, is found to be around 0.03 cm.

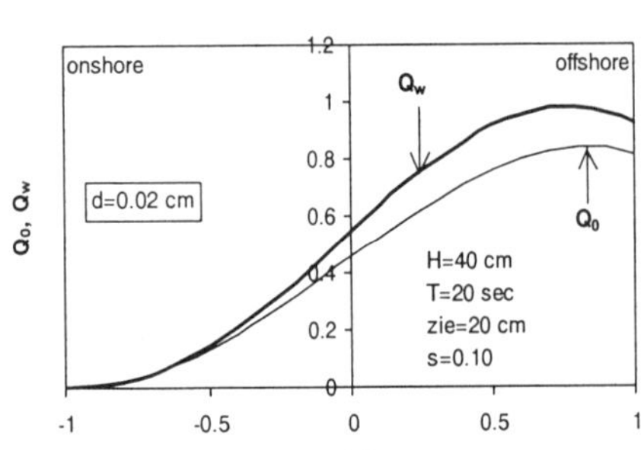

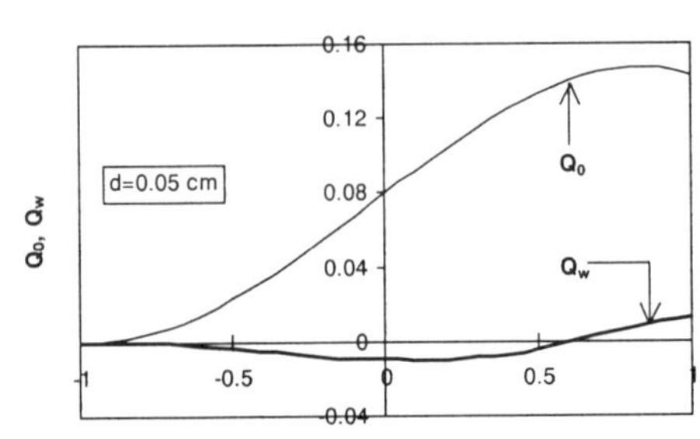

Fig.3: Spatial distribution of time averaged transport rate with (Q_0) and without (Q_w) filtration

TIDAL FLOW NUMERICAL SIMULATION AND SILTATION CALCULATION OF SEAWARD CHANNEL AND HARBOR BASIN OF THE ARTIFICIAL LAKE IN THE NEW HARBOR CITY OF SHANGHAI INTERNATIONAL SHIPPING CENTER

Tang Shifang
(Headquarter of Shanghai Deep Water Port Sub-headquarter of New Harbor City, Development Bank Tower,500 Pudong Road(South),Shanghai, 200120 China
Tel:+86 21 58876443, Fax: +86 21 58880280, Email:sft@shharborcity.com)

Abstract

The first phase construction of the Yangshan Deep-water Port (Shanghai International Shipping Center) is composed of three parts as harbor district, the Donghai bridge and land supporting base of the New Harbor City. The project of harbor district concludes 1600-meters quay, five berths of deep-water container ships, whose annual design handling capacity is 2.2 million TEUs and where the 5th and 6th generation container ships as well as the 8000 TEU container ship can berth. Till 2020 about 30 post-Panamax container ship berths should be finished, whose annual handling capacity is 15 million TEUs at least. The Donghai bridge, which is 31 kilometers long , connects the Luchaogang town in the north and the Xiaoyangshan Harbor District in the south. The part of the bridge over the sea is 25 kilometers in length and 31.5 meters in width, which is designed according to standard of the two-direction six-driveway highway. The New Harbor City, whose planning construction area is 95 square kilometers and residential population is 300000, is the land supporting base of the Yangshan Deep-water Port. At last there'll be an ecotype, intellectualized and modern new harbor city with developed port industry, international transit system and unique style and features.

In the plan the 5 square kilometers artificial lake will be the center of the city and then the construction will diffuse around it annularly. The semi-island connected with the lake shore will stretch into the lake, where there'll be theater, the Marine Museum and the Ocean Aquarium. The administrative and commercial area will be located in the first ring around the center of the lake, which includes the administrative building, the port administration building, the Shipping Exchange and large-scale public square. The next ring is the city park area and then the various "urban island" residential blocks surrounded by the graceful natural scene.

The Luchao Lake is circular, whose diameter is 2.5 kilometers. The highest water level is between 3.3 and 3.4 meter, the landscape water level is between 2.5 and 2.7 meter, the shock pre-reducing water level is 2.2 meter, the elevation of the lake bottom is between 1.0 and –1.0 meter and the side slope is 1:15. There'll be 7 radial drainages connected with the lake. Between the drainages and the lake there'll be a floodgate. Three chain drainages extend from inside to the outside around the lake. The floodgate and lock will be set between the lake and the sea to control the water level of the lake and the pass of the ships. The following research work should be carried out after the construction of the lock:

1. The intrusion of the salt water

When ships pass the lock, there will happen the exchange of salt water and fresh water, which causes the intrusion of the seawater and then contaminates the fresh water area. There exist four types about the ways of seawater encroachment:

density flow encroachment, water level difference encroachment, seawater encroachment caused by the abstraction effect of the ships and the leakage of the navigation lock. The paper discusses the mechanics and the procedure of the seawater encroachment listed above, calculates the speed of the density flow salt wedge and the amount of the encroaching salt caused when the ship passes the lock, and brings forward some measures for the anti-salt of the navigation locks.

2. The siltation below the lock and in the navigation channel

The navigation channel is located in the Luchaogang in the north shore of the Hangzhou Gulf, which is an artificial-dug navigation channel. The siltation, which happens with the change of the flow condition in the project area, is caused by the construction of the navigation lock building and downward dig of water depth of the channel. And it will affect the draw-off to the open sea and the navigation of the channel. So the paper will use the method of 2-dimensional current numerical model to simulate the change of the current field before and after the project and then adopt the empirical formula to calculate the amount of the siltation below the lock and in the channel.

Development of Gravity Drainage System for Beach Protection

Shin-ichi Yanagishima*, Kazumasa Katoh, Naoto Iwasa and Yoshiaki Kuriyama
* Marine Environment and Engineering Department, Port and Airport Research Institute
Nagase 3-1-1, Yokosuka, Kanagawa 239-0826, Japan
Facsimile: +81-468-41-9812, E-mail: yanagishima@pari.go.jp

Introduction

Beaches in Japan have chronically suffered from erosion problems for more than forty years owing to the decrease of sediment supply from rivers and interference with longshore sediment transport by coastal structures. As a result, for beach protection, many coastal structures such as groins, detached breakwaters and seawalls have been constructed, and these work effectively from the viewpoint of disaster prevention. However, the structures keep people from accessing beaches even under mild wave conditions, and deteriorate valuable landscape. Therefore we developed a new beach protection system with gravity drainage, which has no structures on a beach.

Outline of the system

An abrupt beach erosion in a storm is caused by the groundwater seepage from around the top of the berm in the foreshore. In a storm, the groundwater level near the shoreline gets high owing to the penetration of run-up waves, and the groundwater seeps out from around the top of the berm with transporting sediments seaward.

Because in the gravity drainage system, schematically shown in Figure 1, the groundwater is naturally drained seaward through a permeable layer installed below the beach surface and the groundwater level is lowered, the system is expected to prevent an abrupt beach erosion in a storm and protect the beach.

Field measurements

To examine the effect of the system on beach protection, field experiments have been conducted at the Hasaki Coast in Japan facing the Pacific Ocean since 1994. A permeable layer of 8 m in width and 88 m in length was installed about 3 m below the surface in the foreshore, and topography and groundwater level were measured in an area 70 m wide in the alongshore direction and 100 m long in the cross-shore direction.

Furthermore, salinity and the concentrations of nutrients were measured in the permeable layer and in the sea to investigate the effect of the layer on water purification.

Results and conclusions

Figure 2 shows the volumetric changes of beach profiles for 5 years along the permeable layer and 35 m away from the layer, where the ground water level was almost equal to that in the adjacent natural beach. The volume of the beach profile with the permeable layer was mostly larger than that without the layer. This result confirmed that the permeable layer is effective to decrease a speed of beach erosion in a storm, and to enhance the recovery in a calm.

In the permeable layer, the concentration of NH_4-N was higher and that of NO_3-N was lower than those in the seawater. This result indicates that oxidization by bacteria took place in the layer, and hence the permeable layer is expected to promote seawater purification.

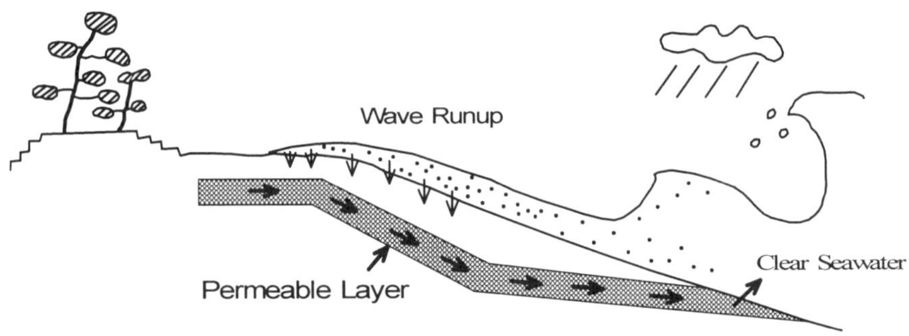

Figure 1 Beach protection system with gravity drainage.

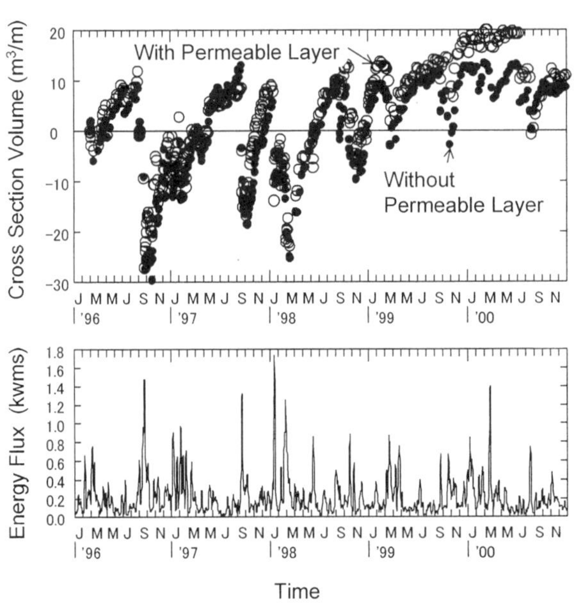

Figure 2 Volumetric changes of beach profiles with and without permeable layer and the offshore wave energy flux. Open circles and solid circles in the upper panel show the volumes with and without the permeable layer.

ESTIMATION OF TURBIDITY USING ADCP AND BEHAVIOR OF SUSPENDED SOLIDS IN THE COASTAL AREA

Toshihiko YAMASHITA (Division of Environmental Resources Engineering, Hokkaido University, N-13, W-8, Sapporo 060-8628, Japan (Fax: +81-11-706-6184, E-mail: y-toshi@eng.hokudai.ac.jp))

Junnya SUMIE (Graduate School of Engineering, Hokkaido University)

Shunsuke NARA (North Japan Port Consultants Co.,Ltd.)

Shin-ichi YAMAZAKI (Civil Engineering Research Institute of Hokkaido)

1. Introduction

Toward clarifying the sediment budget of a river estuary, it is important to understand the process of deposition and movement of sediment of various grain sizes that is discharged from a river in the coastal area. Because such discharged fine-grained components include high content of nutrients such as nitrogen and phosphorus, and form water masses of high turbid, they exert a considerable influence on water quality and biological activities of the sea. Reflection strength, measured along with flow velocity by ADCP, correlates with the concentration of suspended solids in seawater, there is possibility of estimating multi-layered turbidity. Moreover, if the relationship between turbidity and concentration of suspended solids is determined, it is possible to estimate the volume of movement of suspended solids.

2. Procedure

In this study, ADCP and four turbidimeters were installed in the coastal areas of the Mukawa River and the Shiribetsu River in Japan, and on-site observation was conducted to investigate the movement, diffusion, and settling of suspended solids discharged from a river, and the lifting and transport of bottom sediment caused by wave motion. The authors compared data from reflection strength of ACDP and turbidimeters, and examined the possibility of estimating turbidity from reflection strength. Moreover, an attempt was made to calculate transported components of suspended sediment by clarifying the relationship between turbidity and the concentration of suspended solids, and the vertical distribution of velocity by ADCP in the coastal area.

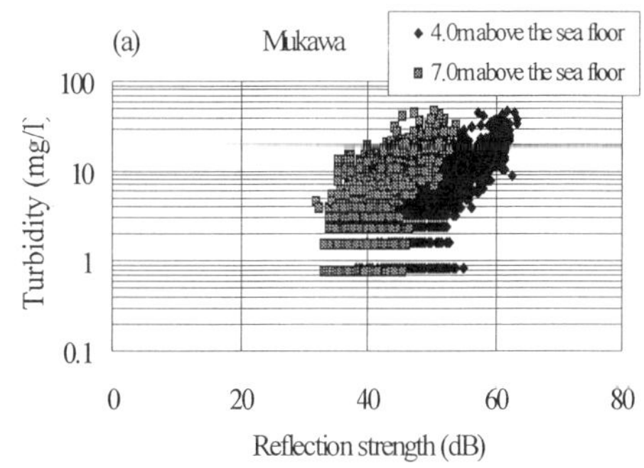

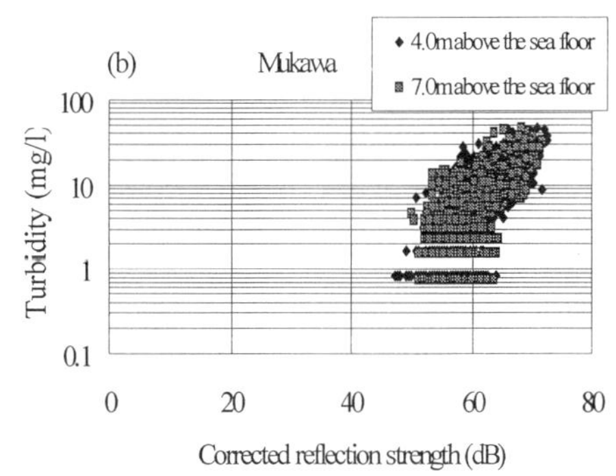

Fig.1 Correlation of turbidity and reflection strength

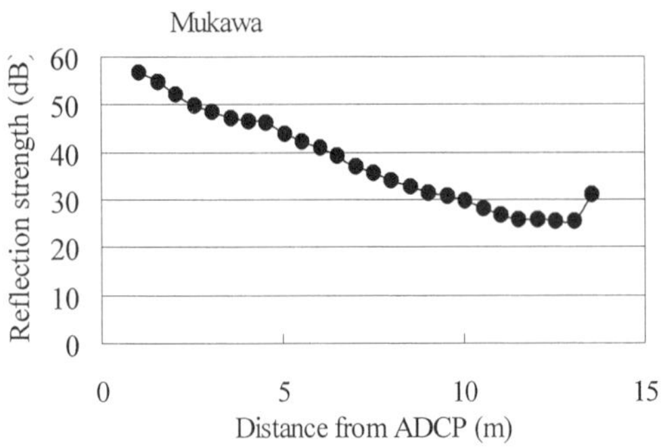

Fig.2 Propagation loss of reflection strength

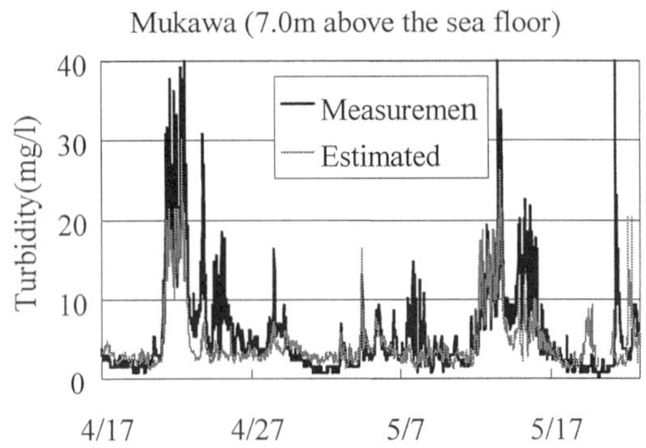

Fig.3 Time series of changes in turbidity

3. Conclusions

The relationship between reflection strength of ADCP and the values measured by turbidity meter showed a systematic difference according to the distance from ADCP (Fig.1(a)). This is known to be caused by energy damping produced when a sound wave transmits through seawater. From the relationship between reflection strength and distance from ADCP when turbidity is uniformly low in the vertical direction, we see that reflection strength decreases as the distance from ADCP increases (Fig.2). Propagation loss was estimated by this damping rate at each point.

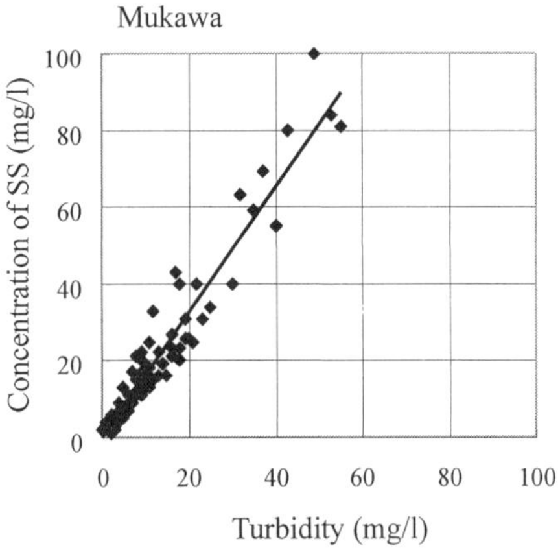

Fig.4 Correlation of turbidity and concentration of SS

A relatively close correlation between the corrected reflection strength adjusted for propagation loss and the turbidity was established at both Mukawa and Shiribetsu coastal areas(Fig.1(b)). The turbidity values estimated from the corrected reflection strength nearly agree with the values obtained by turbidimeter (Fig.3).

The turbidity correlated closely with the concentration of suspended solids (Fig.4). Calculation of the transported suspended sediment at the surveyed points revealed that approximately 22 tons of suspended sediment per 1 km were transported to the northwest during the site observation period (Fig.5).

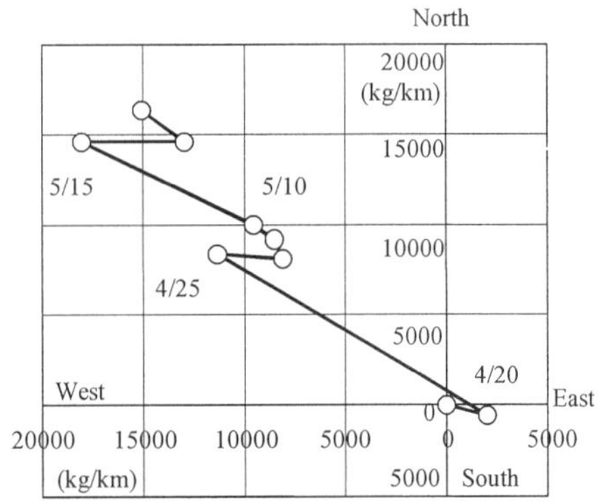

Fig.5 Transport of suspended sediment

Analyzing the Kinetic Characteristics of Flow and Sediment in North Branch of Yangtze Estuary

Minxiong Cao[1], Guozheng Cai[2]
Nanjing Hydraulic Research Institute

By analyzing surveyed data and result of numerical model, some special kinetic characteristic of flow and sediment found at north-branch in Yangtze River.

The paper discusses aspects as follows: (1) Flow characteristic of runoff, tide, storm surge and tide flow. (2) Sediment characteristic of sediment concentration and gradation. (3) Characteristic of salinity. (4) Flow kinetic characteristic (fig).

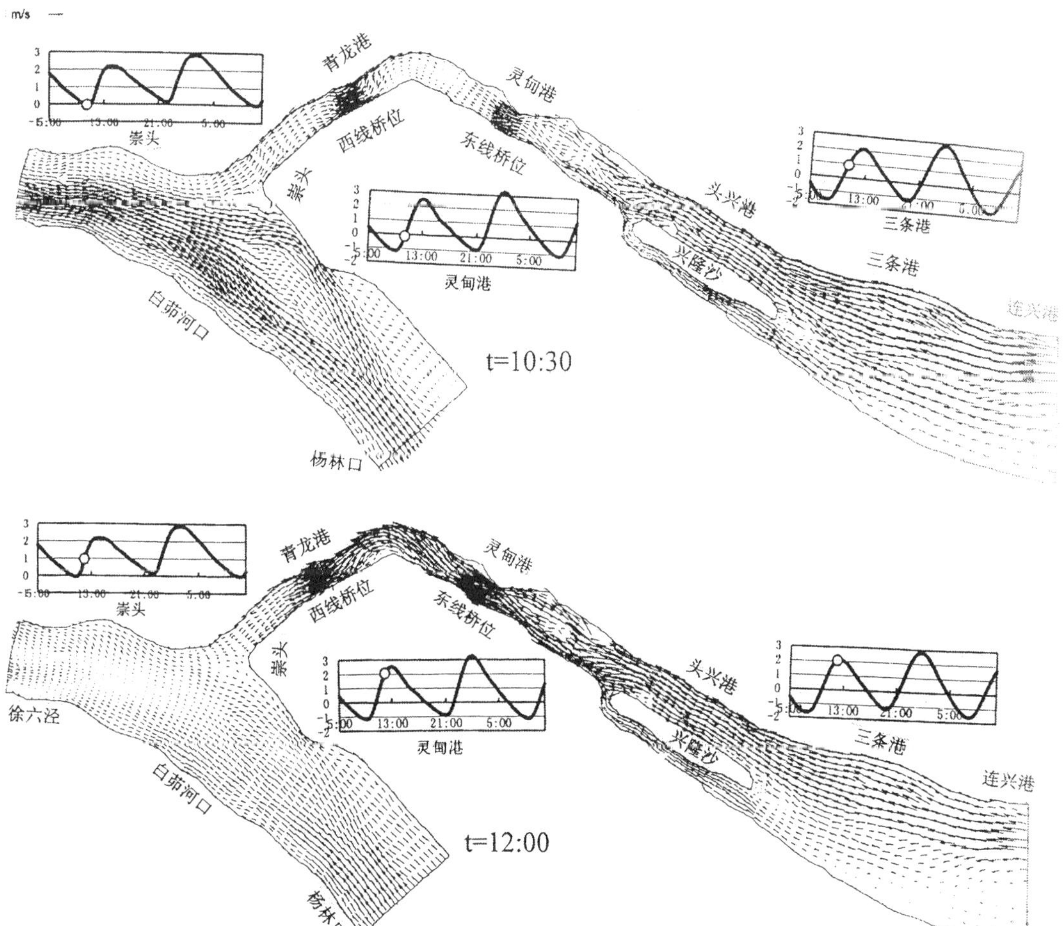

[1]Senior engineer, Master, Doctor Student, Department of River and Harbor Engineering, Nanjing Hydraulic Research Institute, Nanjing 210024 China. Mxcao@njhri.edu.cn or caominxiong@sohu.com.
[2]Senior engineer, Nanjing Hydraulic Research Institute, Nanjing 210024, China.

ANALYSIS ON SILTING AFTER DREDGING IN THE LINGDINGYANG CHANNEL AT DIFFERENT SEASONS

Ying Qiang[1], Xin Wenjie[2] and Li Yunhui[2]
[1] Hohai University, Nanjing 210024 P.R. China
[2]Nanjing Hydraulic Research Institute, Nanjing 210024 P.R. China

Based on eight groups of data of the Lingdingyang channel depth before and after dredging in each season, its characteristics of silting distribution in space and time are investigated. The causes of silting in the channel are analyzed considering the following aspect: topography of Lingdingyang, and change of discharge ratio of runoff and tidal flow, and the conditions of the sediment movement. By the fuzzy mathematical method, a formula is presented, which can determine silting by the different sediment sources.

NUMERICAL STUDY ON WAVE BOTTOM BOUNDARY LAYER OVER RIPPLES

Jiang Changbo[1] Bai Yuchuan[2] Zhao Zidan[3] Lu Hao[4]

1 Associate Prof., Dept. of River & Coast Eng., Changsha Communications University, Changsha, 410076, China. chbjiang@hotmail.com; Dept. of Hydraulic Eng., Tsinghua University, Beijing, 10084, China.
2 Associate Prof., School of Constructional Engineering, Tianjin University, Tianjin, 300072, China, ychbai@tju.edu.cn
3 Prof., School of Constructional Engineering, Tianjin University, Tianjin, 300072, China,
4 Lecturer, Dept. of River & Coast Eng., Changsha Communications University, Changsha, 410076, China.

1. INTRODUCTION

Vortex ripple is widely formed in the coastal region. The vortex ripple enhance the separation of flow and turbulence, it plays a vital role in various coastal process, such as sediment transport, the wave attenuation and the mass transport due to wave. Nevertheless, the flow structure and sediment movement above vortex ripple has not been investigated owing to the complexity of the phenomena.

Du Toit & Sleath (1981) used the Laser-Doppler to measure the flow over a ripple bed, the similar work was conducted by Sato S.J. et al. (1986), very detailed measurements of velocity closed to the bed over fixed smooth and rough bed are carried out using LAD. In resent years, most effort has been focused on obtaining a numerical solution of oscillatory flow over ripple, the first work was described by Lounguet-Higgins (1981), who used the discrete vortex model. Vittorri & Blondeaux (1991) treat a laminar flow situation by solving the vorticity transportation equations. Sato S.J. et al. (1986), Aydin (1987) and Tsujimoto et al (1991) used a two equation turbulent model $k-\varepsilon$ to study the flow around the ripple and the suspended sediment concentration distribution. Recently, Fredsoe, J et al.(1999) have conducted experiments and numerical simulation by $k-\omega$ model to study the flow structure around the vortex ripples under wave action. In the present study, a numerical simulation system based on LES method is developed for analyzing flow structure and dynamic of vortex, the sub-grid-scale turbulent stress is evaluated by the Smagorinsky model (1963). The simplify marker and cell method (SMAC) is used to solve the basic equations in the the curvilinear coordinater system, and a detailed vortex dynamics is discussed.

2 MATHEMATICAL MODELING

2.1 BASIC EQUATION

In the study, the flow is governed by incompressible Naverier-Storkes and continuity equations. By applying the grid filter to Navier-Storkers equations one obtain the filtered governing equations. Then the equationas are transformed from physical space to the curvilinear coordinater computational space, the two dimensional ($i=1,2$) filter continuity equation and momentun equations are:

$$\frac{\partial \bar{u}_1}{\partial \xi_1}+\frac{\partial \bar{u}_2}{\partial \xi_2}-\frac{\partial \bar{u}_1}{\partial \xi_2}\frac{\partial \eta}{\partial \xi_1}=0 \quad (1)$$

$$\frac{\partial \bar{u}_i}{\partial t}+\bar{u}_j\frac{\partial \bar{u}_i}{\partial \xi_j}-\bar{u}_1\frac{\partial \bar{u}_i}{\partial \xi_2}\cdot\frac{\partial \eta}{\partial \xi_1}$$

$$=-\frac{1}{\rho}\frac{\partial \bar{p}}{\partial \xi_i}+\frac{\partial \bar{p}}{\partial \xi_2}\frac{\partial \eta}{\partial \xi_1}-\frac{\partial \tau_{ij}}{\partial \xi_j}+g\delta_{i3} \quad (2)$$

In the LES computations, the effect of small scales appears in the the subgrid-scale (SGS) stress $\tau_{ij}=u_iu_j-\bar{u}_i\bar{u}_j$, which must be modeled., the eddy-viscosity concept model was used in this paper, where:

$$\tau_{ij}=-2\nu_t\bar{S}_{ij}=-2C^2\bar{\Delta}^2|\bar{S}|\bar{S}_{ij} \quad (3)$$

where ν is the fluid viscosity, ν_t is the sub-grid-scale eddy viscosity, C is the model parameter, $\bar{\Delta}$ is the length scale related to the filter width, and the $|\bar{S}|=(2\bar{S}_{ij}\bar{S}_{ij})^{1/2}$ is the magnitude of the resolved-scale-rate tensor.The parameter C can be either constants or functions of the time and space. So they can be evaluated by using Smargorinsky model and dynamic eddy-viscosity model. In this paper, as the first step in modeling wave boundary layers by using LES, only the constant parameter values ($C=0.35$) of Smargrionsky model are considered.

2.2 NUMERICAL METHODS AND BOUNDARY CONDITIONS

The governing equations have been solved by SMAC method. The forward difference scheme is used for the time dependence, and the centered difference scheme is used in the pace for the diffusive terms. The differencing of the convective terms used the donor-cell (second-order upstream) method. The periodic condition is used on the right and left side of the computation domain, the symmetric condition is used at the top, and the no-slip condition is used on the bottom. A comparison between the measurement of the flow over plane bed (Jensen 1989) and ripples bed (Fredsoe et al. 1999) with numerical results is conducted, which shows the simulation results are good, more details can be found in (Jiang 2001).

3. RESULTS AND ANALYSIS

The most important characteristic of the flow around vortex ripples is the dynamic of the vortices, it can be visualized from the velocity field. Using LES method can obtain the instantaneous velocity, so it can easily reveal the dynamic of the vortices over the vortex ripple. Fig.1 clearly shows two typical velocity field. A very strong rotation is produced after the flow pass over the ripple crest, the separation bubble grows quickly from the ripple crest with the increasing of the free stream velocity, which reaches its maximum value shortly after the free stream velocity begin to decelerate. The phase is around $\omega t=150^0$. Even if the outer flow velocity begin to decelerate, the separation bubble still expend and transport to the next ripple. At the same time, the bubble is curling up into the main flow, but it still maintains the strong rotation. As the outer flow reverses, the separation bubble lifts over the ripple crest and eject into the outer flow, became the free vortex. The phase is around $\omega t = 20^0$. Then the next separation bubble begin to form at lee back of the ripple, and at the same time, the former ejected vortex breaks up gradually in the outer flow. The numerical experiments with different value l/a are studied (l is the sand ripple length, a represents the amplitude of the motion of fluid). In addition, the bottom frication velocity and the turbulent velocity distribution are discussed in this paper.

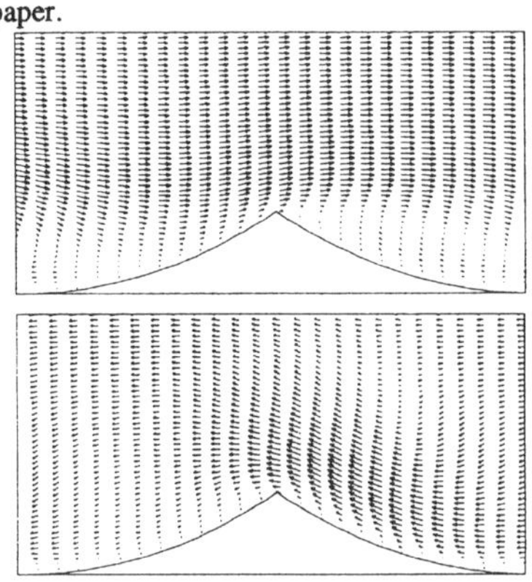

Figure 1: The typical flow velocity field at different phase (Up $\sigma t = 120^0$, Down $\sigma t = 210^0$)

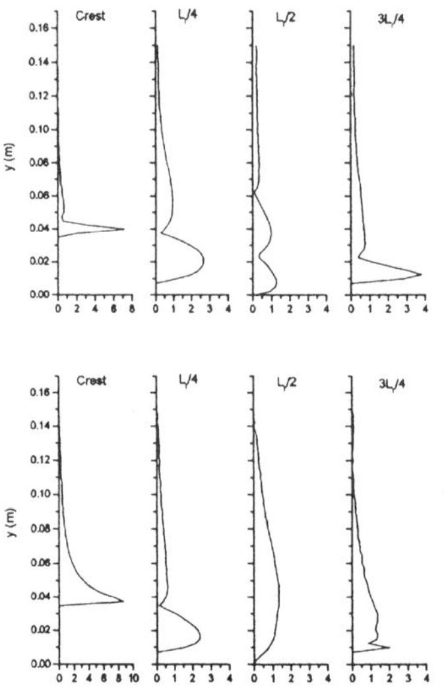

Figure2:The turbulent velocity distribution($\sigma t = 90^0$) (Up $(\overline{u'^2})^{\frac{1}{2}}$, Down$(\overline{w'^2})^{\frac{1}{2}}$)

4. CONCLUSION AND DISCUSSION

Numerical model based on 2D-LES is presented to simulate the vortex dynamics in the wave boundary layer over ripple beds. The vortex characteristics over one period and the main factors influenced the vortex formation are discussed, more over, the bottom frication velocity and the turbulent velocity distribution are simulated. It shows that 2D LES model proves to be a good tool to predict the large scales of quasi-2D turbulent flows as well as that of period unsteady flows. The vortex dynamics and sediment transport under the action of combination wave and current over ripples will be reported on in a future publication.

ACKNOWLEDGEMENTS

This research is supported by the National Natural Science Foundation of China under contracts No. 10202003, and No.59809006.

REFERENCES

Aydin, I., Shuto, N., 1988. An application of the $k-\varepsilon$ model to oscillatory boundary layers. Coastal Eng. Japan.. 30 , 11–24.

Vittori, G. and Blondeaux, P.1991 : Vorticity dynamics in an oscillatory flow over a rippled bed. J. Fluid Mech. 239, 23, p257-289

Du Toit,C., Sleath,J.F.A. 1981. Velocity measurements close to ripple beds in oscillatory flow. J. Fluid Mech. 112, 71-96.

Fredsoe, J., Andersen, K. H.,& Sumer, M. B., 1999 Wave plus current over a ripple-covered bed. Coastal Eng. 38,177-221.

Jiang ,C.B. 2001 Oscillatory bottom layer flow with sand movement and mechanism of sand ripple formation in the open channel. Phd. Thesis, Tianjin University,

Jensen, B.L., Sumer, B.M., Fredsoe, J., 1989. Turbulent oscillatory boundary layers at high Reynolds numbers.J. Fluid Mech. 206, 265–297.

Longuet-Higgins, M.S., 1981. Oscillatory flow over steep sand ripples. J. Fluid Mech. 107, 1-35.

Sato, S., et al., Coastal Eng. In Japan., 29: 65-78 (1986.

Smagorinsky, J., 1963. General circulation experiments with the primitive equations. I. The Basic Experiment. Mon. Weather . Rev., 91, 99-104

Tsujimoto,G et al. 1991 A study on suspended sediment concentration and sediment transport mechanism over rippleed sand bed using a turbulence model. Coastal Eng. In Japan., 34, 178-189

Wave Runups on a Small-Diameter Pile – From Field Experience –

Hajime Mase[1)] and Tomotsuka Takayama[2)]

[1)] Assoc. Prof. and [2)] Prof., Disaster Prevention Research Institute,
Kyoto University, Gokasho, Uji, Kyoto, 611-0011, Japan

Corresponding author is the fist one whose telephone, fax number and e-mail address are as follows:
Tel. +81-774-38-4146, Fax. +81-774-38-4321, e-mail: mase@kaigan.dpri.kyoto-u.ac.jp

Objects

In the past, wave runups on a small-diameter pile are not considered as one of important design factors, since it is assumed that waves are not disturbed by the pile so that the runup heights are equal to the incident wave crest amplitudes. Main concerns have been focused on wave forces.

A medium-scale bridge was built at a certain place in Yamaguchi Prefecture, Japan. After finishing the piers, there appeared the problem that large runups and severe splash occurred at a specific pier. Such phenomenon does not always happen, but occurs under a specific wave condition. The occurrence of runups on a small-diameter piles was learned from the field experience.

This paper reports the characteristics of random wave runups on a small-diameter circular pile by laboratory experiments under non-breaking and breaking conditions, and proposes a prediction formula for the runup height.

Main Results

Experimental data of the 2% excess and the one-third maximum runup heights were mainly analyzed as representative runup heights. It was found that the change of dimensionless runup height, normalized by the local water depth, against the dimensionless water depth was not strongly influenced by the deepwater wave steepness, and the change was approximated by the exponential function. We proposed a prediction formula for the runup height. The correlation coefficient between the predictions and observations was 0.98. There were linear relations between some representative runup heights.

RANDOM WAVE RUNUP ON SHALLOW FORESHORE SEAWALLS

by Hajime Mase[a] and Terry Hedges[b]

[a] *Disaster Prevention Research Institute, Kyoto University, Gokasho, Uji, Kyoto, 611-0011, Japan (mase@kaigan.dpri.kyoto-u.ac.jp)*

[b] *Department of Civil Engineering, University of Liverpool, Brownlow Street, Liverpool, L69 3GQ, United Kingdom (ec22@liv.ac.uk)*

Introduction

Historically, the main purpose of coastal defence was seen as preventing or reducing coastal disasters through the construction of seawalls and other 'hard' structures. More recently, 'soft' countermeasures involving beach management, together with effective systems of population evacuation have become increasingly important. Where structures are regarded as necessary, an integrated system of shore protection may be employed linking a combination of features such as seawalls, submerged breakwaters, artificial reefs, beach nourishment and so on.

There have been many studies on wave transformation, set-up, runup and overtopping. Various prediction formulae and design diagrams have been proposed. Some have been based on tests involving regular waves; others involved random waves. And sometimes the particular application falls outside the range of the available data. The present study aims to extend previous work and to establish the characteristics of random wave runup and representative runup values for seawalls located with their toes close to shorelines and involving some cases in which the seawalls are fronted by artificial reefs.

Experiments on Random Wave Runup

Adopting a scale of 1:45, models of artificial reefs and seawalls were made and installed in a wave flume. The water depth in the uniform section of the wave flume was 45cm, 42.5cm and 40cm, depending on conditions. At the landward end of the 1:20 beach slope, the model seawalls had front slopes of either 1:0.5 or 1:3. The artificial reefs could either be included in the model arrangement or could be removed altogether.

Random waves used in the experiments had JONSWAP type spectra (with a peak enhancement factor, γ=3.3). The target significant wave height was 7cm, and the significant wave periods were 0.9, 1.1, 1.3, 1.5, 1.7, 1.9 and 2.1s. The deepwater wave steepnesses H_0'/L_0 ranged from 0.01 to 0.06. The number of individual waves in each experiment was about 800.

Main Results

The main results of the study are as follows:

1) The Rayleigh distribution provides a generally satisfactory representation of random wave runup;

2) The runup formulae reviewed in this study generally predict smaller values than the measured representative runups;

3) The prediction model based on the imaginary slope and associated curves derived from data on regular waves gives an approximate upper limit to the measured mean of all runups from random waves and an approximate lower limit to the measured average of the highest one-third of the runup values, as shown in Figure 1;

4) If the maximum runup height is estimated, it may be incorporated in a wave overtopping discharge formula, which will be presented at the conference.

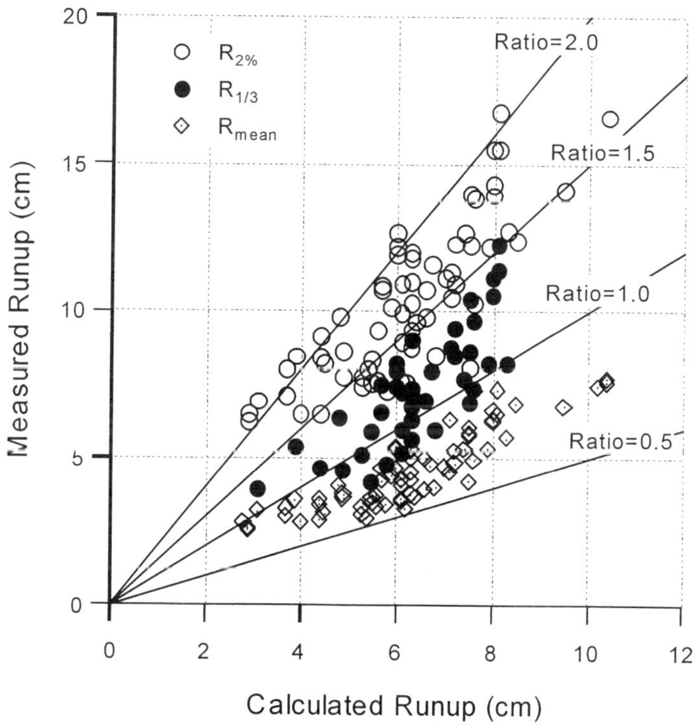

Fig. 1 Comparison of Runup Predictions (by using revised imaginary slope method) and Representative Runup Values

NUMERICAL ANALYSIS OF WAVE OVERTOPPING RATE ON SEAWALL HAVING WEDGE-SHAPED CORNER

Yoichi Moriya[2], Ryo Fujita[1], Tsunehiro Sekimoto[1]

[1]Penta-Ocean Construction Co., Ltd., Institute of Technology, Yonkucho 1534-1, Nishinasuno-machi, Nasu-gun, Tochigi-Pre., 329-2746, JAPAN., Phone: +81-287-39-2107, Fax: +81-287-39-2132 ,

Email: Ryou.Fujita@mail.penta-ocean.co.jp, Tsunehiro.Sekimoto@mail.penta-ocean.co.jp

[2]National Institute for Land and Infrastructure Management, Nagase 3-1-1, Yokosuka, Kanagawa-Pre., 239-0826, JAPAN., Phone: +81-468-44-5029, Email: moriya-y92y2@ysk.nilim.go.jp

1. Introduction

Recently, the problem of wave overtopping has been studied experimentally in horizontally two-dimensional wave field. The properties of wave-overtopping rate for the multi-directional random waves were shown by Sakakiyama and Kajima(1996) and Tomita et al.(2000). On the other hand, in the numerical approach, Moriya and Mizuguchi(1996) calculated wave-overtopping rate for the obliquely incident regular waves along a semi-infinite seawall by utilizing the Boussinesq equations and the wave-overtopping formula of Kikkawa et al.(1967). However, the validity of that model is not confirmed for random waves. In designing drainage facilities behind the seawall, we are required to estimate maximum wave-overtopping rate as well as time-averaged one, considering multi-directional random wave field.

In this study, we propose a numerical model of wave-overtopping rate on a seawall having a wedge-shaped corner, around which wave diffraction and Mach reflection occur, for multi-directional random incident wave.

2. Proposed Model

To evaluate surface elevation along the seawall with a corner, we use the improved Boussinesq equations of Madsen et al.(1991). By using such a non-linear wave equations, diffracted waves from the corner and stem waves along the seawall can be estimated. Then, as a boundary condition in front of seawall, we applied overtopping formula by Kikkawa et al. (equation (1)):

$$P,Q = \frac{2\sqrt{2g}}{3} m(\eta_e - H_c)^{3/2} \qquad \eta_e \geq H_c \qquad (1)$$

$$P,Q = 0 \qquad \eta_e < H_c \qquad (2)$$

Where P and Q are depth-integrated velocity components in the x- and y- direction, g is the gravitational acceleration, η_e is the surface elevation in front of the wall, H_c is the crown height of the wall, and m is the discharge coefficient. The discharge coefficient was set at 0.5 by Kikkawa et al., and 1.0 by Moriya and Mizuguchi. In this study, we set the discharge coefficient at both 0.5 and 1.0, and investigated its effect by comparison with experimental results.

3. Experimental Confirmation

Experimental confirmation of this model is undertaken by using wave basin with multi-directional random wave generator. Fig.1 shows the schematic experimental facilities. Water depth h=10.0 cm is kept constant after 1/10 slope in front of the wave maker, and vertical seawall having a wedge-shaped corner is placed, leaned 22.5 degrees against the wave makers. A significant wave height $H_{1/3}$=2.0 cm and a

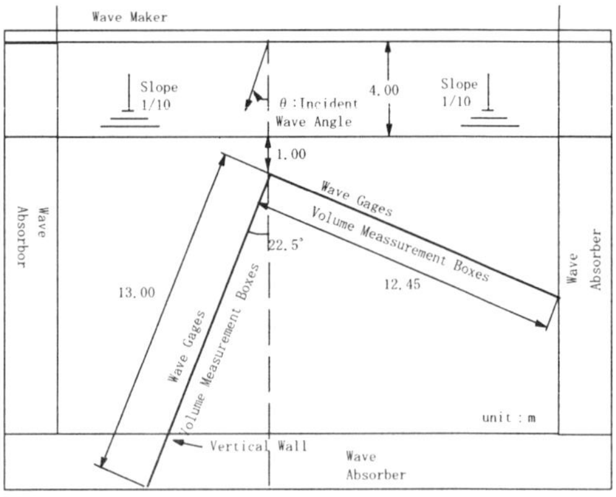

Figure 1: Experimental facilities

significant wave period $T_{1/3}$=1.0 s are kept constant. The crown height of the wall H_c is 2.0 cm. Non-overtopping experiment was also carried out by changing H_c=10.0cm, in order to check the accuracy of the numerical results of the wave field in front of the vertical wall. On the principal wave direction of uni- and multi-directional random incident waves, two conditions of +7.5 and –7.5 degrees are given to make the angles of the incident waves to the seawall, (15deg., 75deg.) and (30deg., 60deg.) respectively. Surface elevation and wave-overtopping rate are measured by the wave gages in front of the wall and volume-measurement boxes behind the wall respectively. In some of volume-measurement boxes, wave gages are set, and time series of wave-overtopping rate are also measured.

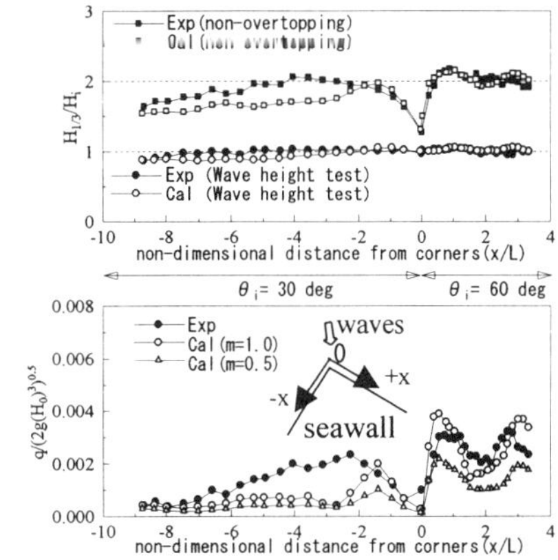

Figure 2: Distribution of wave height in front of the wall and of wave overtopping rate(θ_I: angle against the wall)

4.Conclusions

Fig.2 and Fig.3 show the examples of the comparison between calculated and experimental results(case: principal wave direction –7.5 deg., multi-directional random waves).

In Fig.2, the calculated and the experimental results are in good agreement on the distribution of wave height. Also, the distributions of wave-overtopping rate are similar in shape, though there is a little difference on magnitude, which is caused by slight error in estimating wave height in front of the seawall.

In Fig.3, ratio of calculated maximum wave-overtopping rate to averaged wave-overtopping rate is in good agreement with the experimental results.

From the comparison between calculated and experimental results, it is confirmed that the numerical model proposed in this study can evaluate the distribution of wave overtopping rate along a seawall having a wedge-shaped corner in multi-directional wave field.

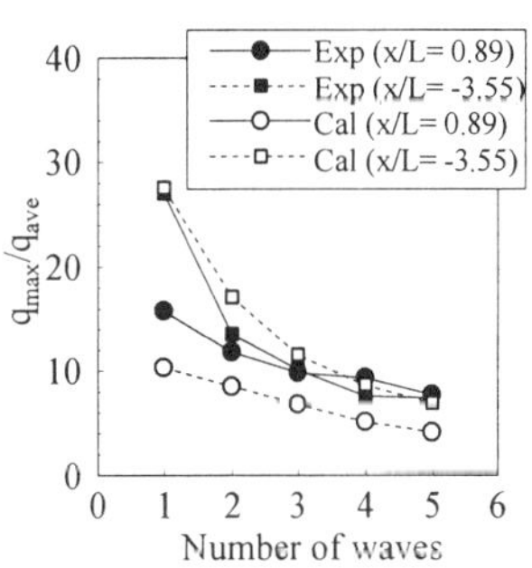

Figure3: Maximum wave overtopping rate

5.Reference

(1)Kikkawa, H., H. Shiigai and T. Kohno (1967) : Basic study on wave overtopping for coastal embankment (1), Proc,14th Japanese Conf. on Coastal Eng., JSCE, pp.335-338. (in Japanese)

(2)Madsen, P.A., Murray, R. and Sorensen, O.R. (1991) : A new form of the Boussinesq equations with improved linear dispersion characteristics (Part 1). Coastal Eng., Vol.15, pp.371-381.

(3)Moriya, Y., and M. Mizuguchi (1996) : Evaluation of wave overtopping considering diffracted waves, Proc. 43rd, Coastal Eng., JSCE, pp.711-715. (in Japanese)

(4)Sakakiyama, T., and R. Kajima (1996) : Wave Overtopping and Stability of Armor Units under Multidirectional Waves, Proc. 25th, Conf. Coastal Eng., pp.1862-1875.

(5)Tomita, T., T. Kawai, T. Umihara, and T. Hiraishi (2000) : Wave Overtopping from Wave-Dissipating Seawall with Corners and Stability of Blocks Covering the Seawall in Multi-directional Waves, Proc, 47th, Coastal Eng., JSCE, pp.886-890. (in Japanese)

LABORATORY EXPERIMENTS ON WAVE OVERTOPPING OVER SMOOTH AND STEPPED GENTLE SLOPE SEAWALLS

Takayuki SUZUKI[1], Masashi TANAKA[2] and Akio OKAYASU[3]

1) Graduate Student, Dept. of Civil Eng., Yokohama National Univ., 79-5 Tokiwadai Hodogaya-ku, Yokohama 240-8501, Japan, E-mail: tsuzuki@cape.ce.jks.ynu.ac.jp, Phone: +81-45-339-4035, Fax: +81-45-348-4565

2) Graduate Student, ditto, E-mail: d02gc113@ynu.ac.jp

3) Associate Professor, Dept. of Marine Science and Tech., Tokyo Univ. of Fisheries, E-mail: okayasu@tokyo-u-fish.ac.jp, Phone: +81-3-5463-0473, Fax: +81-3-5463-0517

1. Introduction

Recently in Japan, coastal structures are constructed with consideration in various aspects such as environment and utilization in ordinary conditions as well as disaster prevention. In that context, stepped gentle slope seawalls that give good accessibility to waterfront are now often used as the near-end protective facility in Japan's Complex Coastal Protection Concept. It is, however, important to estimate accurate wave overtopping rate in the storm conditions, because disaster prevention is the fundamental function of them.

Goda and Kishira (1976) measured overtopping rate over gentle slope seawalls and provided charts for smooth gentle slope seawalls. They conducted experiments also for stepped seawalls, but description on them is very limited. Tamada *et al.* (2002) performed experiments for various smooth slopes, but most of recent gentle slope seawalls are stepped in Japan.

In the present study, experiments on wave overtopping over both smooth and stepped gentle seawalls were performed in laboratory. The slope of seawall surface was 1/3, that is considered to be the typical slope for the current gentle slope seawall design. Overtopping rates between smooth and stepped seawalls were compared. Velocity at the top of seawalls was measured. Reflection coefficients that affect the scalar at the toe of seawalls were also investigated.

2. Experimental Conditions and Data Analysis

Experiments were performed in a wave flume that is 17m long, 0.6m wide and 0.55m high with 1/20 seabed. A 1/3 gentle slope seawall was placed at the on-shore end of the seabed. Both smooth and stepped surfaces were used for the slope of seawalls. Table 1 shows the design parameters of the seawalls. Six wave gages and two Acoustic Doppler Velocimeters (ADV) were installed in the flume to measure incident and reflected waves. Velocity of overtopping water at 2mm above the top of the seawall was measured by a Laser Doppler Velocimeter (LDV). Overtopping rate was obtained from elevation of water surface in a tank placed behind the seawall with a wave gage. The incident and reflected waves were separated by a method proposed by Kubota *et al.*, (1989). The overtopping rate was measured for about 200 waves after the wave field had become steady state condition. The data were taken three times for each of four different wave conditions.

3. Experimental Results

(1) Difference of overtopping rate between smooth and stepped seawalls

Figure 1 shows relationship between ratio of overtopping rate for stepped seawalls against smooth seawalls and crest elevation to offshore wave height ratio. In the figure, overtopping rates of stepped seawalls are less than 45% of those of smooth seawalls. It can be said that stepped seawalls can reduce the overtopping rate considerably.

(2) Relationship between crest elevation and overtopping rate

Figure 2 shows relationship between ratios of crest elevation of seawalls to offshore wave height and overtopping rates, sorted by the water depth at the toe of seawall *h*. For the same overtopping rate, crest elevations of stepped seawalls are about 70% of these of smooth seawalls.

(3) Wave reflection coefficients by seawalls

Figure 3 shows relationship between wave reflection coefficient by seawalls and wave steepness. The top figure (a), in which the water depth at the toe of seawalls is equal to zero, shows almost same values for both smooth and stepped seawalls. In the cases that the water depth at the toe of seawalls exists, reflection coefficients show larger values, and the reflection coefficients for stepped seawalls are smaller than those of smooth seawalls. It can be considered that energy dissipation rate is increased by the steps on the seawalls.

4. Conclusion

Overtopping rate on gentle slope seawalls was investigated in an experimental wave flume. From the experimental results, it can be concluded that: 1) Stepped seawalls reduce overtopping rate in comparison with smooth seawalls. 2) When water depth at the toe of seawall exists, wave reflection coefficients of stepped seawall are smaller than those of smooth seawalls.

In this experiment, velocity and water depth on the top of seawalls were measured. The difference of these values for between smooth and stepped seawalls will also be presented.

Table1 Design parameters of seawalls

	Offshore water depth (*ho*)	Crest elevation from SWL (*hc*)	Water depth at the toe of seawall (*h*)
case(a)	30.0	8.0	0.0
case(b)	35.0	8.0	5.0
case(c)	35.0	11.0	5.0

unit: cm

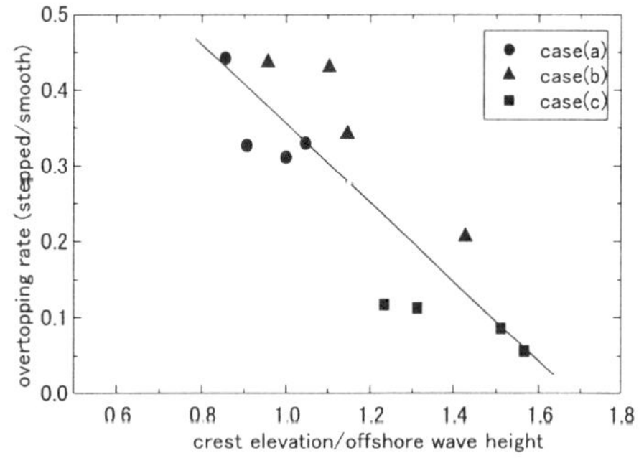

Fig.1 Relationship between overtopping rate and crest elevation to offshore wave height ratio

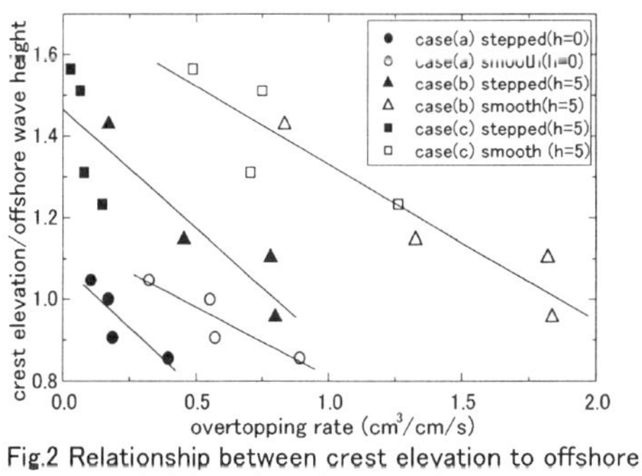

Fig.2 Relationship between crest elevation to offshore wave height ratio and overtopping rate

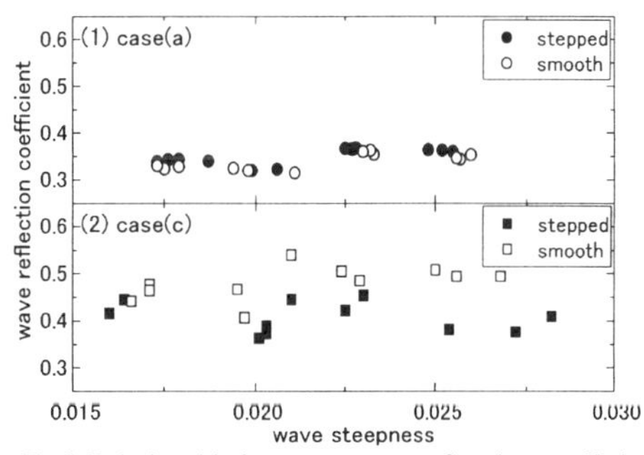

Fig.3 Relationship between wave reflection coefficient and wave steepness

References

Goda, Y. and Y. Kishira (1976): Experiments on Irregular Wave Overtopping Characteristics of Seawalls of Low Crest Types, Technical Note of PARI, No.242, p28. (In Japanese)

Kubota S., M. Mizuguchi, S. Hotta and M. Takezawa (1989): Field Observation of Wave Reflection and Swash Oscillation, Proc. of Coastal Eng., JSCE, Vol. 36, pp119-123. (In Japanese)

Tamada T., M. Inoue and T. Tezuka (2002): Experimental Studies on Diagrams for the Estimation of Wave Overtopping Rate on Gentle Slope-Type Seawalls and These Reduction Effect on Wave Overtopping, Proc. of Coastal Eng., JSCE, Vol. 49 (1), pp641-645. (In Japanese)

Failure probability of breakwater based on neural network

Dong Hyawn Kim[1,*] and Woo Sun Park[2]

[1,2]Coastal & Harbor Engineering Laboratory, Korea Ocean Research & Development Institute, 1270, Sa-dong, Ansan, Kyonggi 425-744, Republic of Korea

Abstract

Some neural network stability models for rubble mound breakwaters are proposed and analyzed. The proposed models give the more reliable results than the well known van der Meer's formula. Among them, the neural network model having the slope angle and the wave steepness as independent inputs shows the best performance. But the neural network model having independent input parameter for significant wave height, significant period is found not to be useful for the design of rubble mound breakwater because the design parameters exceed the training data ranges. In addition, a reliability analysis technique using the trained neural network model is proposed. Based on two analysis examples, it was found that van der Meer's formula gives the larger or smaller failure probabilities than the neural network models. Therefore, one should be very careful in making a decision whether the designed armor units are safe or not if he/she uses only empirical formula. To avoid this situation, it is heavily recommended that more advanced models such as the neural network model proposed here should be simultaneously considered.

Keywords: Breakwater; Stability; Neural network; Reliability based design; Monte Carlo simulation.

**Correspondence to:* Dr. Dong Hyawn Kim, Coastal & Harbor Engineering Laboratory, Korea Ocean Research & Development Institute, 1270, Sa-dong, Ansan, Kyonggi 425-744, Republic of Korea; PH(82)31-400-6345; FAX(82)31-408-5823; email: eastlite@kordi.re.kr

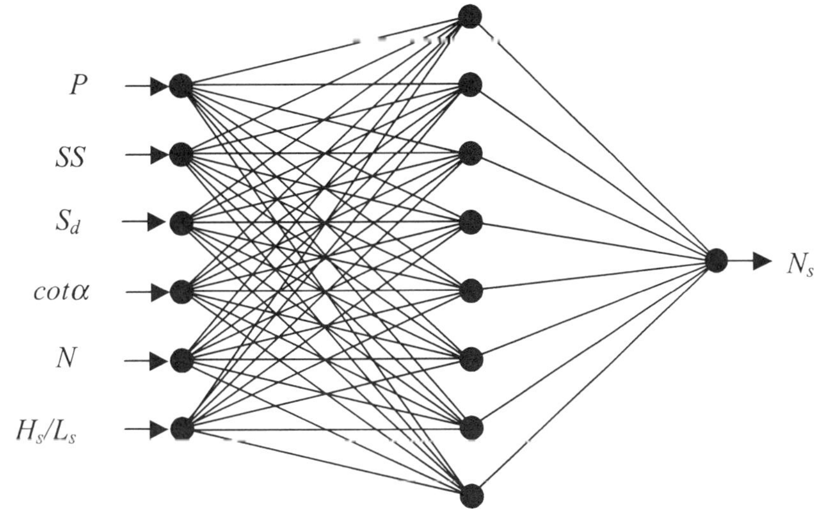

Figure 1. A neural network model for stability number prediction

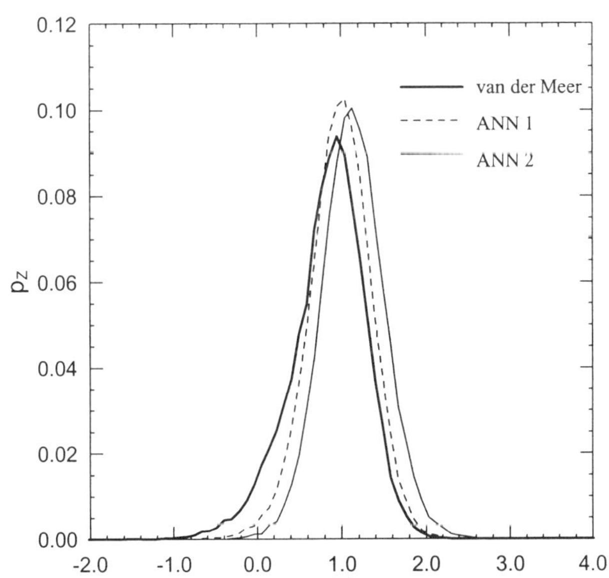

Figure 2. Probability density function for breakwater stability

Analysis of chloride ion penetration in marine concrete structure

Sang-Hun Han[a*] and Woo-Sun Park[a]

[a]Coastal & Harbour Engineering Research Center, Korea Ocean Research & Development Institute, 1270 Sadong Ansan, Kyunggido 425-744, Republic of Korea

Abstract

A new diffusion coefficient model, which considers the effect of chloride binding and evaporable water in concrete, is proposed. FEM program developed on the basis of the new diffusion coefficient model provides an estimation of chloride concentration with concrete depth and aging, and predicted chloride concentrations are compared with experimental values.

The apparent diffusion coefficient, which is a function of chloride binding, evaporable water, and diffusion coefficient, varied according to water-cement ratio. But, curing temperature did not affect the ratio of apparent diffusion coefficient to ordinary diffusion coefficient. The FEM program using the apparent diffusion coefficient estimated the experimental chloride concentration in a little error range. On the contrary, the chloride penetration analysis using ordinary diffusion coefficient overestimated the diffusion rate. Therefore, it is recommended to use the apparent diffusion coefficient proposed in this paper for predicting the initiation age of chloride-induced steel corrosion instead of the ordinary diffusion coefficient.

Keywords : Chloride; Diffusion ; FEM; Evaporable water

*Corresponding author.
(Ph: +82-31-400-6345; Fax: +82-31-408-5823)
Email: shhan@kordi.re.kr

DEFORMATION-BASED RELIABILITY DESIGN OF BREAKWATER CAISSON CONSIDERING VARIABILITY IN WAVE DIRECTION

Su Young Hong[1], Kyung-Duck Suh[2] and Hyuck-Min Kweon[3]

Recently Shimosako and Takahashi (1999) developed a Level 3 reliability design method to calculate the expected sliding distance of the concrete caisson of a vertical breakwater. This method includes stochastic variability in deepwater wave height, tide level, and shallow water wave transformation, and calculates the expected value of the accumulated sliding distance during the lifetime of the breakwater. Monte Carlo simulations were used to take into account the uncertainties of the above-mentioned design factors.

In order to calculate the wave transformation from deepwater to the location of the breakwater, they used Goda's (1975) model, which assumes unidirectional random waves being normally incident to a straight coast with parallel depth contours. In real situations, however, directional random waves with variable principal wave directions are incident to the shore, undergoing refraction as well as shoaling and breaking. Very recently Suh et al. (2002) developed a Level 3 reliability design method to calculate the expected relative damage of the breakwater armor blocks considering the variability in wave directions.

In the present study, we extend the reliability design method of Shimosako and Takahashi (1999) so as to take into account the variability in wave directions by closely following the Suh et al.'s (2002) approach. The directional variability includes directional spreading of random directional waves, obliquity of the design principal wave direction from the shore-normal direction, and its variation about the design value. To calculate the transformation of random directional waves over an arbitrary bathymetry including surf zone, we used Kweon et al.'s (1997) model, which was developed by combining Kweon and Goda's (1996) breaking wave model and Karlsson's (1969) spectral wave refraction model.

Fig. 1 shows the effect of the directional variability on the expected sliding distance. The expected sliding distance considering the variability in wave directions is much smaller than that calculated under the assumption of normal incidence, especially in deeper water. Fig. 2 shows a comparison of the caisson width between the conventional design and the reliability

[1] Grad. Student, Sch. of Civil, Urban, and Geosystem Engrg., Seoul Nat. Univ., Seoul 151-742, Korea. E-mail: hee1008@snu.ac.kr, Tel: +82-2-880-8836, Fax: +82-2-887-0349

[2] Prof., Sch. of Civil, Urban, and Geosystem Engrg., Seoul Nat. Univ., Seoul 151-742, Korea. E-mail: kdsuh@snu.ac.kr, Tel: +82-2-880-8760

[3] Asst. Prof., Dept. of Civil Engrg., Kyongju Univ., Kyongju 780-210, Korea. E-mail: hmkweon@kyongju.ac.kr, Tel: +82-54-279-4915

design. In the figure, B_d is the caisson width calculated by the conventional design method with the safety factor of 1.2, while B_{30} is the caisson width calculated by the reliability design method for the expected sliding distance to be 30 cm. A smaller caisson width is required when the variability in wave directions is taken into account. In conclusion, an over-design of the caisson is expected when the directional variability is neglected.

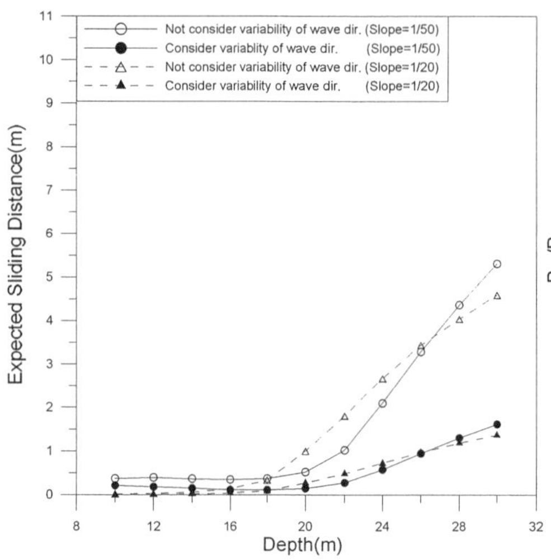

Fig. 1 Comparison of the expected sliding distance(Bed Slope=1/50 ,1/20)

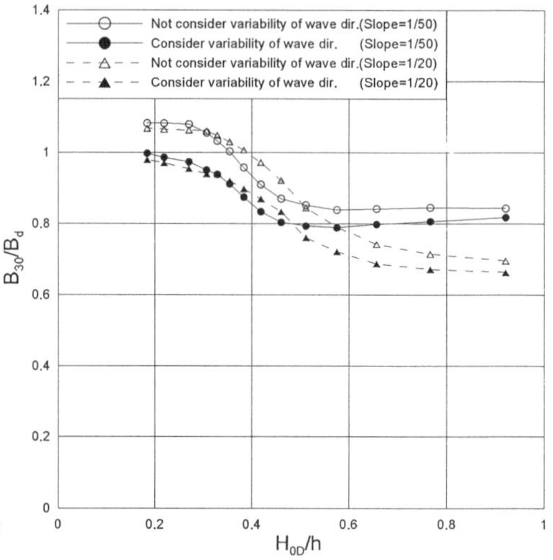

Fig. 2 Comparison of relative cassion width by new design method and conventional method (Bed Slope=1/50 ,1/20)

4. References

Goda, Y. (1975). Irregular wave deformation in the surf zone, *Coastal Engineering in Japan*, 18, 13-26.

Karlsson, T. (1969). Refraction of continuous ocean wave spectra, *J. Waterway Harbors Div.*, ASCE, 95(WW4), 471-490.

Kweon, H.-M., and Goda, Y. (1996). A parabolic model for random wave deformation by breaking on arbitrary beach profiles, *Proc. 25th Coastal Engineering Conference,* ASCE, Orlando, 261-274.

Kweon, H.-M., Sato, K., and Goda, Y. (1997). A 3-D random breaking model for directional spectral waves, *Proc. 3rd Int. Symp. Ocean Wave Measurement and Analysis,* ASCE, Norfolk, 416-430.

Shimosako, K., and Takahashi, S. (1999). Application of deformation-based reliability design for coastal structures – Expected sliding distance method of composite breakwaters, *Proc. Coastal Structures '99*, 363-371.

Suh, K. D., Kweon, H.-M., and Yoon, H. D. (2002). Reliability design of breakwater armor blocks considering wave direction in computation of wave transformation, *Coastal Engineering Journal*, 44(4), 321-341.

IMPROVED EVALUATION OF THE EXPECTED SLIDING DISTANCE OF A CAISSON AND PRACTICAL PARAMETERS OF UNCERTAIN FACTORS

TAE-MIN KIM

Graduate student, Department of Civil Engineering, Kyoto University Graduate School

Gokasho, Uji, Kyoto 611-0011, Japan

e-mail: tmkim@kaigan.dpri.kyoto-u.ac.jp

TOMOTSUKA TAKAYAMA

Professor, Disaster Prevention Research Institute, Kyoto University

Gokasho, Uji, Kyoto 611-0011, Japan

e-mail: takayama@kaigan.dpri.kyoto-u.ac.jp

Many studies on the application of the reliability design method, as an alternative of the conventional one, to caisson-type breakwater have been carried out (e.g. Shimosako and Takahashi, 1998, 1999; Takayama et al., 2000; Goda and Takagi, 2000; Goda, 2001). Especially, the computation of the expected sliding distance proposed by Shimosako and Takahashi quantitatively clarifies the risk of sliding failure, and the expected sliding distance is a useful stability index against sliding of a caisson in the reliability design. However, detail studies for accurate estimation of the expected sliding distance are not sufficient until now. Therefore, this fact motivated this paper.

The objectives of this paper are classified into two items. One is to compare simulation results of the expected sliding distance due to the existing parameters of uncertain factors (e.g. estimation errors of deepwater wave height, calculation errors of wave transformation, wave period and wave force, uncertainties of friction factor), which is considered in reliability design procedures, and the other is to propose alternatives for improved estimation of the expected sliding distance in the reliability calculations. As alternatives, a doubly-truncated normal distribution (Kim and Takayama, 2003) and average expected sliding distance are proposed with a validity through the simulation results. Additionally, new parameters of uncertain factors are presented through analysis of the existing studies (Takayama and Ikeda, 1993; Takenaka et al, 1999), and their characteristics are investigated through case study.

Corresponding author. Tel : +81-774-38-4143, Fax: +81-774-38-4321

E-mail address: tmkim@kaigan.dpri.kyoto-u.ac.jp (Tae-Min Kim)

Major conclusions drawn from this paper are as follows.

1) Through the simulation results computed by the existing probability distributions of uncertain factors, it is confirmed that the probability distributions of uncertain factors affect largely the expected sliding distance.
2) In the Monte-Carlo simulation, the seed of random variable largely affects the expected sliding distance. The influence of the seed on expected sliding distance increases as the water depth becomes deep. Therefore, the average expected sliding distance proposed in this paper is a better index for sliding failure of caisson-type breakwater than the expected sliding distance.
3) The simulation results calculated by the employment of the original normal distribution cause abnormal sliding distance because the random variables of wave force and friction factor in means of Monte-Carlo simulation have some values outside the region where experimental data are not valid. For effective considerations of the uncertain factors in the reliability design of caisson-type breakwater, a doubly-truncated normal distribution is more realistic than the original normal one.

References

Goda, Y. and H. Takagi (2000). A reliability design method of caisson breakwaters with optimal wave heights, *Coastal Engineering Journal*, 42 (4), pp. 357-387.

Goda, Y. (2001). Extreme wave statistics for reliability-based design of caisson breakwater, *Proc. of Int. workshop on Advanced Design of Maritime Structures in the 21st century, PHRI & MLIT,* Japan, pp.1-13.

Kim, T.M. and T. Takayama (2003) A proposal for effective consideration of uncertain factors for reliability-based design of caisson-type breakwaters, *Coastal structures conference 2003*, ASCE, Manuscript in preparation.

Shimosako, K. and S. Takahashi (1998). Reliability design method of composite breakwater using expected sliding distance, *Report of the Port and Harbour Research Institute* 37, 3, pp. 3-30 (in Japanese).

Shimosako, K. and S. Takahashi (1999). Application of deformation-based reliability design for coastal structures — Expected sliding distance method of composite breakwaters —, *Coastal Structures '99,* ed. I. J. Losada, Spain, Balkemare, pp. 363-371.

Takayama, T. and N. Ikeda (1993). Estimation of sliding failure probability of present breakwaters for probabilistic design, *Report of the Port and Harbour Research Institute* 31, 5, pp. 3-32 (in Japanese).

Takenaka, H., M. Nishida., H. Sakakibara., K. Tonomo and H. Sato (1999) Accuracy of calculation method for wave transformation and wave force based on field, *Proc. Coastal Engineering*, JSCE, Vol. 46, pp. 826-830 (in Japanese).

THEORETICAL STUDY OF THE REFLECTION OF OBLIQUE INCIDENT WAVES BY PARTIALLY-PERFORATED BREAKWATERS

Yucheng LI, Hongjie LIU and Dapeng SUN

State Key Laboratory of Costal and Offshore Engineering

Dalian University of Technology, Dalian(116023), China

In the recent years, breakwaters with perforated front walls have been widely used. it can reduce the wave reflection and wave run-up in front of the structures. Then, the construction cost may decrease.

Several researchers have investigated the interaction of waves with perforated structures. Some are experimental examination of the reflection characteristics of a perforated-wall breakwater; others are analytical models for predicting the reflection coefficient. But most analytical models deal with the case in which the perforated wall (or plate) is fully opened. And most researchers pay their attentions to the interaction of normal incident wave with breakwater. In engineering practice, owing to the stability requirement of the structures, the front wall of breakwater is partial-perforated rather than fully opened.

In the present paper, the reflection of oblique incident waves from breakwaters with a partial-perforated front wall is investigated. The fluid domain is divided into two sub-domains and eigenfunction expansion method is applied to expand velocity potentials in each domain. In the eigen-expansion of the velocity potential, evanescent waves are included. Numerical results of the present model are compared with experimental data and they are in good agreement. The effect on reflection coefficient of porosity, the relative chamber width, the relative water depth in wave absorbing chamber and the water depth in front of structure are discussed.

Numerical Simulation of Deformation Process of Wave-Dissipating Blocks by 3D-DEM

Eiji HARADA[1], Hitoshi GOTOH[2], Tetsuo SAKAI[3] and Masahiro OHNO[4]

[1] JSPS Doctoral Research Fellow, Dept. of Civil Engrg., Kyoto Univ.,
Yoshida Honmachi, Sakyo-ku, Kyoto, 606-8501, Japan
tel: +81-753-5099, fax: +81-761-0646, e-mail: harada@coast.kuciv.kyoto-u.ac.jp (correspondence)
[2] Associate Professor, Dept. of Civil Engrg., Kyoto Univ.
[3] Professor, Dept. of Civil Engrg., Kyoto Univ.
[4] Graduate Student, School of Civil Engrg., Kyoto Univ.

Objectives

Although the breakwater covered with wave-dissipating blocks is one of the most popular coastal structures, collapse of the wave-dissipating block mound due to the direct wave attack have been reported. The required condition of the breakwater covered with wave-dissipating blocks is to keep the arrangement of the blocks with the suitable gap between blocks for absorbing wave enegy. And also the engagement between blocks is required to keep sufficient resistance to the wave attack. Flexibility in deformation fitting itself to the change of ground level, such as land subsidence due to the seabed liquefaction or scouring, is an advantage of the wave-dissipating block mound. Therefore,the change of engagement between blocks after the land subsidence is important in the design of wave-dissipating block mound. However, a large deformation of mound brings extremely high stress over the yeild condition of the concrete blocks, under which the part of the block-leg is destroyed. That failure of the block-leg leads to the lose of the engagement between blocks, and resultant lift off the single block unit. Consequently lose of the engagement between blocks could lead to the catastrophic destruction of the wave-dissipating block mound. In this paper, to understand the disaster mechanism of the wave-dissipating block mound under the sudden land subsidence, the numerical simulator based on the Distinct Element Method was proposed.

Numerical Model

The model for the deformation of the wave-dissipating block mound was described by the full 3D-DEM to calculate the interactions between contacting blocks explicitly. Hence the motion of each individual block is tracked clearly. In DEM, block/block and block/wall interactions are modeled by the spring and dashpot system assumed at the contacting points with other blocks or wall. The element of the fixed boudary wall and moving wall is a spherical particle the diameter of which is 0.5m. While the wave-dissipating block is formed by connecting four small spheres with one large sphere rigidly to keep the gravity centers of four small spheres on vertex of a regular tetrahedron at the gravity center of which, the center of large sphere locates. And also the diameter of large one is 1.0m while the that of small is 0.5m. The motion of the blocks are tracked by the rigid body model which keeps the geometrical relation of those five spheres.

Results

Figure 1 shows the typical snapshots and force vector plots of solution. Initially, In the top layer 56, middle layer 56, bottom layer 72, totally, 184 blocks are arranged almost uniformly on the fixed wall. The center part of the fixed bottom wall suddenly goes down by 1.0m, then blocks rush into the hollow space. The force between contcting blocks is shown by the cylinder, the direction of which indicates the direction of the force vector, the strength of force is proportional to the diameter of the cylinder. The remarkable force concentration between contacting blocks is shown at the bottom of the moving wall and the right edge of hollow space. These information can be a help of prediction of block-leg failure.

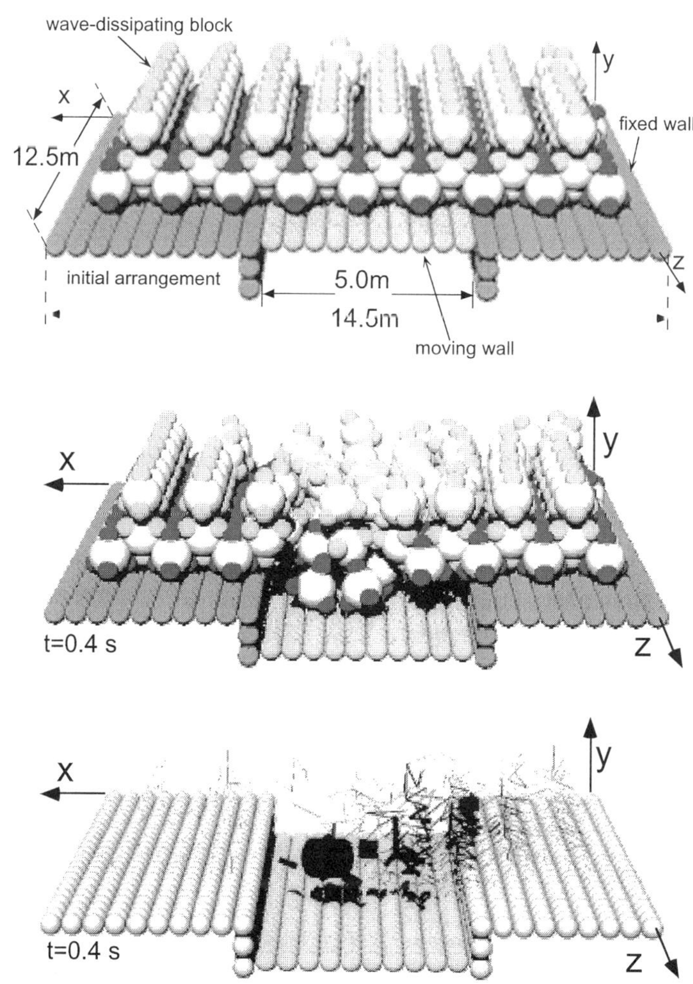

Fig.1 Snapshots and Force vector plots

REFLECTION OF WAVES OVER VARIOUS-SHAPED SUBMERGED BREAKWATERS UNDER REGULAR WAVE ACTION

Yong-Sik Cho
Associate Professor, Department of Civil Engineering, Hanyang University,
Seoul 133-791, Korea. ysc59@hanyang.ac.kr
Tel: 82-2-2290-0393; Fax: 82-2-2293-9977

Young-Taek Kim
Researcher, Water Resources and Environmental Div., Korea Institute of Construction Technology,
2311, Daehwa-dong, Ilsan-gu, Goyang, Gyonggi 411-712, Korea. ytkim@kict.re.kr
(Graduate Student, Department of Civil Engineering, Hanyang University, Seoul, Korea 133-791)
Tel: 82-31-910-0654; Fax: 82-31-910-0251

Jong-In Lee
Senior Researcher, Water Resources and Environmental Div., Korea Institute of Construction Technology
2311, Daehwa-dong, Ilsan-gu, Goyang, Gyonggi 411-712, Korea. jilee@kict.re.kr
Tel: 82-31-910-0359; Fax: 82-31-910-0251

Ocean surface water waves traveling in water of varying depths may experience a lot of physical phenomena such as reflection, diffraction, shoaling and refraction due to variation of bottom topography, nonlinear interaction among different harmonics and interferences with man-made structures. The present study mainly concerns with reflection of water waves propagating over submerged breakwaters.

Wave reflection of various-shaped submerged breakwater is analyzed by using physical and numerical modelings. Shapes of submerged breakwaters are rectangles, triangles and trapezoids.

The physical modeling is performed in a two-dimensional wave channel which has the dimension of 2.0m high, 1.0m wide and 56m long, and an electric-servo typed wave generator. Laboratory experiments are performed for various cases, that is, variations of water depths, widths of breakwaters, arrays of breakwaters and so forth.

Measured results from the physical modeling are compared with those of the numerical modelings. The numerical models are based on the boundary element method, the eigenfunction expansion method and the hyperbolic typed wave equations that contain the fast varying terms. Very reasonable agreements are observed. .

Wave-induced Currents in front of a Vertical Breakwater

ZUO Qihua DING Bingcan WANG Dengting

The mass transport flow and the radiation stress flow induced by water wave play an important role on the dynamic research of coastal engineering. On scouring around a large-scale circular cylinder, Kaitsui[1] considered not only sediment transportation of the radiation stress flow but also Eularian flow and the mass transport flow. Saito[2] even thought that the mass transport flow near the bottom boundary layers plays a key role. But till now, few studies have been made on currents in front of a vertical breakwater caused by different incident waves, especially on those along breakwaters.

Based on the nonlinear parabolic approximation of mild slope equation, the wave field in front of a vertical breakwater with different wave incident angles, that is, $10^\circ, 20^\circ$ and 30° respectively is simulated firstly for flat or varying bottom and then based on the depth intergraded average N-S equation, the Eulerian flow, the mass transport flow and the radiation stress flow induced by wave are computed. The computational domain is 2000m × 1000m, different time and space step are adapted. The bottom frictional coefficient f and eddy viscosity factor A_h are taken as constant. Table 1 shows the velocities and the corresponding positions of the maximum values of Eulerian flow u_e, the mass transport flow u_m and the radiation stress flow u_r for typical wave period, wave height and water depth. In the table, the coordinates (x', y') are the ratio of space length and wave length of deep waves. From the table it is seen that wave fields, water depth, topography and incident angles have great effects on flow velocities and flow fields. No matter the flat or sloping topography, the velocities of the mass transport flow and the radiation stress flow have almost the same magnitude. Therefore, for currents induced by waves, not only the radiation stress flow but also the mass transport flow should be taken into account.

The major conclusions are:

1. The wave fields, water depth, bottom topography and incident angles have a great effect on current induced by waves in front of the breakwater, especially on the sloping topography and around the wave broken zone, therefore, enough attention should be paid.
2. For currents induced by waves in front of the breakwater, not only the radiation stress flow but also the mass transport flow should be considere

Reference:

1. Kaisui H, Toue T. Bottom shear stress in coexistent field of superimposed waves and current and scouring around a large-scale circular cylinder [J]. Coastal Eng (in Japanese), 1992, 35(1).
2. Saito E, Soto and T. Shibayama. Study on local scour around a large diameter circular

cylinder in wave field [A]. Proc of 35th Japanese Conf on Coastal Eng [C], JSCE (in Japanese), 1988.

Table 1 Calculated results of Eulerian flow、mass transport flow and radial stress flow

Topography	Water depth (m)	Incident angle (°)	u_e ($m \cdot s^{-1}$)	(x', y')	u_m ($m \cdot s^{-1}$)	(x', y')	u_r ($m \cdot s^{-1}$)	(x', y')
Flat bottom	5	10	0.18	(22.0,22.6)	0.29	(22.0,22.6)	0.13	(21.7,22.6)
		20	0.21	(14.1,23.1)	0.32	(14.1,23.1)	0.40	(13.4,23.3)
		30	0.25	(10.6,12.7)	0.36	(10.6,12.7)	0.52	(10.1,17.4)
	10	10	0.04	(17.2,18.2)	0.07	(17.2,18.2)	0.06	(17.1,17.8)
		20	0.05	(11.7,17.6)	0.08	(11.7,17.6)	0.21	(11.2,17.8)
		30	0.07	(5.2,17.6)	0.11	(5.2,17.6)	0.35	(5.0,17.2)
Varying bottom (1:100)	5	10	0.46	(24.6,7.7)	0.76	(24.6,7.7)	0.56	(23.0,9.9)
		20	0.61	(22.3,8.0)	0.91	(22.3,8.0)	1.26	(21.9,8.6)
		30	0.91	(20.8,6.9)	1.26	(20.8,6.9)	1.09	(19.2,8.2)
	10	10	0.48	(17.6,15.9)	0.79	(17.6,15.9)	0.73	(15.4,17.0)
		20	0.65	(13.1,15.5)	1.00	(13.1,15.5)	1.44	(12.6,17.5)
		30	0.89	(9.3,14.5)	1.28	(9.3,14.5)	1.14	(11.1,15.4)

(1) ZUO Qihua
Title: Professor; Senior engineer
Affiliation: One of members of International Steering Committee of APAC in 2003. His main research direction is coast dynamics. He has published several dozen papers.
Address: Nanjing Hydraulic Research Institute, Guangzhou Road 223, Nanjing, 210029, China
Tel: +86 25 5828008 Fax: +86 25 5829333 E-mail: qhzuo@njhri.edu.cn

(2) DING Bingcan
Title: Senior engineer
Affiliation: His main research direction is coast dynamics. He has settled in Canada.
E-mail: bcding888@hotmail.com

(3) WANG Dengting:
Title: Master; Assistant engineer
Affiliation: His main research direction is the interaction of wave and structure. He has published 3 papers.
Address: River and Harbor Department, Nanjing Hydraulic Research Institute, Hujuguan 34, Nanjing, 210024, China
Tel: +86 25 5829341 Fax: +86 25 5829333 E-mail: dtwang@njhri.edu.cn

Subsidence of Rubble Stones due to Wave-Induced Seabed Liquefaction

Tetsuo SAKAI[1], Hitoshi GOTOH[2], Eiji HARADA[3] and Yasufumi IMOTO[4]

[1] *Professor, Dept. of Civil Engrg., Kyoto Univ.,*
Yoshida Honmachi, Sakyo-ku, Kyoto, 606-8501, Japan
tel:+81-75-753-5097, fax: +81-75-761-0646, e-mail: sakai@coast.kuciv.kyoto-u.ac.jp
[2] *Associate Professor, Dept. of Civil Engrg., Kyoto Univ.,*
Yoshida Honmachi, Sakyo-ku, Kyoto, 606-8501, Japan
tel:+81-75-753-5098, fax: +81-75-761-0646, e-mail: gotoh@coast.kuciv.kyoto-u.ac.jp (correspondence)
[3] *JSPS Doctoral Research Fellow, Dept. of Civil Engrg., Kyoto Univ., Japan*
[4] *Graduate Student, School of Civil Engrg., Kyoto Univ., Japan*

Objectives

Under the existence of momentum liquefaction of seabed due to the action of high waves, a subsidence of armor blocks and rubble stones, which weaken the protection against scouring, occurs. In the maintenance planning of the coastal structures, the prediction of the subsidence of armor blocks and rubble stones is one of important items. The key factor of the subsidence of armor blocks and rubble stones is the block-block or stone-stone interaction, while most of the previous studies have treated the subsidence of single block or stone. A behavior of rubble stones at the foot of seawall under the momentum seabed liquefaction is experimentally and numerically investigated. Especially in the numerical model, the stone-stone interactive structure of rubble stone mound is explicitly taken into account.

Experiment and Simulation Model

The experiment is conducted in the steel water tank, which has rectangular cross section. A behavior of glass beads mound, which is D=25.0 mm in diameter and 2.33 in specific density, on the sand layer, which is d=0.25 mm in diameter and 2.65 in specific density, under the sinusoidal water pressure change is observed through a glass side wall of the tank. The period and amplitude of water pressure oscillation are T=4.0 s and 2.5 mAq, respectively. Under these hydraulic and seabed conditions, the momentum liquefac tion of sea bed clearly occurs.

The motion of rubble stones and sand are described by the 3D-DEM to calculate the stone/stone, sand/sand and stone/sand interactions explicitly. The present model is the extension of the MBS, or Movable Bed Simulator, proposed by Gotoh and Sakai(1997), which is the Lagrangian model of motion of sediment particle in turbulent flow with considering interparticle collision. The driving force of liquefaction is modeled as the apparent lift force, which cancel the gravity force for a sort moment. Short-term loading and rather long-term unloading are alternatively repeated, as a simplified model of momentum liquefaction.

Results

Figure 1 shows the typical behavior of glass beads mound under the action of momentum liquefaction of seabed. The backside of the glass beads mound is a vertical wall, hence the initial cross section of the glass beads mound is rough triangular. The foot of the mound slides fore direction and the height of the mound gradually decreases under the repetition of the momentum liquefaction. The result of numerical simulation is shown in Fig. 2. Fundamental tendencies, such as the fore sliding and the height decreasing of mound, are reproduced clearly by the present model, at least qualitatively. Further, the refinements on the present model to improve the quantitative agreement with the experiment are discussed.

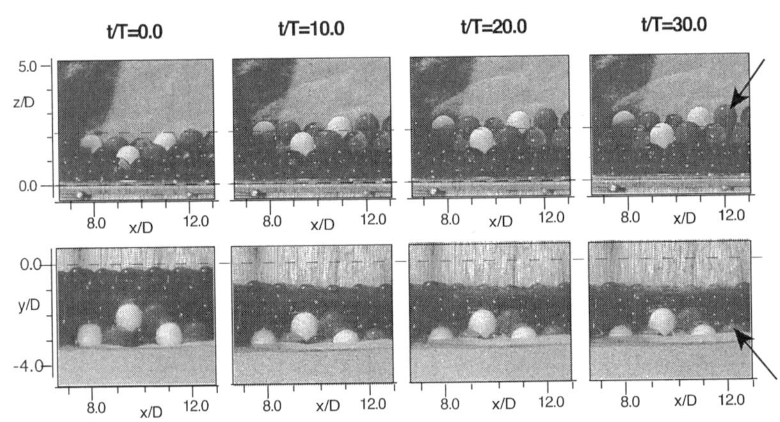

Fig. 1 Subsidence of glass beads mound (experiment; upper: top view, lower: side view)

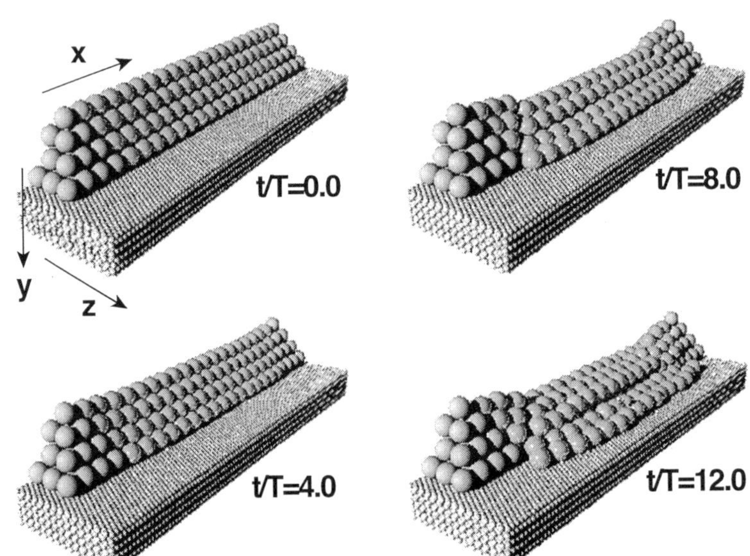

Fig. 2 Results of numerical simulation

RESEARCH ON BEARING CAPACITY CALCULATION METHOD OF LARGE DIAMETER CYLINDER IN SOFT CLAY

Wu Meng-Xi
Institute of Mechanics, Chinese Academy of Science, No. 15, North Forth Ring West Road,
Beijing, P.R. China 100080
Ph: 8610-62554188
Email: wumengxi2000@263.net; hytlx@imech.ac.cn

Abstract

Large-diameter cylinder inserted in soils is a new structure style in nearshore and port engineering. It has been used as retaining quay walls and other kinds of port project structures such as dikes. It has wide potential application for its efficiency. The bearing and failure mechanism of big diameter cylinder in soft clay under wave action is very complicated. The design theory and calculating method is still under development. It is very important to do some research on the calculating method of the bearing capacity.

The main cause of the capsizing failure of the big diameter cylinder structures in the testing segment of the Second Grade River-Route Controlling Project in the Bayou of Yangtse River during the 'wei-ma-xun' typhoon, is the bearing capacity deficiency. The improper calculation of the bearing capacity caused the insufficiency of the cylinder inserted in the soft clay.

Undrained shear strength (vane strength) in the field is 7kPa at the mudline with an average increasing rate of 1.33kPa/m. The average sensitivity of the soils is about 3. The elastic modulus is 1500kPa gained from the confined compression tests or 500 kPa from the unconsolidated-undrained shear tests on pushed samples. The diameter of the concrete cylinder is 12m, and the thickness of the wall is 0.23m above the mudline and 0.21m under the mudline. The total height of the cyclinder is 22.2m with a length of 15.2m under the mudline. There is a concrete cap which is 3.5m high and 4800 kN weight over the cylinder.

The wave height of 1% cumulate frequency during the typhoon is 7.38m, wave periods 7.4s and wave wavelength is 67.7m. The total lateral wave force is 7066kN, total vertical force is 3561kN, and total bending moment at the mudline is 32858 kN·m.

Based on the plastic limit analysis, a method for the ultimate bearing calculation of the cylinder structure inserted in soft soils is proposed in this paper. It is assumed that a gap has formed at the opposite side of the cylinder at the ultimate state. The factor of safety of the structure by this method is 0.82, and the corresponding ultimate lateral capacity is 5818kPa.

Abaqus 6.20 is used to simulate the soil cylinder behavior. The ideal elastic-plastic (Mises) model is used. Fig. 1 is the load-displacement curves of the structure.

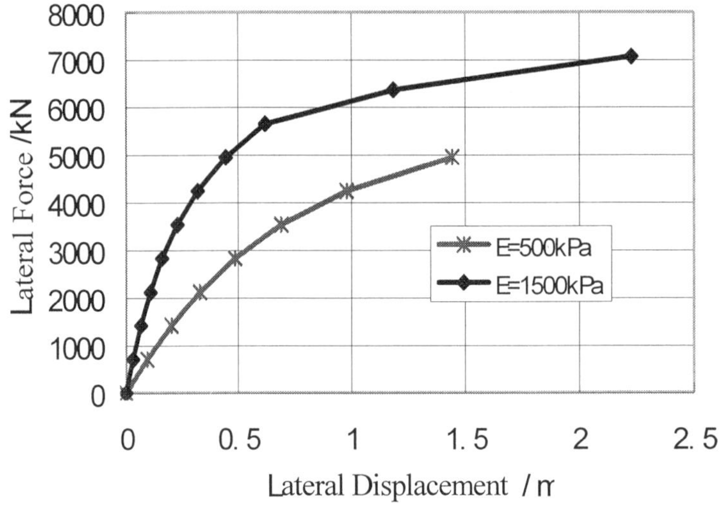

Fig. 1 Lateral load vs Lateral displacement at mudline of the cylinder

It indicated that the ultimate lateral loading capacity is about 7000kPa and 5000kPa while the elastic modulus is 1500kPa and 500kPa respectively. The average value is 6000kPa, consistent with the capacity calculated by the method proposed above.

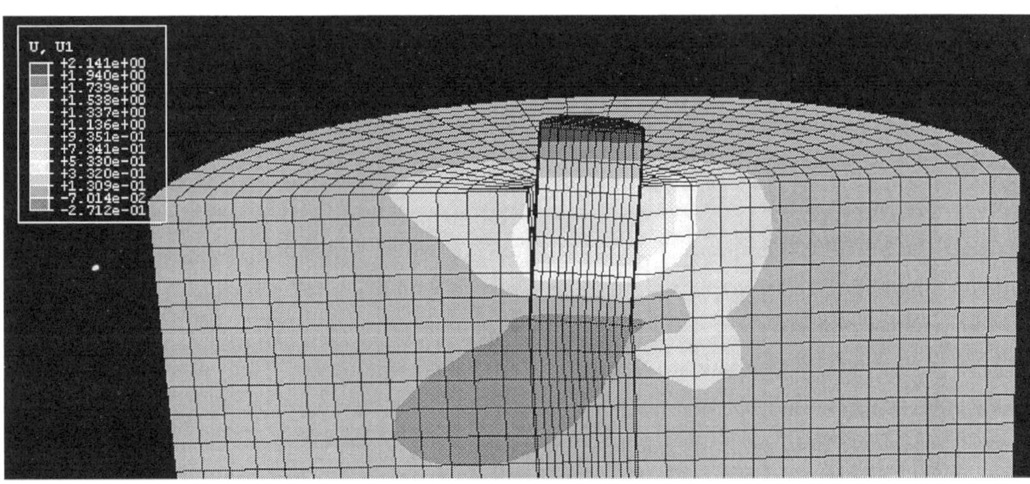

Fig. 2 Lateral displacement contour of the cylinder soil system

Fig. 2 is the Lateral displacement contour of the cylinder soil system under the last loads step in fig.1 while the elastic modulus is the lower. A gap was formed in the opposite side of the cylinder. It also agrees with the assumption used in the ultimate analysis.

Experimental Study of Breakwater with Penetrating Box Foundation

LI Wei[a*], BIE Shean[b], CHEN Wenshu[b] and REN Kai[b]

[a]Planning and Construction Department, Port of Tianjin Authority, Tanggu 300456, China
[b]Department of Port and Coastal Engineering, Tianjin University, Tianjin 300072, China
*Email: lw_99@sina.com

1. Introduction

The new type of breakwater with box foundations consists of box foundation penetrated into soft seabed soil and upper structures in various forms. The structure as a whole can be installed on the site conveniently, be transported over water by means of pneumatic floating and can also be reused if necessary.

This paper will study the mechanical properties of breakwater with penetrating box foundation by means of static model test and wave test, discuss the relations of the vertical bearing capacity and horizontal resistant capacity of box foundation as well as the instability failure forms to their dimensions and external loads, in an attempt to explore the method of calculating and analyzing the stability of penetrating box foundation.

2. Static Model Test

The static model test was conducted in a 6m long, 3m wide and 1.8m high steel flume filled with soil. Five steel model structures with three different widths and three different embedded depths were employed. The scale between the models and their prototypes is about 1:5 to 1:8.

2.1 Vertical Bearing Capacity Tests

2.2 Vertical Uplift Tests

2.3 Horizontal Sliding and Overturning Resistant Tests

3. Stability Tests in Wave Flume

The purpose of testing with wave flume is to check the stability of box foundation under the action of dynamic loads and observe the forms of instability failure of the structure when wave loads are higher than the designed limit loads.

The model box is 0.8m wide, 0.94m long (width of wave flume is 1.0m) and 0.32m high. The steel box foundation model weighing 171kg is placed in the mud pond of wave flume, on the top of box body is a 0.8m high upright wall. The soils in the mud pond are the same as those used in the static stability tests.

4. Conclusions

(1) The mechanical properties of the stability of box foundations are similar to those of rigid foundations with embedment i.e. the vertical earth pressures at the bottom of the foundation are distributed in a straight line distribution pattern under the action of vertical loads and the passive and active earth pressures are produced at the front and back sides of box under the action of horizontal loads.

(2) The sliding and tilting stability, the vertical bearing capacity and earth pressures at the foundation bottom of penetrating box foundations can be calculated by the methods for the rigid foundations with embedment.

(3) When the penetrating box foundations in clay move upward, a suction force will be introduced in the soils at the bottom of box, which will attenuate with time. The suction force and its attenuation are related to the water permeability and rheological features of soils, as well as the movement speed of box, etc.

(4) Under the action of waves, the instability failure of box foundations mainly takes the form of large horizontal displacement .When the structure experiences a reciprocating motion, the wave force acting on the structure will be lowered.

Experimental and Numerical Evaluation on Consolidation Characteristics of Soft Ground Breakwater

Woo-Sun Park[1], In-Sung Jang, O-Soon Kwon, and Ki-Dai Yum
Coastal and Harbor Engineering Research Laboratory, Korea Ocean Research and Development Institute, Ansan P.O. Box 29, Seoul 425-600, Korea

In constructing breakwaters located on a soft ground, considerable expenditure and time consumption are required for a ground improvement. The conventional breakwaters such as a rubble mound breakwater and a caisson breakwater constructed on the soft ground reveal several problems, i.e., an economical low-efficiency, a difficulty in construction, a guarantee of safety, and a seawater pollution due to heavy dredging and dumping materials. Therefore, many attempts have been made to develop new type of breakwaters that can be applicable for soft ground efficiently, for example, a floating type breakwater, membrane type breakwater, pile type breakwater, curtain type breakwater, and so on. Those breakwaters may be applicable for soft ground, however, as not a main breakwater but a secondary assistant breakwater.

The authors have developed a new type of soft ground breakwater (called below *soft ground breakwater*) as shown in Fig. 1, which can endure the unfavorable conditions of conventional breakwaters. The soft ground breakwater is supposed to prevent the excessive consolidation settlement due to the self-weight with an aid of the buoyant box, and to ensure the sufficient lateral resistance by use of bottom walls.

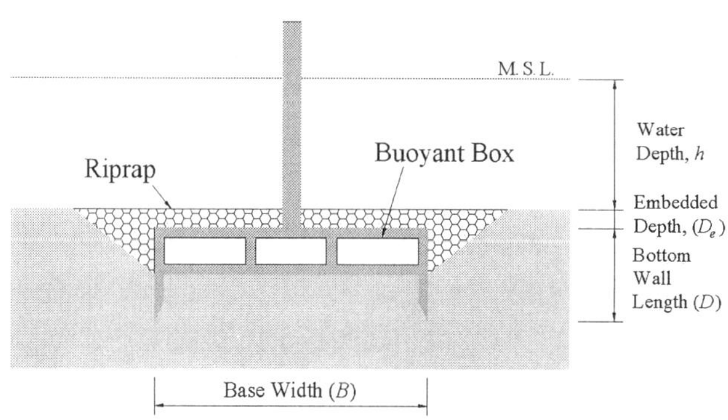

Fig. 1 Conceptual diagram of the soft ground breakwater.

In this paper, the evaluations on consolidation behavior of the soft ground breakwater were performed from the laboratory model test and finite element analysis for various types of breakwater and ground conditions. And, the applicability of the soft ground breakwater was also examined by comparing with consolidation settlement based on Terzaghi's one dimensional consolidation theory.

The model test results reveal that for some cases punching shear failure mode happened

[1] Corresponding author, Phone:82-31-400-6325, FAX:82-31-408-5823, E-mail:wspark@kordi.re.kr

when the applied load was higher than the yield stress of the soil, and caused the abrupt settlement. In order to consider the effect of such a punching shear failure, square root-t method, which is used in the evaluation of the coefficient of consolidation from the conventional consolidation test, was utilized. From the modified curve of applied pressure and settlement a validity of square root-t method, which was utilized to eliminate the effect of initial settlement from total settlement, was verified. Both model test and numerical results also showed that the bottom wall is more efficient to reduce the consolidation settlement than the base width due to the decrease in the thickness of the consolidation layer and the increase in the resistance, and that the effect of the base width is negligible if the bottom wall is long enough.

The consolidation settlements of the conventional breakwater have been generally computed by the Terzaghi's one dimensional consolidation theory. However, the soft ground breakwater proposed herein has a complicated bottom wall system, it is difficult to utilize directly the theory. For a convenience in a practical design, the applicability of the Terzaghi's theory to the soft ground breakwater was examined by comparing the consolidation settlements obtained from numerical analysis and model test with the computed values from the Terzaghi's theory. As shown in Fig. 2, which compares the model test and numerical results with the Terzaghi's solutions, the Terzaghi's theory can conservatively be adopted with some accuracy if assumed the location of the consolidation loading is an end of the bottom wall.

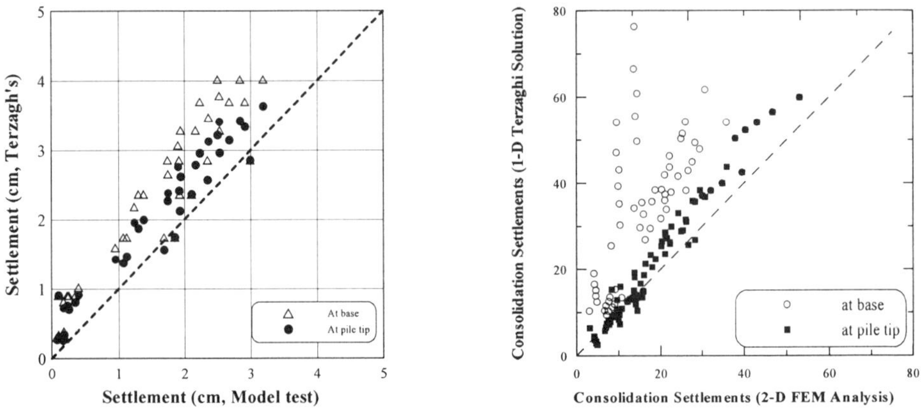

(a) Comparison with model test results (b) Comparison with numerical results

Fig. 2 Comparion of the model test and numerical results with the Terzaghi's solutions

KEYWORDS : soft ground breakwater, laboratory model test, finite element analysis, consolidation settlement, Terzaghi's one dimensional consolidation theory, square root-t method.

EXPERIMENTAL STUDY OF IRREGULAR WAVE IMPACT ON PILED WHARF WITH PERMEABLE SLOPE SHORE CONNECTING

Bing Ren[1] *and Yongxue Wang*[2]

Abstract

The safety of coastal and offshore structures such as piled wharves, shore trestles, oil drilling platforms, etc, has great relation to the wave impact. Under confused sea condition, the wave impact load on the structure is destructive when waves propagate underneath the structure and surge up to its subface. And the structure's service life may also be threatened by the negative pressure occurred when waves separate from the subface of the structures.

In previous research of regular wave slamming on small horizontal cylinders, the impact force is defined as the rate of change of momentum of associated with mass of water moving past the structure. The slamming force F_i is expressed as $F_i = C_s \frac{1}{2} rU^2 Dl$, where D and l are the diameter and length of the cylinder respectively, and C_s defined as the slamming coefficient, is theoretically equal to p (Kaplan and Silbert, 1976; Faltinsen, 1977; Sarpkaya, 1978). Experimental studies have yielded values of C_s which exhibit a considerable degree of scatter, ranging from 1.0 to 7.79 (Dalton and Nash, 1976; Faltinsen, 1977; Miller, 1977; Sarpkaya, 1978; Campbell and Weynberg, 1980; Isaacson and Prasad, 1992; Isaacson and Prasad, 1993).

For wave slamming against structure members with large dimension in the splash zone. French(1979) studied the solitary wave slamming on the platform by laboratory experiments. Kaplan(1992) presented a analytical solution with the same assumption as for small horizontal cylinder. There were very little research work on irregular wave impact problem.. In order to yield a thorough comprehension of the irregular wave impact mechanics, more research work should be done to understand the kinematics of the irregular wave impact process and the corresponding distribution of impact pressure

This paper presents the experimental investigation of irregular wave slamming on the piled wharf with permeable slope shore connecting. The experiments were conducted in the large wave-current tank in the State Key Laboratory of Coastal and Offshore Engineering, Dalian University of Technology. In the experiment, the significant wave height $H_{1/3}$ is in the range from 0.1m to 0.2m, and peak period of spectrum T_p in the range from 1.0s to 2.0s. The ratio of $s/H_{1/3}$, which refers to the subface level of the structure (s) to the incident wave height ($H_{1/3}$), is between 0.1 and 0.4. The time-domain analysis results of the irregular wave impact pressure on the deck of the piled wharf under various case studies are presented. The distribution of the impact pressure along the deck of the piled wharf is described. The influence of different incident wave parameters and relative clearance $s/H_{1/3}$ on

[1] Lecture, State Key Laboratory of Coastal and Offshore Engineering
Dalian University of Technology, Dalian 116024, China. Bren@dlut.edu.cn
[2] Professor, State Key Laboratory of Coastal and Offshore Engineering
Dalian University of Technology, Dalian 116024, China.. wangyx@dlut.edu.cn

the impact pressure is discussed. The empirical formula of the characteristic peak pressure P_c on the deck of the piled wharf are proposed.

Figure 1 below show the comparison of the test data and the computed value of the peak pressure by empirical formula.

Keywords: Irregular Wave, Wave Slamming, Piled Wharf

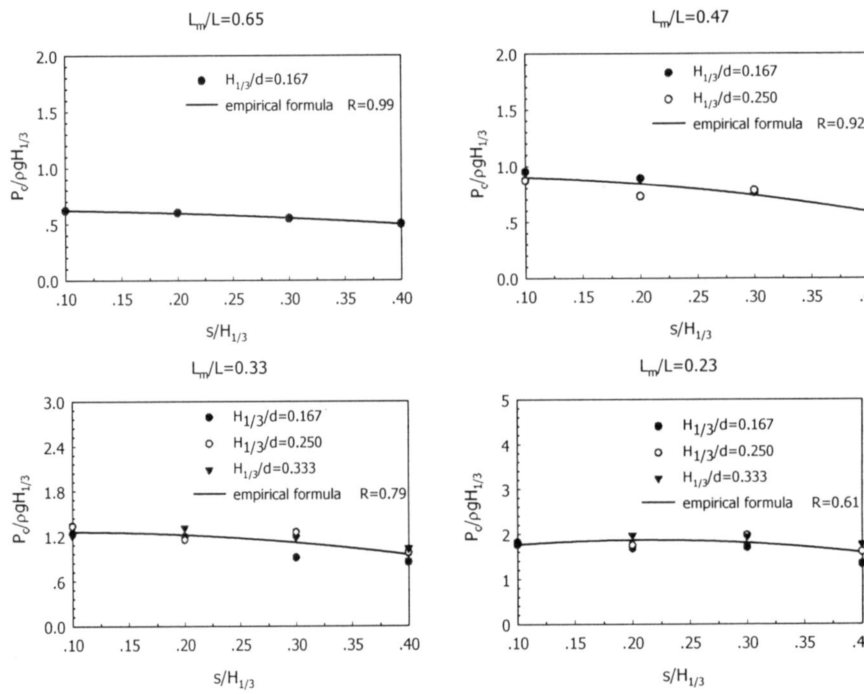

Figure1 comparison of the test data and the computed value of the peak pressure

APPLICATION OF NEURAL NETWORK IN CALCULATION OF WAVE FORCES ON A VERTICAL WALL

WANG Dengting, ZUO Qihua and SHEN Ying

With the development of artificial neural network (ANN) [1], it had been widely used in engineering fields. Mase, et al.[2] established a model utilizing ANN to judge impact of waves on the superstructures of composite breakwaters. In this paper, on this basis, the BP (back-propagation) network with one hidden layer is modified and a neural network is established to calculate wave forces on a vertical wall by use of the additional momentum method and auto-adjustive learning rate. The comparison between the experimental values and the calculated results by Goda's formula [3], the Chinese standard formula [4] and the present method shows that the neural network is suitable to predict wave forces on a vertical wall. Figure 1 shows the establishment process of neural network.

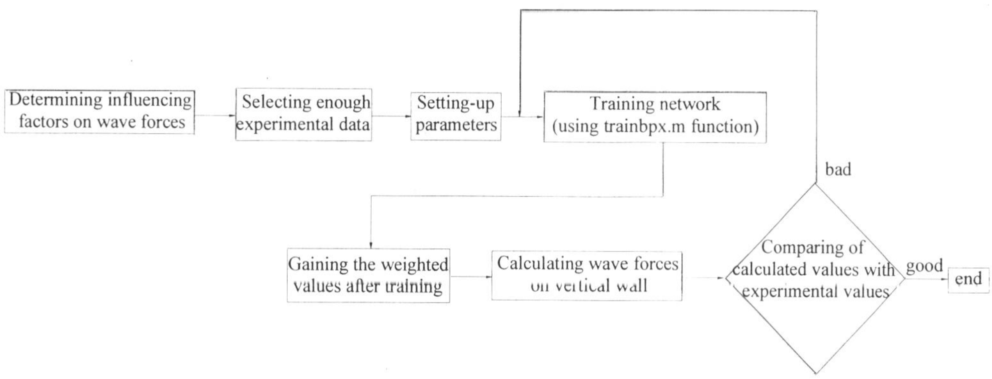

Fig. 1 Establishment process of neural network

Owing to restriction of the selected experimental data, the input layer neuron numbers are taken to be d/H, H/L, d_1/H, B/H, respectively, and the output layer neuron to be dimensionless number $F/\gamma H^2$ (H is the incident wave height where the vertical wall is located; L is the wavelength; d is the depth in front of the vertical wall; d_1 is the depth on the foundation-bed; B is the shoulder width of foundation-bed; F is wave forces). In this paper, the hidden layer number is taken to be 1 and the hidden layer neuron number to be 9. 100 groups of experimental data are selected to establish a neural network model and 45 groups of experimental data to examine the suitability of the model. All the selected data do not consider the condition of wave overtopping. Initial learning rate lr=0.02; increasing multiplier factor of auto-adjustive learning rate lr_inc=1.05; decreasing multiplier factor of auto-adjustive learning rate lr_dec=0.7; error speed err_ratio=1.04; momentum factor mc=0.95; expected error err_goal=10.0.

Figure 2 is the comparison between the experimental and calculated values by the neural network model. (correlation coefficient = 0.878). It is seen from the figure that the calculated results of the neural network model are superior to those of Goda's formula and Chinese standard formula. However, for an actual project, if only the magnitude of the correlation coefficient is considered, it may result in certain danger. So in this paper, it is suggested that the value of $F/\gamma H^2$ for establishing the model should be multiplied by a suitable coefficient and the setting-up of other parameters does not change, satisfactory results can be obtained. Figure 3 gives the comparison between experimental and calculated values by the neural network model considering that the factor is 1.1.

If after carrying out systematic and complete experiments, more perfect data are added to the neural network model, it will be further optimized. The present model can be popularized and applied to other respects of coastal and ocean engineering such as prediction of tides, wave running-up etc.

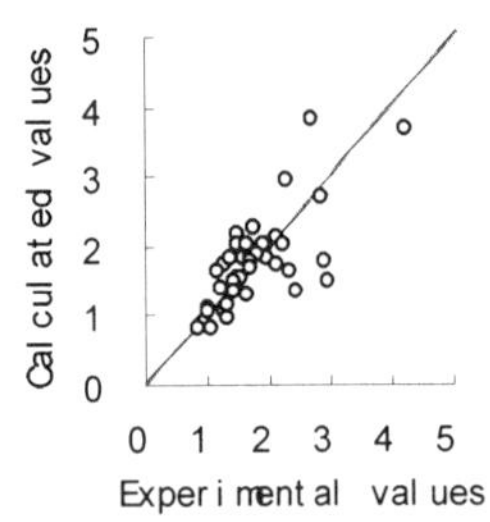

Fig 2 Comparison between experimental and calculated values by the neural network model

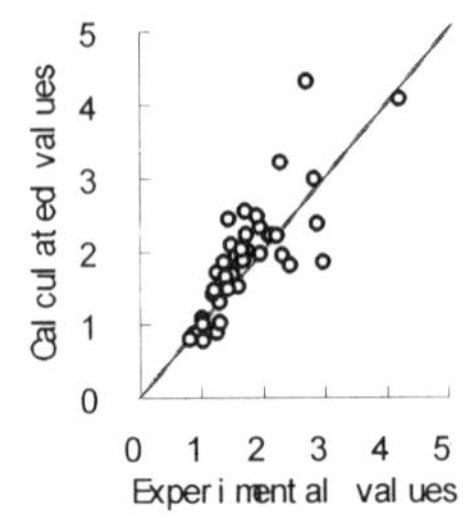

Fig 3 Comparison between experimental and calculated values by the neural network model considering suitable factors

Reference:

1 Cong, S. (1998). "The theory and application of neural network facing MATLAB toolbox" China University of Science and Technology Press, Hefei, China.

2 Mase, H. and Kitano, T. (1999). "Prediction model for occurrence of impact wave force" *Ocean Eng.*, Vol. 26: pp 949-961.

3 Goda, Y. (1985). "Random seas and design of marine structure" Univ. of Tokyo press, Tokyo, Japan.

4 Code of Hydrology for Sea Harbor. (JTJ213-98). People's Communications Publishing House, Beijing, China.

The authors' Information:

(1) WANG Dengting:

Title: Master; Assistant Engineer

Affiliation: His main research interest is interaction of waves and structures, and he has published 3 papers.

Address: River and Harbor Department, Nanjing Hydraulic Research Institute, Hujuguan 34, Nanjing 210024, China

Tel: +86 25 5829341 Fax: +86 25 5829333 E-mail: dtwang@njhri.edu.cn

(2) ZUO Qihua

Title: Professor; Senior Engineer

Affiliation: Member of International Steering Committee of Second International Conference on APAC in 2003. His main research interest is coastal dynamics, and he has published dozen of papers.

Address: Nanjing Hydraulic Research Institute, Guangzhou Road 223, Nanjing 210029, China

Tel: +86 25 5828008 Fax: +86 25 5829333 E-mail: qhzuo@njhri.edu.cn

Direct measurement of wave force acting on a rubble mound breakwater in hydraulic experiment

Dal-Soo Lee[a*], Changil Kim[b], Yong Min Oh[c] and Kwang Soo Lee[d]

[a]Principal Research Scientist, Coastal and Harbor Engineering Laboratory, Korea Ocean Research and Development Institute, Ansan P.O. Box 29, Seoul 425-600 South Korea
Ph: +82-31-400-6336; Email: dslee@kordi.re.kr

[b]Graduate Student, Department of Civil Engineering, Konkuk University, 1 Hwayang-Dong, Kwangjin-Ku, Seoul 143-701 South Korea
Ph: +82-31-400-6323; Email: cikim@kordi.re.kr

[c]Principal Research Scientist, Coastal and Harbor Engineering Laboratory, Korea Ocean Research and Development Institute, Ansan P.O.Box 29, Seoul 425-600, South Korea
Ph: +82-400-6312; Email: ymoh@kordi.re.kr

[d]Principal Research Scientist, Korea Ocean Research and Development Institute, 1270 Sadong, Ansan, Seoul 425-600 South Korea
Email: kslee@kordi.re.kr

For the direct measurement of the wave forces acting on a rubble mound breakwater, a breakwater frame model was devised. The core part in the model was filled with approximately 40 thousands of specially shaped empty polyhedrons instead of natural stone to reduce total weight. The specific gravity and pore ratio of this alternative core material are 0.55 and 0.43, respectively. Natural armor stones were adopted for the cover layer and filter layer on the front slope of this model.

Two cases of experiments were performed in a wave flume of 53.2m(L)x1.25m(H)x1.0m(W); (1) for the breakwater model hung at the load cell meter fixed on the upper part of the wave flume (1.5cm above the bed), (2) for the model installed on the bed of the wave flume. Twelve regular waves were applied and water depth was 30cm.

In both cases, wave run-up and run-down were measured using the inclined wave gage installed along the front slope of the model. Water elevations at two points on the front slope and wave pressures along the front slope and along the model bottom were measured. Incident, reflected and transmitted wave data were also collected. For case (1), additional items of the total horizontal and vertical forces were observed using the load cell meter.

Integrated wave forces from pressure gages were compared with those measured by the load cell meter. Results show that horizontal components evaluated from the two different methods coincide fairly well. However, vertical components of the wave forces

* Corresponding author. Phone: +82-31-400-6336; E-mail: dslee@kordi.re.kr

measured directly with the load cell meter are much smaller than those integrated using pressure values.

Measured wave forces will also be compared with those calculated using the linear equations of partially reflected wave-form on the slope suggested by Goda (2002, personal communication). In the equations, it is assumed that the wave form is the same with that in front of a virtual vertical wall whose reflection coefficient is equal to that of the rubble mound breakwater considered.

Fig. 1 Setup of a rubble mound breakwater model

A SIMPLIFIED THEORY ON TOTAL WAVE PRESSURE EXERTED ON A RUBBLE MOUND BREAKWATER

Yoshimi Goda [1] and Akira Matsumoto [2]

[1] ECOH Corporation, 2-6-4, Kita Ueno, Taito, Tokyo, 110-0014, Japan,
Tel:+81-3-5828-2184, Fax:+81-3-5828-2176, goda@ecoh.co.jp

[2] TETRA Co., Ltd., 2-7, Higashi Nakanuki, Tsuchiura, Ibaraki, 300-0006, Japan,
Tel:+81-298-31-7411, Fax: +81-298-31-7693, matsumoto@tetra.co.jp (Corresponding Author)

1. Introduction

A rubble mound breakwater is the most fundamental structure to protect coastal area from wave actions. Although numerous studies on the stability of armor blocks have been conducted, very little knowledge is available on the wave pressure exerted on the rubble mound breakwater (for example, Troch and De Rouck, 2002). Where the bearing capacity of the subsoil is large enough to support the breakwater, precise evaluation of total wave pressure may not be necessary. However, if the geotechnical property of subsoil is not appropriate, careful examination on the total wave pressure must be conducted to assure the stability of the breakwater against a slip failure of the subsoil caused by the weight of breakwater and the wave pressure acting on the breakwater.

In the present paper, a simplified theory to calculate the total wave pressure exerted on a breakwater slope is proposed. Its applicability is examined by comparing the theoretical and numerical results. Discussion is also made on the properties of the pressure attenuation inside the breakwater.

2. Theory

Let us consider a system of partial standing waves on the breakwater slope (Figure 1). Suppose that a virtual reflecting wall RQ is located on $x=x_1$. Maximum runup with the height $\eta_{\max}$ occurs at this position. The water surface elevation of the standing waves $\eta(x,t)$, which are the superposition of incident waves and reflected waves stemming from the virtual reflecting wall, is given by the following equation:

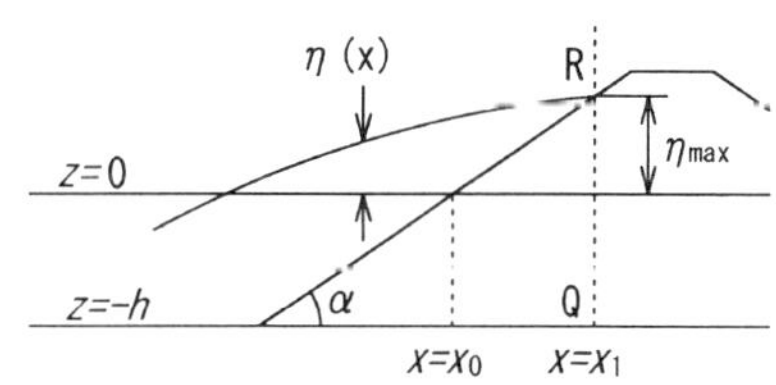

Figure 1: A partial standing wave.

$$\eta(x,t) = \frac{1}{2} H \sqrt{1 + 2K_R \cos 2k(x - x_1) + K_R^2} \cos\left[\frac{2\pi}{T}(t - \tau)\right] \tag{1}$$

in which τ denotes the time lag expressed by Eq. 2:

$$\tau = \frac{T}{2\pi} \tan^{-1}\left[\frac{(1 - K_R)\sin k'(x - x_1)}{(1 + K_R)\cos k'(x - x_1)}\right] \tag{2}$$

where H is the incident wave height, K_R is the reflection coefficient, T is the wave period and k is the wave number corresponding to the water depth outside of the breakwater. The term k' represents the modified wave number on the breakwater slope that accounts for the reduction in the wave celerity due to decreasing water depth. Wave pressure intensity can then be calculated by substituting Eq. 1 into Eq. 3. Finally, total wave pressure is obtained by integrating Eq. 3 over the range $z=-h$ to $\eta(t)$.

$$p(x,z)=\begin{cases}\rho g[\eta(x)-(x-x_0)\tan\alpha] & : \ 0\le z\le \eta_{\max} \\ \rho g\eta(x)\dfrac{\cosh k(h+z)}{\cosh kh} & : \ -h\le z<0\end{cases} \qquad (3)$$

3. Calculated and Theoretical Total Wave Pressure

Figure 2 shows the pressure distributions and associated water surface elevations obtained by the numerical computation and the proposed theory when the total pressure on the breakwater slope takes its maximum value. A numerical wave flume CADMAS-SURF (Isobe et *al.*, 1999) was employed for the numerical computation. A modified wave number of $k'=1.6k$ was introduced to approximately estimate the time lag τ. The agreement between numerical and theoretical results is good. The error in total wave pressure is less than 10%. Slight disagreement of pressure intensity in the range of $z>0$ is due to the wave nonlinearlity. Numerical computation is based on nonlinear theory, while the proposed method is based on the linear wave representation. Therefore, maximum runup height $\eta_{\max}$ obtained by proposed theory is smaller than numerical one. To improve the accuracy of theoretical total wave pressure, a higher order solution will have to be applied to the pressure calculation procedure.

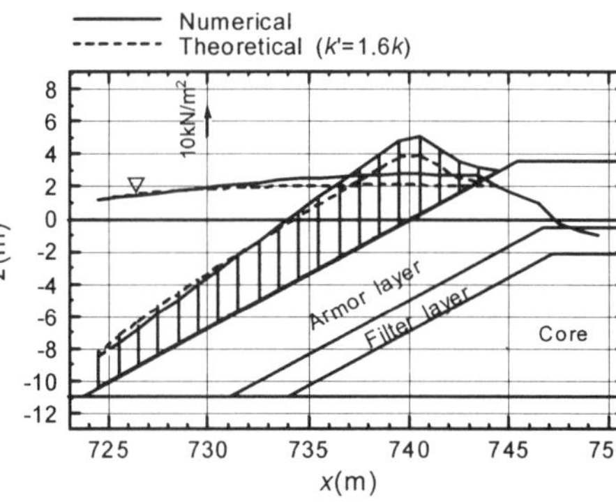

Figure 2: Pressure and water surface elevations.

4. Pressure Attenuation Inside the Breakwater

Figure 3 shows maximum pressure distribution together with maximum water surface elevation obtained by numerical computation. Wave pressure is not sufficiently absorbed inside the breakwater but a significant intensity of pressure remains and act on the subsoil. This implies the necessity of taking the total wave pressure into account when the bearing capacity of the subsoil is poor.

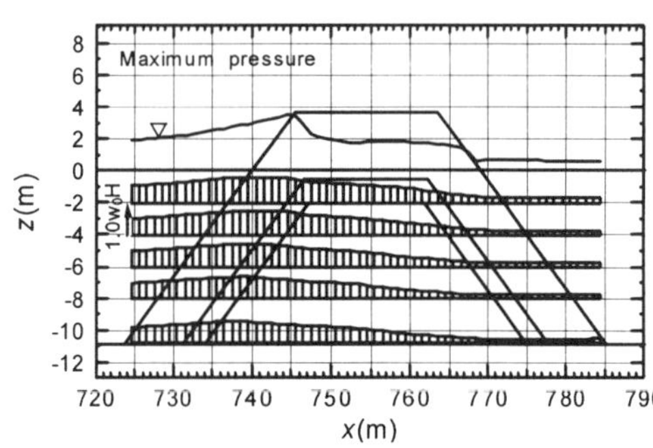

Figure 3: Maximum wave pressure distribution.

5. Conclusions

The proposed simplified theory can evaluate the total wave pressure exerted on the rubble mound breakwater. The theory will assist the design of a rubble mound breakwater to be built on a soft subsoil by providing an estimate of wave force on the breakwater.

References

Isobe, M. et *al*. (1999): *Proc. of Civil Engrg in the Ocean*, Vol.15, JSCE, 321-326 (in Japanese).

Troch, P. and De Rouck, J. (2002): Abstract, 28th ICCE, Paper No.181.

INFLUENCE OF PROXIMITY OF THE SEABED ON WAVE FORCES ON SUBMARINE PARALLEL PIPELINES

Vijayakumari, P.L[*] and Sundar, V[**]

[*]Senior Lecturer, Department of Civil engineering, N.S.S College of Engineering, Palakkad
Presently Research scholar, Department of Ocean Engineering, Indian Institute of Technology Madras, India.
Email: vijayapl@hotmail.com
[**]Professor, Department of Ocean Engineering, Indian Institute of Technology Madras, India.
Email: vallamsundar@hotmail.com

Abstract

Submarine pipelines are usually employed for the conveyance of oil and gas, seawater for cooling systems & watersupply projects, disposal of drainage water etc. In the case of pipelines close to the seabed, the proximity of the seabed influences the forces on the pipeline in the coastal zone. The flow behavior and the forces on the pipeline are affected due to the presence of another pipeline adjacent to it. In addition, the beach slope also alters the flow field around the pipelines and the distance between the pipelines as well as the clearance between the pipe and the sea bed governs the wave induced forces on them. For an efficient design of parallel submarine pipelines near a sloping bed, a proper understanding of the effect of vertical clear spacing between the seabed and the bottom of the pipe and the effect of the horizontal clear spacing between the adjacent pipes is essential. This prompted the investigators to take up a detailed experimental program on the present topic.

The experimental investigations were carried out in a 72.5m long, 2m wide wave flume in a constant water depth of 1.0m in the Department of Ocean Engineering, Indian Institute of Technology Madras, India. A computer controlled wave maker operating either in piston or in hinged mode is installed at the one end of the flume. It is capable generating both regular waves and random waves of predefined spectral characteristics. A sloping rigid bed with a slope of 1:10 was fabricated and erected in the other end of the flume. The slope started from a distance of 40 m from the wave maker. Two pipes of same of diameter, D= 0.315m and length of 1.99m were used for the tests. One of the pipes, being the test pipe was housed with eight pressure transducers with measuring range of 0.2 bar around its circumference to register the time history of the dynamic pressures. The water depth at the location of the test pipe was maintained at 0.827m. The experiments were conducted for w/D=1 (w is the horizontal clear spacing between the test pipe and the dummy pipe on seaside or leeside). The pipelines were placed normal to wave direction near the sloping bed and subjected to the action of regular waves. The tests covered a range of wave period, T of 1.0 to 2.0 sec with an increment of 0.2 sec. For each of the T, at least three wave heights, H were employed for the tests. The time histories of dynamic pressures around the test pipe and wave elevations in phase of the cylinder axis were simultaneously acquired by personal computer with a sampling interval of 0.025 sec for duration of 1 min. Further, the tests were done with the dummy pipe on the seaside as well as on lee side of the test pipe. For each of the test condition said above, the tests were repeated for different e/D ratios, where, 'e' is the clear distance between the pipe and the sloping bed. The experimental setup for a dummy pipe placed on the lee side of the test pipe is shown in Fig.1.

---

[1] Corresponding author: P.L. Vijayakumari, Senior Lecturer, Department of Civil engineering, N.S.S College of Engineering, Palakkad, Kerala, India and Presently Research scholar, Department of Ocean Engineering, Indian Institute of Technology Madras, India. **Email: vijayapl@hotmail.com**

The pressures were integrated to obtain the force time history from which the peak horizontal and vertical forces are evaluated. The circumferential variation of the dynamic pressures in a non-dimensional form is reported as a function of scattering parameter, ka, where, 'k' is the wave number and 'a' is the radius of the pipe. The dimensionless vertical and horizontal forces are also presented as a function as a relative depth (d'/L', where d' is the water depth at the test pipe and L' is the wavelength over the test pipe). The effect of the vertical gap ratio on the forces have been studied and reported. The details of the model set up, experimental procedure, results and discussion are presented in this paper.

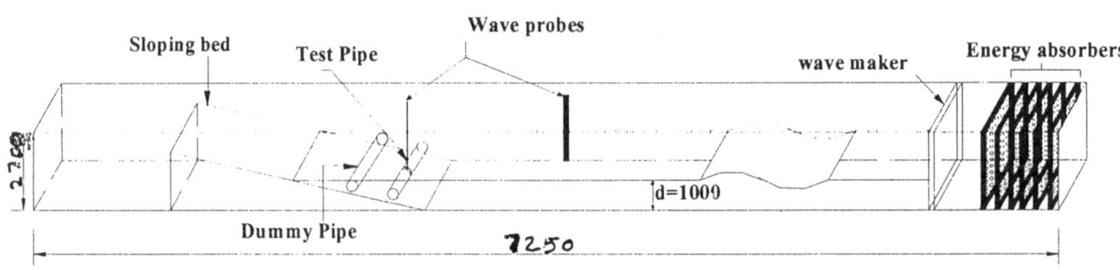

DETAILS OF THE WAVE FLUME WITH THE DUMMY PIPE ON THE LEESIDE

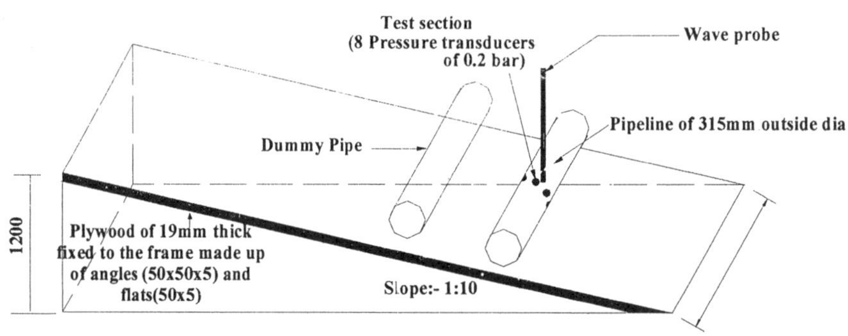

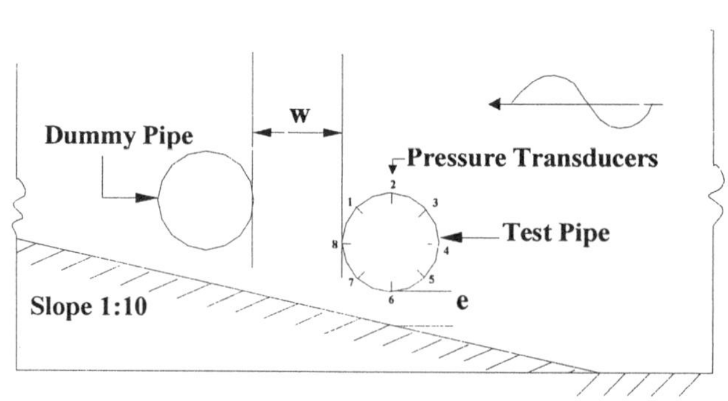

SECTIONAL ELEVATION OF THE DOUBLE PIPE SETUP

FIG.1.EXPERIMENTAL SETUP WITH THE DUMMY PIPE ON THE LEEWARD SIDE OF THE TEST PIPE

APPLICATION OF A COMPOUND MODEL METHOD IN THE MODEL TEST OF SHANGHAI YANGSHAN DEEPWATER PORT

Li Bei

(Tianjin Research Institute of Water Transport Engineering,Tanggu, Tianjin, 300456 China)

Tel:+86 22 25707168-382, Fax: +86 22 25795125, Email: libei2001@sina.com

Abstract

In the model test of Yanshan Deepwater Harbor of Shanghai International Shipping Center, 30 kilometers away from shoreline outside the mouth of the Hangzhou Bay, a "composite simulation type" research method of compound models is used to simulate and study the tidal flow variation characteristics and sediment movement variation process in the project sea area before and after the construction schemes are put into effect.

The physical model system includes a flashboard-type tide-generating system in the east boundary to simulate the tide from out sea and a flashboard-type discharge check system in the west side to simulate the tidal flow moving back and forth from the top of the Hangzhou Bay. The whole model is designed approximately along the tidal flow direction. Because the simulated sea area is larger and there exits the Coriolis effect, there exits not only phase difference but also transverse water flow in the south and north boundaries, therefore, the discharge flowing into and out of the south and north boundaries is controlled by double-way pumps.

Making use of the characteristics of mathematical models that can simulate the Coriolis effect caused by the rotation of the eatrth and integrating the boundary position, the computation grid, data output, and so on, are designed. The computed tidal level and discharge are given to the physical model as its conditions of four open boundaries. Through the tidal level and discharge distribution of the model boundaries, the water is forced to flow into and out of the boundaries to remedy errors caused by the physical model which cannot simulate the Coriolis effect, so as to attain the aim of simulating the practical tide and tidal flow well.

To provide the physical model with accurate boundary conditions, two sets of mathematical models are set up. The first set is a one of a large sea area including the Changjiang River estuary and the Hangzhou Bay. The whole computation domain is about 232.5km long in the east-west direction and about 225.5km wide in the south-north direction. In order to finely simulate the tidal flow movement in the project sea area with many islands and passages, a mathematical model for the project sea area is set up, with the first set mathematical model providing boundary conditions. The computation domain is about 72km long in the east-west direction and about 50km wide in the south-north direction and the area is 3600km^2, which stretches out 12km respectively from the four boundaries of the physical model as transition area. In view of the characteristics of computation sea area with many islands and irregular shoreline and the run of the physical model's boundaries, the mathematical models adopt arbitrary triangular grids, with the minimum spatial step 30m and the maximum spatial step 1230m. In the computation domain, there are 35 islands being taken into account.

WAVE DIFFRACTION FROM A UNIFORM CYLINDER IN FRONT OF VERTICAL WALLS

Bin TENG, De-zhi NING

The State Key Laboratory of Coastal and Offshore Engineering,

Dalian University of Technology, Dalian, 116024, China

Fax: 86-411-4708526 Email: bteng@dlut.edu.cn

Abstract

Many researches have been carried out on wave diffraction from structures in an open water area. For a uniform cylinder MacCamy and Fuchs (1954) developed an analytic solution. For arbitrary 3D bodies, many numerical methods are available. However, for wave diffraction from bodies in front of vertical walls, less research has been seen.

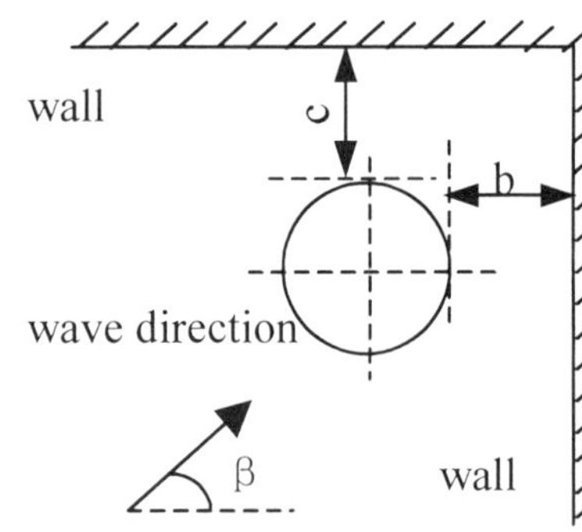

Fig. 1 Definition sketch

The purpose of the paper is to extend the researching scope of wave diffraction from bodies in open water into the case in front of walls intersecting normally. Based on the image theory, the wave diffraction from a cylinder in front of the two vertical walls is transformed into a problem of diffraction of four symmetric incident waves from four symmetrically arranged cylinders. The transformed problem is solved by an analytical method based on eigenfunction expansion, in which Graf addition theorem is applied in the transformation of coordinates systems.

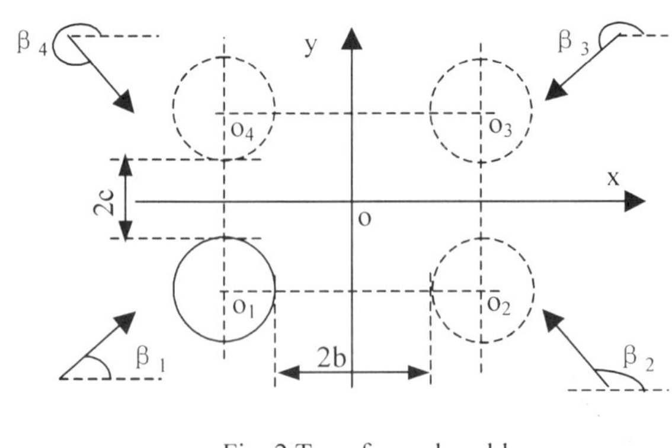

Fig. 2 Transformed problem

By comparison with the results calculated by a boundary element method, the validity of the present analytical solution is proved. Numerical experiment has been performed to examine the effects of the distances between the cylinder and the vertical walls and wave incident direction on the wave force on the cylinder. The calculation shows that the wave force on the cylinder in front of the two vertical walls is several times larger than that on the cylinder in the open water. The force on the cylinder in front of the two walls oscillates with wave number, and the oscillating frequency increases with the increase of the distance between the cylinder and the walls. The maximum wave force on the cylinder in the x-direction increases with the decrease of the wave incident angle, and vise verse for the wave force in the y-direction. Figures 3 and 4 show the effect of the distance between the cylinder and the vertical wall on the wave force in the x-direction at $c/a=0.2$, $\beta=30^0$ and $\beta=45^0$ respectively. OW denotes the case in the open water.

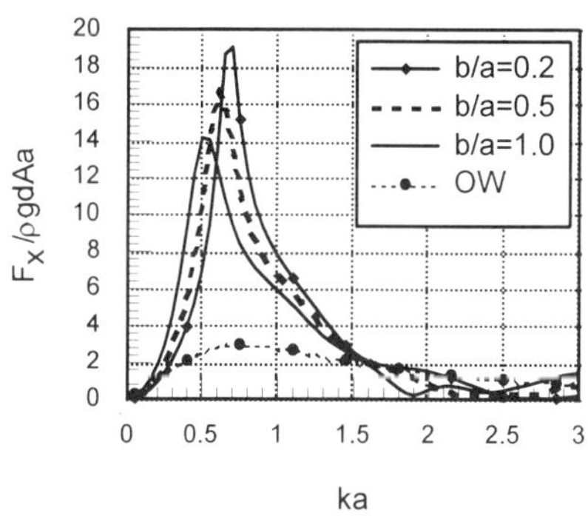

Fig. 3 Wave force in the x-direction at $\beta=30^0$, $c/a=0.2$

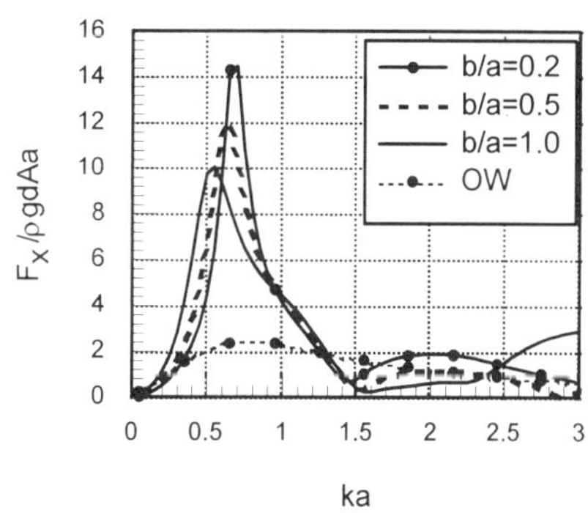

Fig. 4 Wave force in the x-direction at $\beta=45^0$, $c/a=0.2$

RESEARCH ON INFLUENCE OF HANGZHOU BAY MAJOR BRIDGE TO QIANTANG BORE

Shaolong Xiong[1] Jian Zeng[2] Haiqian Han[3]

Abstract: At Qiantang estuary, when the tidal range is large while the water in the river is relatively shallow with a decreasing depth toward upstream, the back slope of the tidal wave catches up with the front gradually due to increasing nonlinearity and riverbed resistance. The front slope of the tidal wave becomes more and more steep and finally a violent bore is formed.

The Qiantang River runs into the East China Sea through the funnel-shaped Hangzhou Bay. Because the ratio of river runoff and the tidal flow is rather small, a large amount of sediment originating from the Changjiang Estuary has been carried into the bay by flood tidal current and has deposited to form a large longitudinal sand bar which is known to be 130 km in length from Zhapu to Wenyan. As the tidal wave propagates into the funnel-shaped Hangzhou Bay, the tidal range increases steadily owing to narrowing width of the bay. When it climbs in the shallow slope upstream of Ganpu, the wave deforms violently and the wave front becomes steep and then breaks to form the well-known Qiantang Bore as shown in Fig. 1.

Hangzhou-Bay Major Bridge from Zhengjiadai to Fengshou Sluice is 36 km in length and it is the longest sea-crossing bridge in the world. A physical model is used to investigate the effects of the Hangzhou-Bay Major Bridge on the Qiantang Bore, the tidal level, the tidal velocity and the tidal volume. Laoyancang and Jinshan are selected as upstream and downstream boundaries of the model respectively, as shown in Fig. 2. The model is governed by similarity laws for gravity, resistance and flow continuity. It is molded according to the underwater topography measured in September 2000 and verified well with hydrometric data at the same period. Table 1 gives the calculated and model-adopted similarity scales. The research indicates that during spring tide with the frequency of tidal range p=1%, the high tidal level will decrease by 0.06m and the low level increase by 0.03m and the tidal range decrease by 0.09m at Ganpu, which is less than two percent of the tidal range of the spring tide at Ganpu; the height of the tidal head will decrease by 0.02m at Yanguan, which is less than one percent of the original height of the tidal head after the construction of the bridge. It is then concluded that the influence of the Hangzhou-Bay Major Bridge on the Qiantang Bore is negligibly small.

Table 1. Calculated and model-adopted scales

Scale	Equation	Calculated	Adopted
Horizontal	λ_l		1000
Vertical	λ_h		100
Velocity	$\lambda_u = \lambda_v = \lambda_h^{1/2}$	10	10
Current time	$\lambda_t = \lambda_l / \lambda_u$	100	100
Roughness coefficient	$\lambda_{n_b} = \lambda_h^{2/3} \big/ \lambda_l^{1/2}$	0.68	0.64

1 Professor, Senior Engineer , Zhejiang Institute of Hydraulics and Estuary,
Add: 50 East Fongqi Road, 310020 Hangzhou, China, Tel: 86-571-86516945, Fax: 86-571-86960283, E-mail:slxiong@163.com

2 Engineer, Zhejiang Institute of Hydraulics and Estuary, E-mail: zengj163@163.com

3 Engineer, Zhejiang Institute of Hydraulics and Estuary, E-mail: qtjhhq@btamail.net.cn

Fig 1. World –famous Qiantang Bore

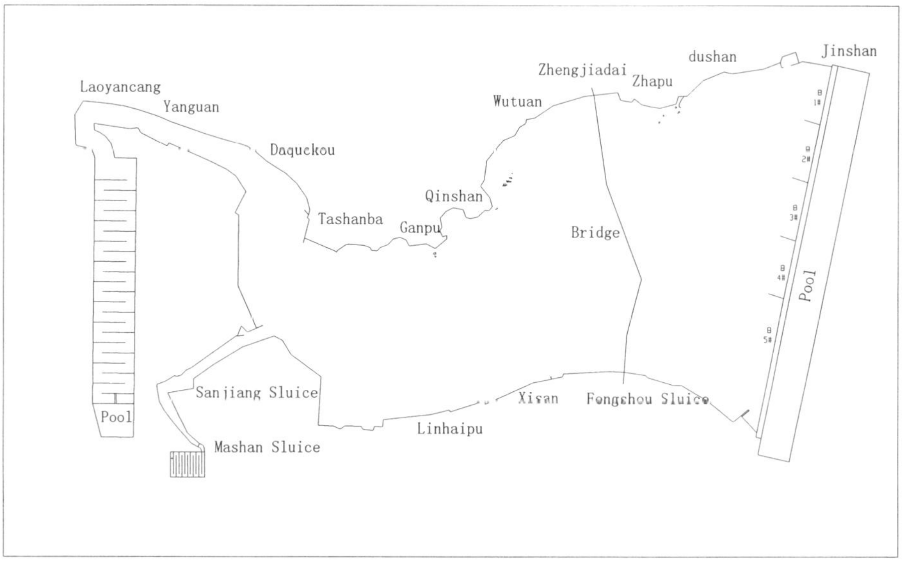

Fig 2. Layout of the Model for Hangzhou-Bay Major Bridge

References:

Shaolong Xiong etc.(2001). "Feasible Research of Hangzhou Bay Access Project, Experimental Report for Physical Model", Technique Report, Zhejiang Institute of Hydraulics and Estuary, Hangzhou, China.

Shaolong Xiong etc.(2002). "Physical Model for Hangzhou Bay Major Bridge and Research on Local Scour of bridge Span", Technique Report, Zhejiang Institute of Hydraulics and Estuary, Hangzhou, China.

Shaolong Xiong etc.(2002). "Design and Verification of Physical Model for Hangzhou Bay Major Bridge", *Donghai Marine Science*, 2002 No.4.

EXPERIMENTAL AND NUMERICAL STUDY OF WAVES IN POROUS STRUCTURE

Mohammed Fazlul KARIM
Ph.D Student, Dept. of Civil & Env. Engg., Saitama Univ., 338-8570 Japan
Email: sgd3051@post.saitama-u.ac.jp, fazlulk@hotmail.com,
Phone: 090-6189-4709, Fax: 81-48-858-3561

Katsutoshi TANIMOTO
Professor, Dept. of Civil & Env. Engg., Saitama Univ., 338-8570 Japan
Email: **tanimoto@post.saitama-u.ac.jp**
Phone/Fax: 81-48-858-3561

Abstract

Many coastal structures constructed for the purpose of harbor securing and shore protection are porous. It is well known that wave properties such as height, energy and force are changed considerably inside the porous structure. However, all these changes depend on the several structural parameters including width, material size and porosity of the structure. Therefore, it is essential to investigate the influence of all these parameters. Wave motion in permeable structures is often studied experimentally. However, in the recent years several numerical modeling are also done. Losada et al. (1995), Hsu et al. (2002) and Liu et al. (1999) are the few examples of previous studies on wave structure interaction in porous media. However, all of these studies are related with breakwaters where part of the wave energy is transmitted by overtopping or passes through the structure. In case of seawall there is no transmission of wave energy. Hence, different characteristics of reflection, energy dissipation and wave damping are expected.

This study is to investigate the wave damping, energy dissipation and reflection properties of vertical permeable structure. A series of physical experiments have been conducted for different wave and structural conditions. Width of the wall, material size and porosity are the variables in the physical experiments. The experiments were conducted in the hydraulic laboratory of Saitama University. A total of 60 tests were performed for the incident waves of heights 3.68 to 12.30 cm and period 1.2, 1.6 and 2.0s (2 structure width × 2 material types × 3 periods × 5 wave heights). However, it is also understood that physical models have their natural limitations. Small-scale physical modeling is influenced by scale effect while large-scale modeling is relatively expensive. Moreover, the measurement during wave breaking itself is highly challenging. Therefore, numerical models are the valuable research tools for studying non-linear wave transformation and wave structure interaction in porous media.

Based on the latest development in the computational fluid dynamics, a numerical model using direct simulation of Navier-Stokes equation is developed. Modified Navier-Stokes equations are used to describe the wave motion in porous structure. The SMAC method is employed to solve the time evolution of the velocity and pressure field. For the free surface treatment volume of fluid (VOF) function is introduced and its advection equation is solved using Harvie and Fletcher (2001) scheme. To avoid extrapolation or interpolation of prime variables (u,w and p) at the free surface, a two-phase model proposed by Hieu and Tanimoto (2002) is used. The turbulence model is incorporated using Smagorinsky sub-grid scale model. Schematic view of numerical model setup is shown in Fig.1. The results of numerical model are compared with the experimental data. Comparison of spatial wave height distributions inside and outside the porous structure is shown in Fig. 2 where 'L' represents the wavelength. The results of numerical model fairly agree with the experimental data. Fig. 3 shows time profiles of free surface for one wave period with experimental crest and trough levels. It also shows good agreement with experimental data. The model is then applied to investigate the effect of different wave and structural parameters influencing reflection, wave force, energy dissipation and wave damping. A typical example of wave damping pattern inside the structure is shown in Fig. 4 for different input wave conditions. Since, porous armor layers are used for many practical structures, i.e., seawalls or caisson breakwaters, present model would be very useful for detail analysis of non-linear wave structure interactions.

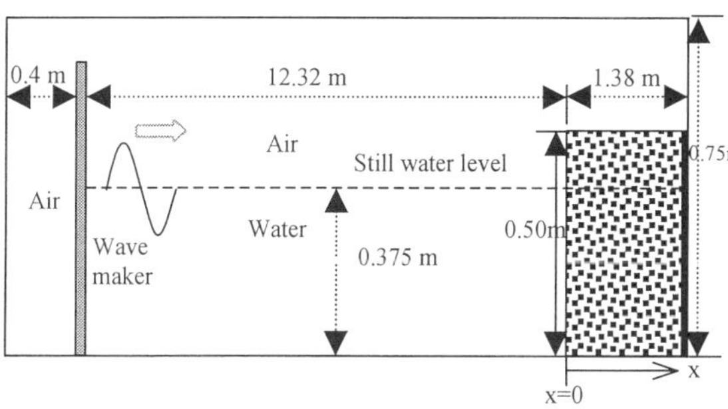

Fig.1.: Schematic view of numerical model setup

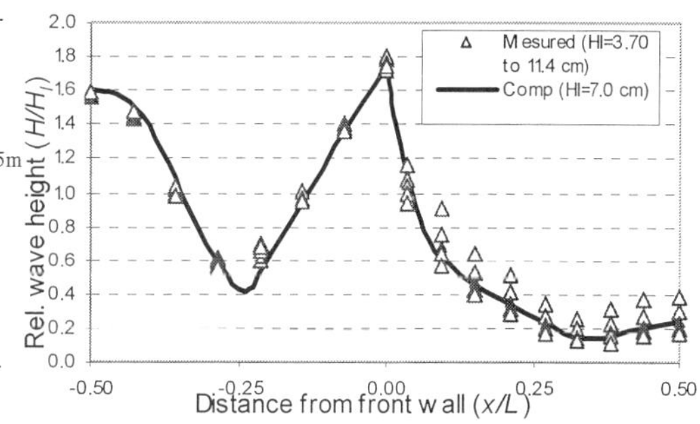

Fig.2.: Comparison of spatial wave heights

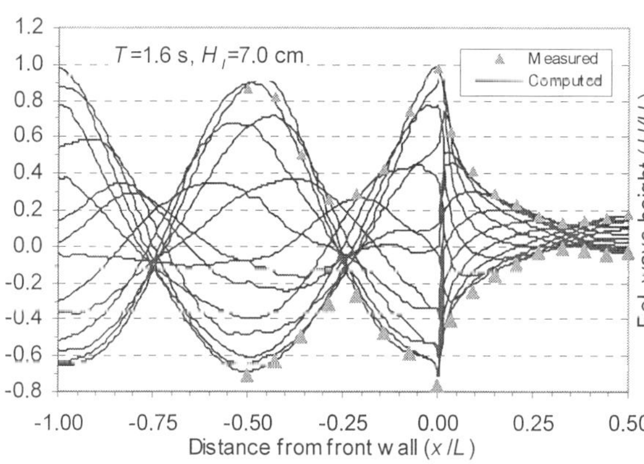

Fig.3.: Surface profile inside and outside porous structure

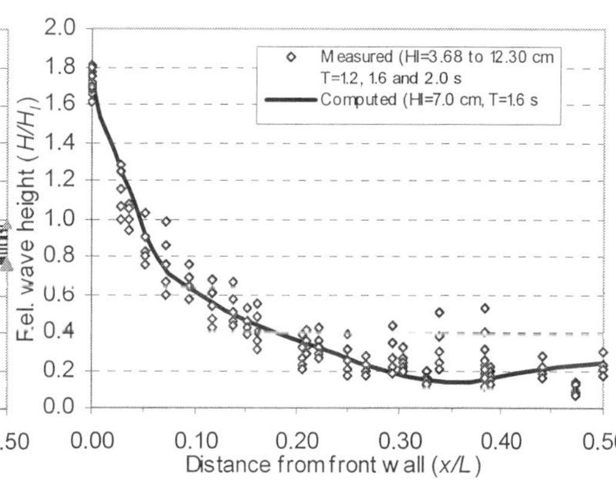

Fig 4.: Wave damping inside the porous structure

Reference:

Harvie, D.J.E. and Fletcher, D.F. (2001), A new volume of fluid advection algorithm: the defined donating region scheme, Int. Journal of Numerical Methods in Fluids, Vol. 35, pp.151-172

Hieu, P.D. and Tanimoto (2002), A two-phase flow model for simulation of wave transformation in shallow water, Proc. 4th International summer symposium, JSCE, pp. 179-182.

Hsu, T.J., Sakakiyama, T. and Liu, P.L.F (2002), A numerical model for wave motions and turbulence flows in front of a composite breakwater, Coastal Engg., Elsevier, Vol. 46, pp.25-50

Liu, P.L.F., Lin, P., Chang, K., and Sakakiyama, T. (1999), Numerical modeling of wave interaction with porous structures, Journal of waterway, Port, Coastal and Ocean Engineering, ASCE, Vol. 125, No. 6, pp. 322-330

Losada, I.J., Losada, M.A. and Martin, F.L. (1995), Experimental study of wave induced flow in a porous structure, Coastal Eng., Elsevier, Vol.26, pp.77-98

WAVE TRANSMISSION THROUGH DOUBLE VERTICAL SCREEN BREAKWATERS

Balaji R[*] and Sundar V[**]

[*]MS Scholar, e-mail: bala_bink@hotmail.com
[**] Professor, e-mail: vallamsundar@hotmail.com[!]
Department of Ocean Engineering, Indian Institute of Technology Madras, India

Abstract

For the formation of artificial harbours and boat marinas, breakwaters are constructed aiming to maintain required tranquility. Among various types of breakwaters, vertical breakwaters require less space, less construction material and time saving. However, the vertical impermeable front face on its seaward side experiences excessive dynamic loading due to reflections of the incident waves. There are several methods adopted, such as dissipating chamber systems to reduce the loads as well as the reflection from the structures. The vertical wave screen are mainly used as breakwaters in small boat harbours and marinas, where less space for breakwater construction, less reflection requirement, and wave transmission is permitted to a limited level. They are also used as wave dampers, in order to reduce the reflection of wave from existing vertical breakwaters. The permeable screen may be made of rectangular, triangular or circular elements, the configuration of which depends on the degree of allowable transmission of waves on its leeward side. The present study is to investigate the hydrodynamic characteristics of twin vertical wave screen breakwaters through an experimental investigation.

The present wave screen model consists of series of horizontally spaced pipes with 160mm (outer) diameter as wave intercepting elements. Two different seaward side screen porosity (23.8 and 5.9%) and a constant leeward side screen porosity of 5.9%, by varying the spacing between the pipes (refer Fig. 1) were considered for investigation. To find the effect of chamber width on the transmission and reflection, the spacing between the screens were varied from 0.5m to 2.0m at an interval of 0.5m. The model was placed at 30m from wave paddle in a 72.5m, 2m wide and 2.7m deep wave flume in Department of Ocean Engineering at Indian Institute of Technology Madras, India. Figure 2 shows the cross sectional view of the wave flume with the model setup. The tests were done in a constant water depth of 0.95m and wave screen extended through out the depth and piercing the free surface and overtopping was not allowed. Three wave gauges positioned in front of the model registered the composite wave elevation, which were used for decomposing it into incident and reflected components. Two more wave gauges, out of which one was positioned in between the screens that registered the water level oscillation inside the chamber and other wave gauge was placed after the leeward side screen that measured the transmitted wave elevation.

[!] **Corresponding author:** Prof. V. Sundar, Department of Ocean Engineering
Indian Institute of Technology Madras, Chennai – 600 036, India
Ph: 91-44-257 8629 Fax: 91-44-235 0509

The model was subjected to regular waves with wave periods ranging from 1 to 2 sec at an interval of 0.2 sec. For each wave period, three different wave heights were adopted for the tests. The variations of reflection coefficient (K_r), Transmission coefficient (K_t) and loss coefficient (K_l) are reported as a function of B/L where L is the wavelength and B is the chamber width. The effect of screen porosity and the spacing between screens on the wave reflection and transmission are also discussed in this paper. In addition, the water level oscillation inside the chamber is normalised with the incident wave height and presented in this paper. The details of the model, hydrodynamic testing facility, experimental setup, procedure and analysis along with the interpretation of results are presented and discussed in this paper.

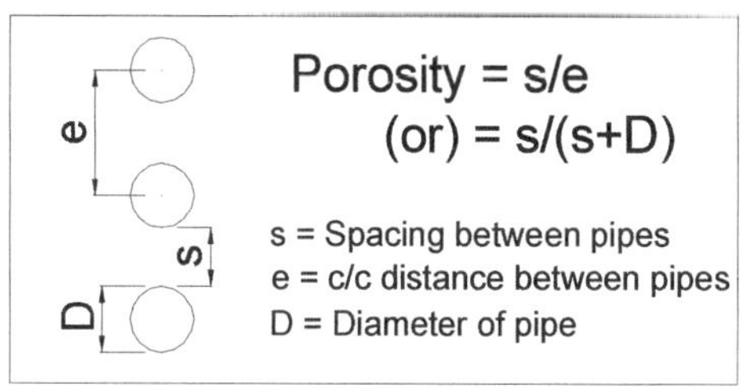

Fig. 1 Definition of porosity

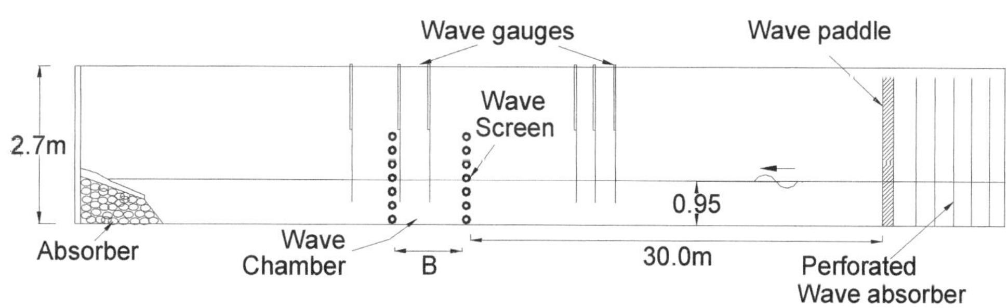

Fig. 2 Cross sectional view of wave flume with wave screen model

WAVE FIELD ANALYSIS IN A HARBOR WITH PERMEABLE BREAKWATERS

Moonsu Kwak[1], Kisang Lee[2] and Chongkun Pyun[3]

1. Civil Engineering, Myongji College, 356 Hongeun, Seoul 120-776, KOREA
2. Harbor & Coastal Dept., Saman Engineering Consultants, Union Building, Kwachun 427-800, KOREA
3. Civil & Environmental Engineering, Myongji Univ., Yongin 449-728, KOREA

1. Introduction

The main purpose of the breakwater, which is the most typical element in harbor plan, is to maintain the calm area in a harbor by preventing waves propagated from offshore. But, because of blocking nature of the impermeable breakwater, which causes fundamental environmental changes in coastal water circulation system, has brought about serious environmental problems. In recent years, the construction of the permeable breakwater, which allows the seawater exchange to improve the water quality in a harbor, is being carried out. Although the permeable breakwater is able to improve the water quality by expediting the seawater exchange, it can create transmitted waves, which could disturb the calmness in a harbor. In this study, the numerical models was established to analyze the wave field in a harbor which is constructed in the permeable breakwater. The time-dependent mild-slope equation was used as the governing equation for the model, and the arbitrary transmitted coefficient condition was used as the boundary condition on the permeable breakwater. For the verification of the model, the two dimensional hydraulic model test was used to analyze the transmitted wave of a permeable breakwater. Also, the applicability of the model was examined for the real harbor with permeable breakwater by comparing the analysis of the wave field, which was obtained by applying the model to the Jeju New Harbor, with the results of the three dimensional hydraulic model test.

2. Governing Equation and Transmission Boundary Condition

The governing equation for the model was the time-dependent mild-slope equation, which integrates the velocity from the bottom of the sea to the water surface and shows it in the form of the line discharge. For the boundary condition of the arbitrary transmitted coefficient, the phase exchange after passing the breakwater was calculated from the line discharge obtained from the coexistence area, the incident and reflected waves of the breakwater. The reflectivity and transmitted coefficient was given to calculate the line discharge for the transmitted waves. The resistance coefficient, which reflects the energy loss of the incident waves by permeable breakwater, was calculated by the equation given by Sollitt(1972), and was used to calculate the transmitted wave velocity in permeable breakwater.

---

1. Assistant professor (T.82-2-300-1290, F.82-2-300-1138, Email : moonsu@mail.mjc.ac.kr)

2. Engineer (T.82-2-509-4848, F.82-2-503-5277, Email : kslee@samaneng.com)

3. Professor (T.82-31-330-6409, F.82-31-336-9705, Email : ckpyun@mju.ac.kr)

3. Verification of Numerical Model

For the verification of the model, a wall type permeable breakwater with length 0.7 m, width 0.4 m, height 0.8 m in the two dimensional wave flume was installed, and a hydraulic model test to the regular wave, which meets non overtopping, a non breaking condition was done and the results were compared. One wave gage was installed on the front of the wave paddle and three on the rear of the permeable breakwater with 0.7 m interval to one another(Fig. 1). Fig. 2 shows the comparison of the water surface elevation of the hydraulic model test and numerical model at each measurement point. The computation result of this model shows the decreasing phenomenon of a water surface elevation by the energy loss caused by the wall, and it good matches with the result of hydraulic model test at all measurement points.

4. Wave Field Analysis in a Harbor

In this study, the model was applied to Jeju New Harbor, where the construction of the permeable breakwater is being considered, and the applicability of the model at the harbor was examined by comparing the computation result with the hydraulic model test. Fig. 3 shows the distribution of the wave height of the computation result when the impermeable breakwater was constructed, and Fig. 4 shows when the permeable breakwater was constructed for the 100 m of the head of breakwater. The result of the permeable breakwater gave an increased wave height at behind of the breakwater when it was compared with that of the impermeable breakwater. Fig. 5 is the bar chart, which compares the result of the wave height ratio in the harbor region to the result of the hydraulic model test and the numerical computation. When the permeable breakwater was constructed, the wave height ratio at behind of the breakwater increased by approximately 0.1 compared to the impermeable breakwater, and the difference between the two decreased as the measurement was taken nearer to the inner harbor. Therefore, the Jeju New Harbor needs to have the quay at behind of the permeable breakwater blanketed with the wave dissipating block so that the transmitted wave could be effectively broken.

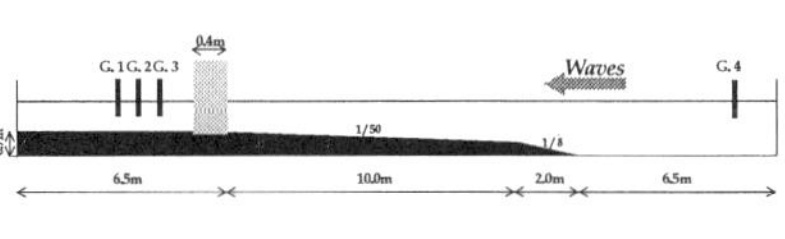

Fig. 1 Sketch of hydraulic model test.

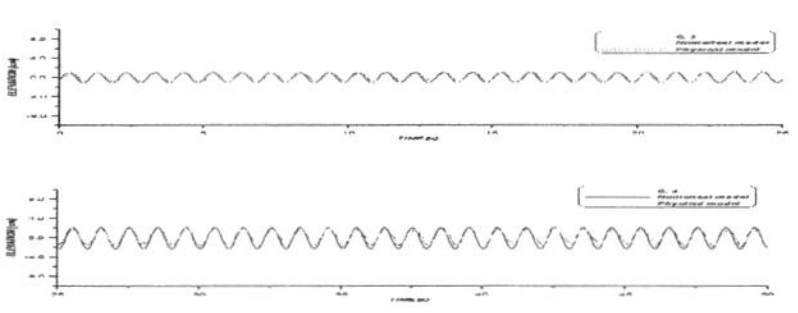

Fig. 2 Comparison of experiment and computation.

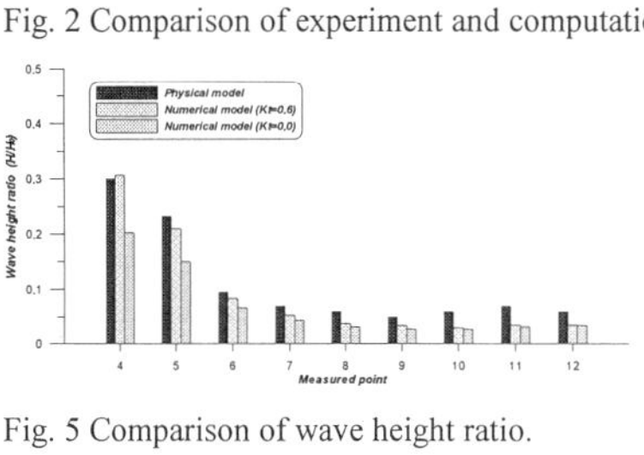

Fig. 5 Comparison of wave height ratio.

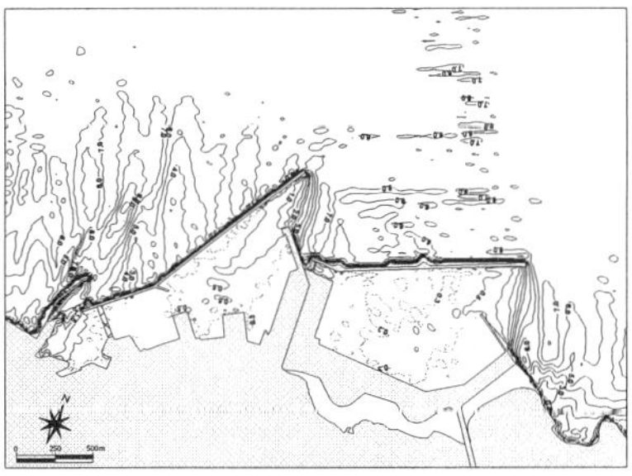

Fig. 3 Distribution of wave height $(K_T = 0.0)$.

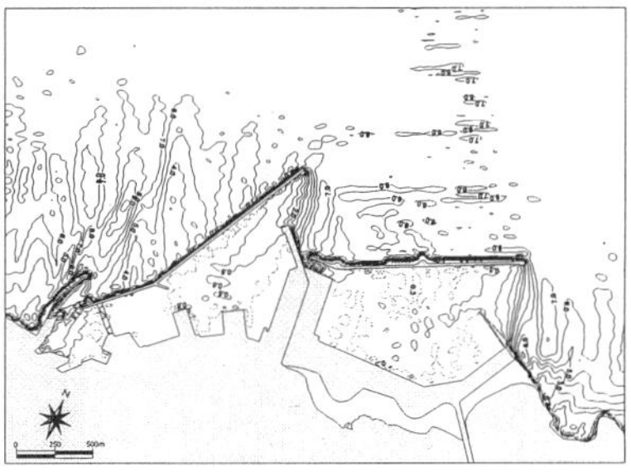

Fig. 4 Distribution of wave height $(K_T = 0.6)$.

WAVE DAMPINGS BY AN ARRAY OF CIRCULAR CYLINDERS AND A GROUP OF MODEL PLANTS

Kenjirou Hayashi[1], Kubota Yuusuke[2] and Tosiyuki Shigemura[3]

[1,2,3] Dept of Civil and Environmental Engineering, The National Defense Academy
1-10-20 Hashirimizu, Yokosuka, 239-8686 Japan
Fax : 0468-44-5913, Email : hayashik@cc.nda.ac.jp

It is important to understand the hydrodynamics of a group of plants for the application of it as a good alternative material for protecting the banks of inland waterways against waves and current compare to the classical materials as concrete block or stones. This paper describes an investigation into the wave damping capacity of plants. The damping of waves transmitting a group of model plants installed in the wave flume are measured. The wave forces acting on an array of circular cylinders, which is a simple model of a group of reeds, and wave damping by it are also measured and evaluated theoretically.

Laboratory experiments were carried out in a wave flume 40m length, and 0.8m width for the study of wave forces and wave damping of a array of circular cylinders. The array of circular cylinders of diameter D=1cm and 1m length with 1.8m length and 0.8 width was installed in the middle of wave flume length. The ratio of circular cylinders in unit bed area in the cylinder array was λ =0.0091. The incident wave height H_i, reflected wave height H_r from the array of cylinders and transmitting wave height H_t are measured by using four wave gages setting just in front of and just behind of the array of cylinders. The wave forces acting on a rigid circular cylinder in the array of cylinders are also measured by using a three components strain gauge type load cell. The experiments were made in regular waves with still water depth d=60cm. The ratios of water dept to wave length L were d/L=0.42, 0.18, 0.14, and 0.10.

The other laboratory experiments were carried out in the same wave flume for the study of wave damping of a group of plants. The group of model plants with 1.1m length and 0.8 width was installed in the middle of wave flume length. Model plant is a bundle of 48 vinyl tubes. The diameter D_p and height S_p of vinyl tubes were 0.1～0.2cm and about 22cm. The ratio of plants in unit bed area was λ_r=0.016. The incident wave height H_i, reflected wave height H_r from the group of model plants and transmitting wave height H_t are measured by using four wave gages. The experiments were made in regular waves with still water depth d=15, 22 and 60cm. The ratios of water depth d to wavelength L were d/L=0.1, 0.2 and 0.3.

Waves transmitting through an array of circular cylinders lose energy due to the works they do on the cylinders [1)]. The following equation is derived by considering the conservation of energy equation and assuming that linear water theory is valid, the drag coefficient C_D is constant over the depth, and the bed is locally flat.

$$K_t=H_t/H_i = [1-(32/3)\lambda(H_i/D)\{dx/(g.T^2)\}(CD/n)\{1/3+1/(\sinh(2\pi d/L))^2\}]^{1/2} \quad \text{---(1)}$$

In which K_t = transmitting coefficient ; Hi = incident wave height for the small partial distance dx ; H_t = transmitting wave height for dx ; g = gravity ; T= wave period (= $1/f_w$) ; n = $1/2+(2\pi d/L)/\sinh(2\pi d/L)$.

The main result obtained in the present works is as follows.

1)The variations of K_t to H_i/L for the case of array of circular cylinders are shown in figure 1 in the case of d/L=0.18. The theoretical values obtained by using equation 1 are also plotted in this figure. The values of experiments are estimated quite well by the theoretical value for the case of C_D=1.5. The value of C_D=1.5 is obtained by using Morison`s equation to the measurement value of in-line forces acting on a rigid circular cylinder in the array of cylinders.

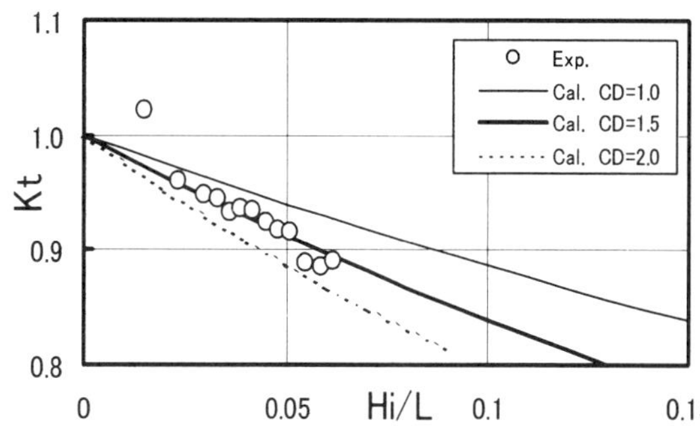

Fig. 1 Kt and Hi/L (d/L=0.18 B/L=0.54)

2)The variations of K_t to H_i/L for the case of a group of model plants are shown in figure 2 in the case of water depth d=22cm, d/L=0.2 and B/L=1.0. The theoretical values obtained by using equation 1 are also plotted in this figure. The values of experiments are estimated well by theoretical values of C_D=1~2. These values of C_D are also obtained by using Morison`s equation to the measurement value of in-line forces acting on a vinyl tube of model plant.

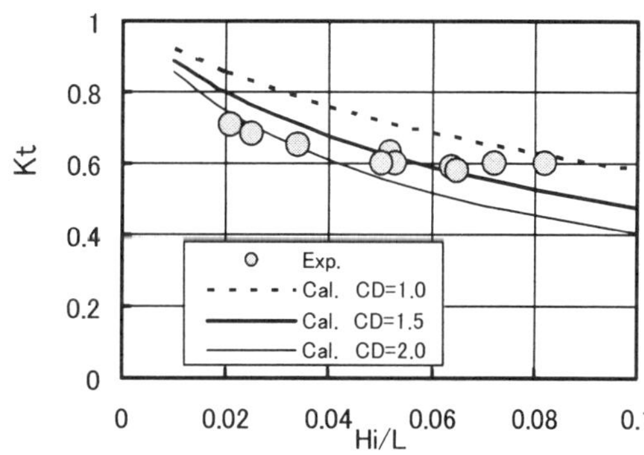

Fig.2 Kt and Hi/L for Model Plants
d=22cm Sp/d=1.0 d/L=0.2 B/L=1.0

References

(1) Dalrymple, R. A , Kirby, J. T. and Hwang, P. A.: Wave diffraction due to area of energy Dissipation, Jour. Waterway, Port, Coastal and Ocean Eng. Vol. 110, No.1, pp.67-79, 1984.

SOLUTIONS TO WAVE TRANSMISSION AND REFLECTION BY BOTTOM MOUNTED WAVE-PERMEABLE STRUCTURE IN SHALLOW WATER

Li Xi[1] Yan Yixin[2]

ABSTRACT

For determination of absorbing coefficient, one-dimensional wave is studied to give the analytical expression of absorbing term, and the analytical predictions are compared to available physical laboratory data in flume for wave propagation through bottom-mounted breakwaters. We consider the problem of wave partial/full reflection and transmission by a wave-absorbing layer with the focus on the understanding of wave reduction by wave-permeable structures. We obtain accurate predictions of reflected and transmitted results around the wave-permeable breakwater for various incident-wave conditions. The results show that wave permeable breakwaters with proper absorbing coefficient can be used as an effective alternative to massive gravity breakwaters in reduction of wave transmission in shallow water.

Key words: wave equations, wave-absorbing coefficient, and wave permeable breakwaters

1. Introductions

For permeable structures such as bottom-mounted breakwaters, the incident waves interact with the porous breakwater causing transmitted and reflected waves to propagate towards the boundaries. The solution of wave permeable breakwaters is expressed by introducing the damping item to the governing equations in accordance with energy dissipation. In numerical study the 'open boundary' or 'fully absorbing boundary' condition is required, which requires the transmitted and reflected waves small. The numerical problem can be approximated by the absorption solutions to weak reflected and fully absorbing of real wave permeable breakwater. Therefore in principle, the same theoretical solutions could be applied in numerical simulation of the wave permeable structures.

2. Mathematical formulation of absorbing coefficient

2.1 Governing equations

The one-dimensional wave equation with an absorbing term read:

Continuity equation

$$n\zeta_t + hU_x = 0 \tag{1}$$

Momentum equation

$$U_t + gn\zeta_x + SU = 0 \tag{2}$$

where h is the constant water depth, η is the free surface elevation, U is the average velocity, n is the porosity coefficient for wave permeable structures, S is the non-linear absorbing term.

We assume time-harmonic motion of ω, introducing $U = n\upsilon(x)e^{i\omega t}, \zeta = \eta(x)e^{i\omega t}$, from (1) and (2) get the equations

$$\eta_{xx} = \frac{\omega i}{h}\frac{S+\omega i}{g}\eta = r^2\eta \tag{3}$$

Introduce phasor constant $r=\alpha+i\beta$, the α is called as the damping coefficient, the β is in accordance to the wave number k.

1 Li Xi, Lecturer, Department of Ocean Engineering, Hohai University, Nanjing 210098, P.R.China., Phone: (25)3787339
2 Yan Yi Xin, Professor, vice president, Hohai University, Nanjing 210098, P.R.China., Phone: (25)3786611

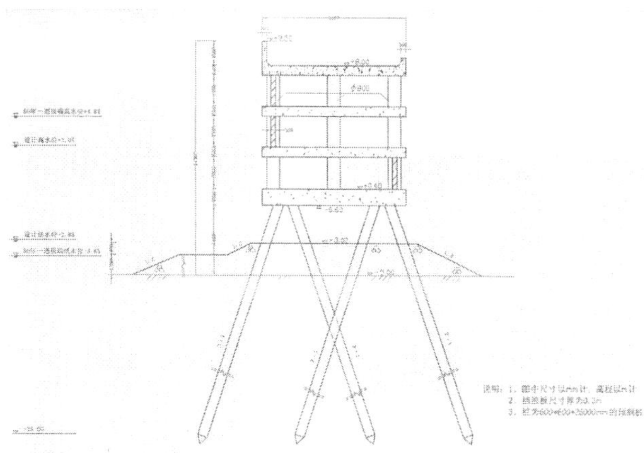

Figure 2 Cross-section configuration of wave-permeable breakwaters as the link bridge (Yan Y., Li X., etc. 2002)

In order to show the usefulness of the model, we applied the model in the feasibility study of the Port of Rizhao, the east coast of China. The port consists of two series of piers with different spans providing berthing and operating facilities, referred to Figure 3, and linked by the wave permeable breakwaters. The input significant wave height and peak spectral period near the site has been extrapolated from the field data collections. The application result of wave climate is the following (Figure 3) with a partial reflection of 25%.

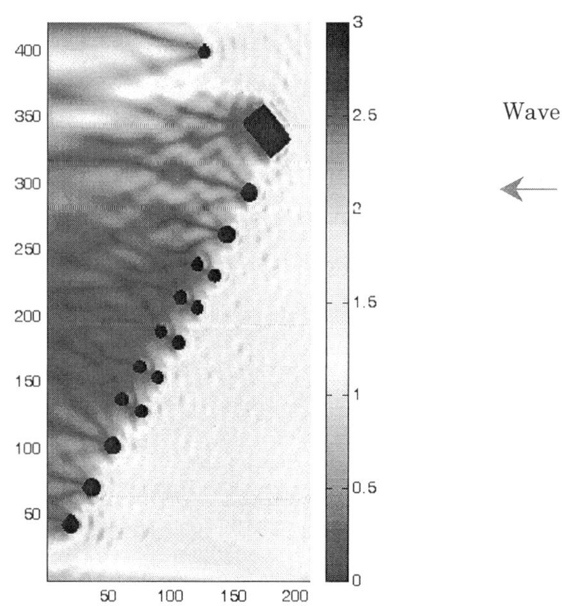

Figure 3 Numerical simulation of wave-permeable breakwaters (Li X. 2002)

In the two-dimensional wave, we directly use the absorbing coefficients from the one-dimensional wave analysis. Combined diffraction-refraction and reflection-transmission effects would be considered.

Numerical Computations for a Sloshing Problem

J. H. Kyoung[1], J. W. Kim[2], Sung P. Cho[1] and K. J. Bai[1]

[1]*Seoul National University*
[2]*American Bureau of Shipping*

ABSTRACT

A sloshing phenomenon in a liquid storage tank is numerically simulated. In nature a sloshing problem in a tank, i.e., a liquid storage tank or a cooling water tank in a nuclear power plant can cause a severe disaster during an earthquake, or an excessive motion of a oil tanker at sea. During excessive sloshing, the sloshing-induced impact load can cause a critical damage on the tank structure. Recently, this problem becomes an important issue in FPSO design. A three-dimensional free surface flow in a tank is formulated in the scope of potential flow formulation with the nonlinear free surface conditions. A finite-element method based on the Hamilton's principle is employed as a numerical scheme. The problem is treated as an initial value problem. The nonlinear computations are made through an iterative method at each time step. In the computations the hydrodynamic loading on the pillar in the tank is obtained. The computed results are compared with previous results obtained by others.

SOUTHWEST MONSOON EFFECT ON PLANKTON OCCURRENCE AND DISTRIBUTION IN PARTS OF BAY OF BENGAL

Mohammad Zafar
Institute of Marine Sciences, University of Chittagong, Bangladesh.
E-mail: zafarimscu@ctgu.edu / zafarims@yahoo.com ,
Fax 880-31-726310 /710347, Tel. 880-31-682654

Abstract

The present investigation was carried out for one year from March 1988 to February 1989 and focuses on hydro-plankton aspects in parts of Bay of Bengal. The hydrological factors were mainly governed by the monsoon and river systems flowing into the Bay. A marked seasonal fluctuation in salinity, dissolved oxygen, water temperature and suspended particles were observed, generally which is a unique feature of northeastern coastal part of the Bay of Bengal.

The higher abundance of phytoplankton was recorded in January (3332 cells/l) during the low turbidity and bright sunshine period. The minimum was found in August (1295 cells/l) when maximum rainfall and fresh water discharge were recorded. Phytoplankton population was shown to have a significant negative correlation with suspended particles in the four stations in the studied area (S_1, $r = -0.887$, $P < 0.1$; S_2, $r = -0.920$, $P < 0.001$; S_3, $P < 0.01$; S_4, $r = -0.920$, $P < 0.001$). Maximum density of zooplankton was recorded in September (326146.36/100 m^3), which indicates a seasonal influence on the distribution pattern of the different group of zooplankton. Hydromedusae, Mysids and *Squilla* are shown as major indicator groups and they represent only <1% of the zooplankton population. The zooplankton abundance could not be significantly correlated with phytoplankton ($r = 0.09$, $P < 0.05$).

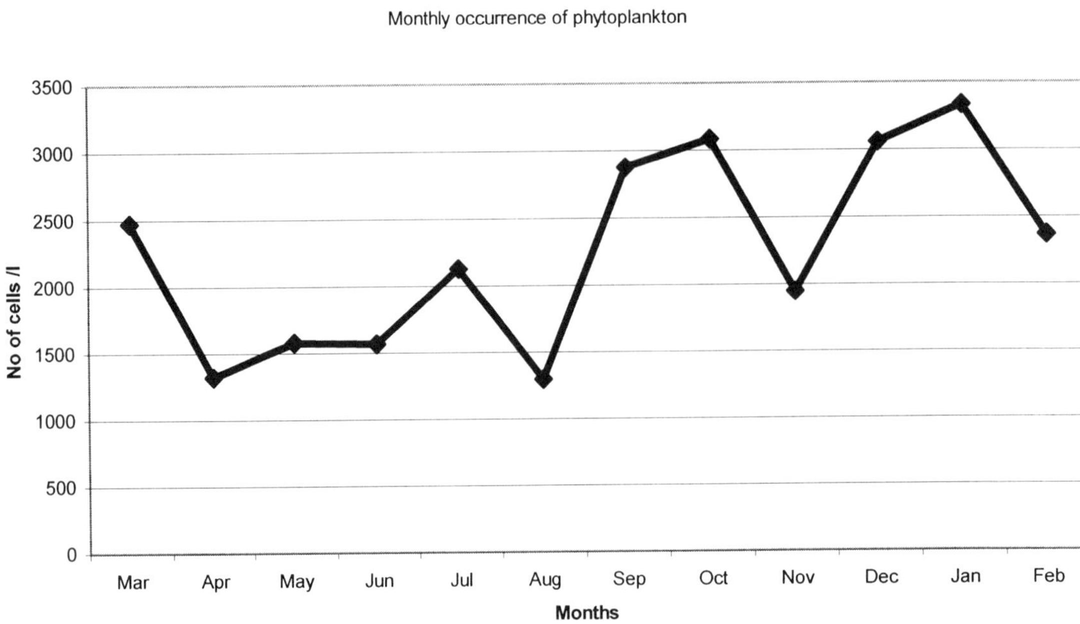

Fig 1. Monthly occurrence of Phytoplankton in the Kutubdia Channel.

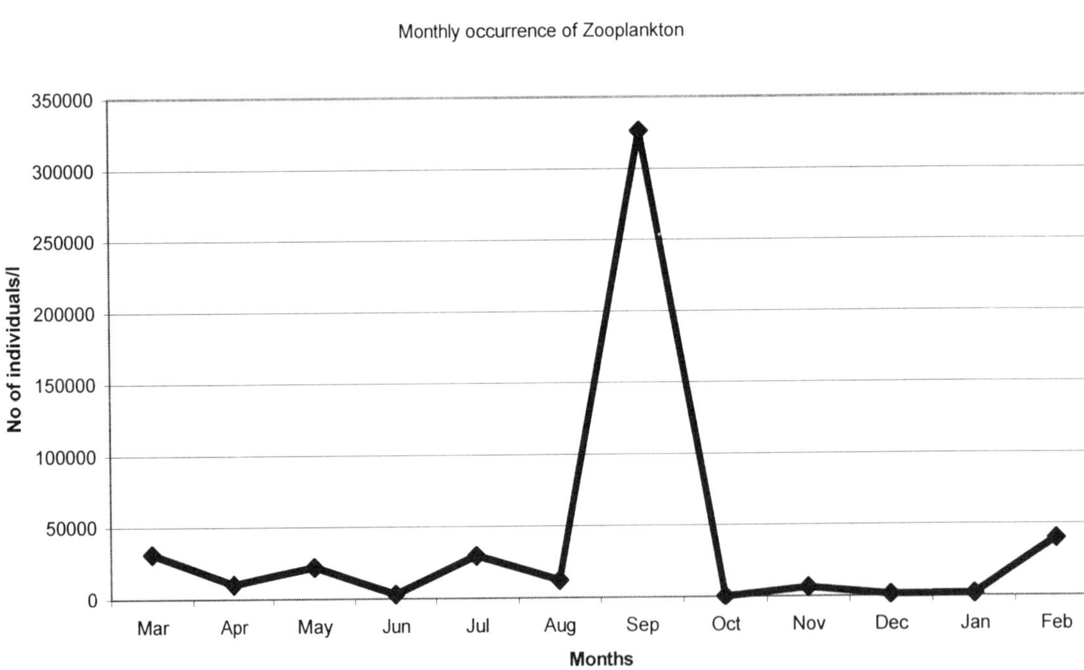

Fig 2. Monthly occurrence of Zooplankton in the Kutubdia Channel.

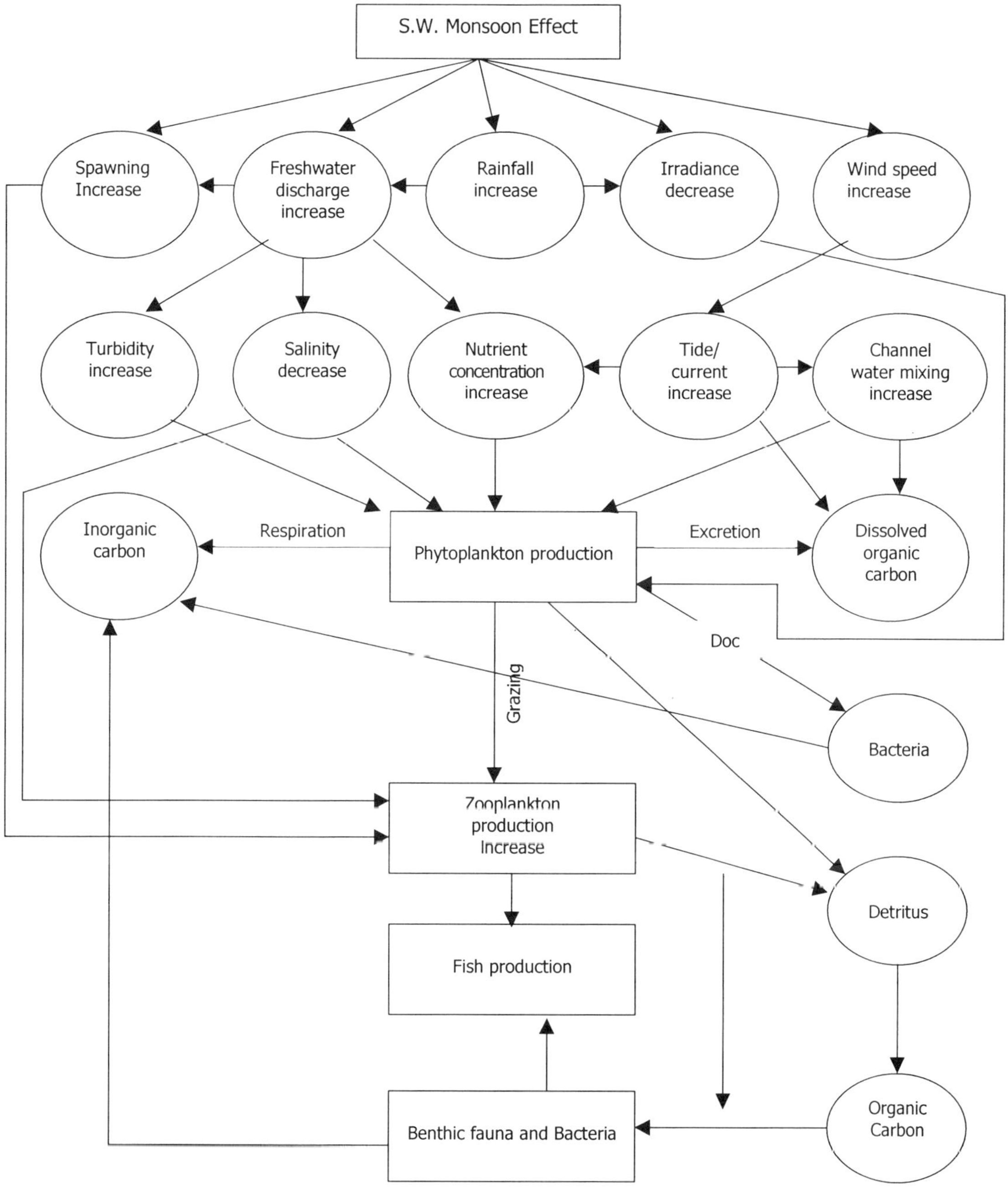

Fig 3: Southwest Monsoon Effect on Plankton

CONSEQUENCES OF NORTH PACIFIC SEA-LEVEL OSCILLATION FOR ESTUARINE WATER QUALITY AROUND THE JAPANESE COAST

Tadashi HIBINO[1)] **and Naoki ITABASHI**[2)]

1) Associate Professor, Department of Social and Environmental Engineering,
Graduate School of Engineering, Hiroshima University
4-1, Kagamiyama 1 chome, Higashi-Hiroshima 739-8527, Japan
TEL/FAX +81 824 24 7816, e-mail hibinot@hiroshima-u.ac.jp

2) Doctoral student, Department of Social and Environmental Engineering,
Graduate School of Engineering, Hiroshima University
4-1, Kagamiyama 1 chome, Higashi-Hiroshima 739-8527, Japan
TEL +81 42 372 7857, FAX +81 42 372 6398, e-mail Naoki.Itabashi@ss.pacific.co.jp

Introduction Many short-period climatic variations over the circum-pacific area can be associated with the abnormal behavior of the Aleutian Low. The Aleutian Low plays an important role in generating weather patterns over the contiguous United States (Garriot, 1904; Kelein, 1965). The synoptic aspects of the teleconnection involve the position and strength of the westerlies as well as the longitude and amplitude of the quasi-stationary, a long-wave train generated downstream from the variable Aleutian Low. The dependence of the Aleutian Low in fall on the sea-water temperatures in the Northern Pacific developed during the antecedent summer is considered (Namias, 1976). Namias suggested that the few cases where the indicated relationship apparently failed could be explained in terms of external influences like the blocking of the Bering Sea, or the occasional existence of unusually strong SST gradients which change atmospheric baroclinicity and sometimes influence the Aleutian Low. It transpires that a 500hPa height fluctuation corresponding to the PNA pattern presented by Wallace and Gutzler (1981) causes the SST anomalies represented by the first EOF. Yamada and Sekine (1997) indicated that a southward shift along the Japanese coast of the subarctic gyre in response to the southward shift of the Aleutian Low results in the negative SST anomalies, which can not be attributed to either the sensible / latent heat fluxes or Ekman transport.

The water quality and physical structures in Japanese estuaries may be strongly affected by the ocean currents around Japan. Variations in salinity inside Ohfunato Bay (39 ° 02'N, 141 ° 44'E, Pacific ocean side, northeastern part of Japan) may not only be influenced by the inflow of river water but also by intrusions of ocean water. In fact, at a 30m depth, river inflows have been found to have only a minor effect on salinity compared with variations in salinity that are correlated with sea level and atmospheric pressure variations (Tsuruya & Hibino, 1999). Variations in sea level distribution around Japan are considered to correlate well with the southward intrusion of the Oyashio due to the eastward shift of the Aleutian Low as indicated by the relationship between sea level variation and bottom salinity in Ohfunato Bay (Hibino, 2002). These facts suggest that outer ocean currents influenced by pressure patterns may play an important role in determining the water quality inside a semi-enclosed bay.

In order to understand the influence of circum-pacific pressure systems on water quality in Japanese estuaries, the relationships between development of the Aleutian Low and patterns of sea level and sea surface temperature distributions from the north to western part of the North Pacific were analyzed. The present study introduces useful procedures for investigating the outer sea conditions to analyze anomalous changes in water temperature and sea level at various ports in the North Pacific.

Conclusions Main conclusions of the present study are summarized as follows:

(1) The annual change of SLH is considered to be mainly controlled by high and low pressure systems. SST around the Japanese coast is well controlled with the North Pacific Sea-level Oscillation (NPSO) influenced by the developed patterns of the Aleutian Low (Fig.1, Fig.3). This fact indicates that the ocean currents around the Japanese coast are strongly influenced by NPSO.

(2) During the last ten years, SLH has been clearly rising around Japan. In comparison, SLH appears to have been falling from Adak eastward (Yaktat in Fig.2). The increasing of equatorward transport of the eastern limb of the subtropical gyre made SLH increase around western Japan from 1988-97. The spread of the Aleutian Low has caused SLH to increase in the west and decrease in the eastern parts of the North Pacific from 1998 (Fig.3).

(3) It may be possible to use meteorological and water level data as historical long-term indicators of SST at the entrance of estuaries. Because of the close relationship between SLP, SLH, and SST, it is hoped that such information can be used as a tool for assessing historical water quality in Japanese estuaries, or even for predicting water quality in these systems in the future.

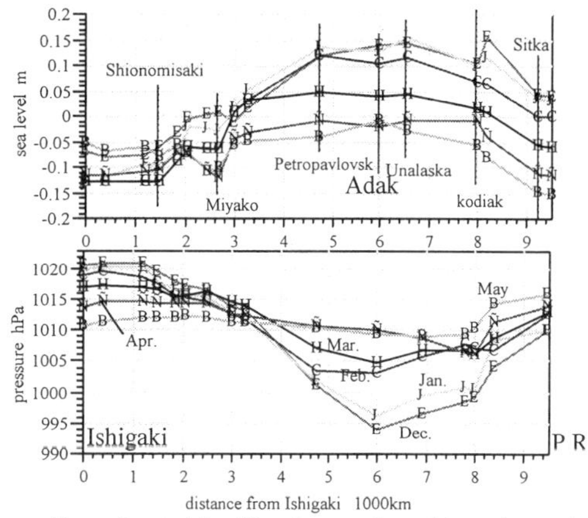

December to May (Oyashio southward intrusion period)

Fig.1 SLH and SLP gradients along West to North Pacific coast (from Ishigaki to Prince Ropert)

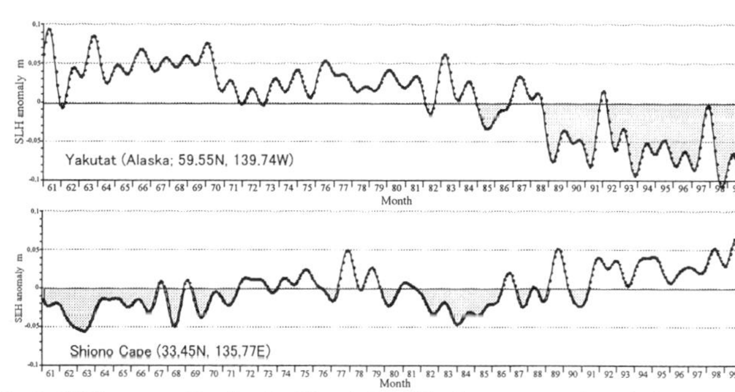

Fig.2 SLH anomalies in the North Pacific (over annual variation)

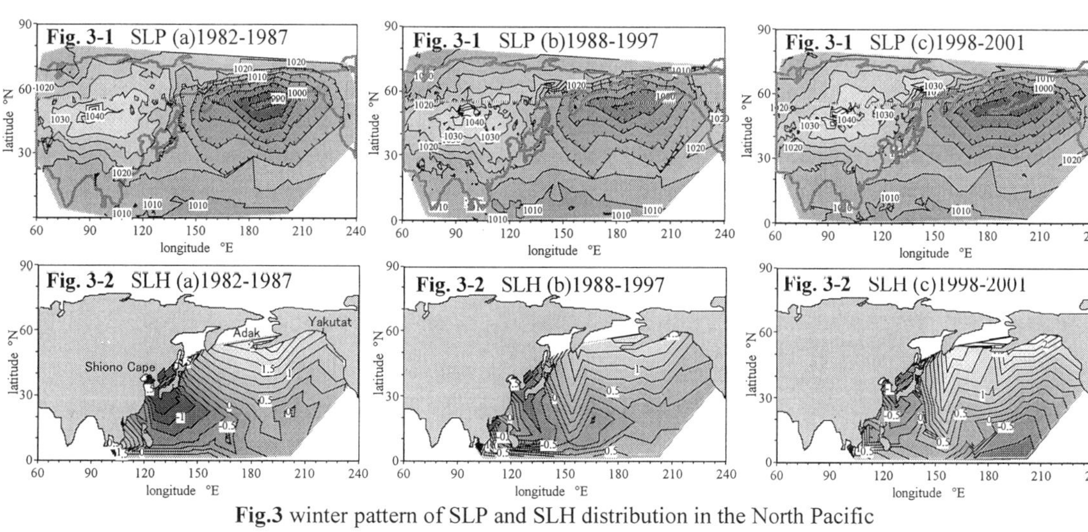

Fig.3 winter pattern of SLP and SLH distribution in the North Pacific

High Nutrient Water Intrusion to the Kashima Coast Induced by Kuroshio Axis Fluctuations

Hiroshi YAGI*, Kumiko ADACHI** and Akira NIHIRA***

*Department of Civil Engineering, Tokyo Institute of Technology, 2-12-1 O-okayama, Meguro-ku, Tokyo ,152-8552, Japan. (Phone 81-35734-2591, E-mail:hyagi@fluid.cv.titech.ac.jp)

**National Research Institute of Fisheries Engineering,

***Ibaragi Prefectural Fisheries Experimental Station

1. INTRODUCTION

In order to understand an ecosystem of nearshore region facing to an open coast, it is important to know a primary productivity and a nutrient budget of it. However, a nutrient cycle in an open coast has not been understood enough yet, because it might be strongly affected from an offshore region but the mechanism and quantity of nutrient supplies from offshore region has not been revealed yet. The objective of the present study is to show the basic seasonal variation of nutrients in an open coast and to reveal the influence from offshore region to near regions, especially an effect of oceanic currents fluctuations through a long term field measurements.

2. FIELD MEASUREMENTS AND RESULTS

Filed study was curried out at the Kashima coast which is a typical open coast in Japan (Fig.3). Observational stations were aligned in the on-offshore direction to measure the cross-shelf spatial distribution of temperature, salinity, dissolved inorganic nutrients(NO_3, NO_2, NH_4, PO_4, SiO_2) and chlorophyll.a. Field survey was performed by 'TOKIWA' (research vessel) once a month from February to November in 2001. Measurement results showed a typical seasonal variation of water properties; high nutrients and low stratification of temperatures and salinities in winter changed to the low nutrient and high stratification in summer. The interesting point is that the nutrients concentration rapidly increased in May and decreased in Jun as shown in Fig.1. This phenomenon was considered to be related to the offshore oceanographic condition because there was no significant river discharge during this period and this high nutrient water corresponded to high salinity. Figure 2 shows the cross-shelf spatial distribution of temperature from April to Jun. Cold water less than 10 degrees(hatched area) moved toward shallow water region in May and returned to the offshore deepre area in Jun. This fact suggests that the upwelling of offshore deeper water occurred during this period and low temperature and high nutrient water intruded to the coastal region. Satellite image data(Fig.3) and the other oceanographic data(sea surface level etc.) indicated that the Kuroshio flowing off Kashima coast approached to the coastal region during this period. These facts suggest that Kuroshio approaching to the Kashima coast induced the upwelling of offshore deeper water and intrusion of high nutrient water to the nearshore region. In the present study, the response of Kashima coastal water in May 2001 was compared with the other Kuroshio approaching event in August 1999(Fig.3). The results of comparison have shown that there are two kinds of response of Kashima coastal water to the Kuroshio approaching event. The response of Kashima coastal water and the upwelling mechanism by the Kuroshio approaching event have been investigated by the CFD model and will be shown in the final paper.

3. CONCLUSIONS

The field measurement results indicated that the Kuroshio approaching to the Kashima coast induced the upwelling of offshore deeper water and low temperature and high nutrients water intruded to the nearshore region. From the comparison with the other Kuroshio approaching period, it was revealed that there are two kinds of response of Kashima coastal water to the Kuroshio approaching event.

REFERENCE

Lee,T.N. Atkinson, L.P. and Legeckis,R.(1981): Observation on a Gulf Stream frontal eddy on the Georgia continental shelf, April 1977, Deep-Sea Res, 28, 347-378.

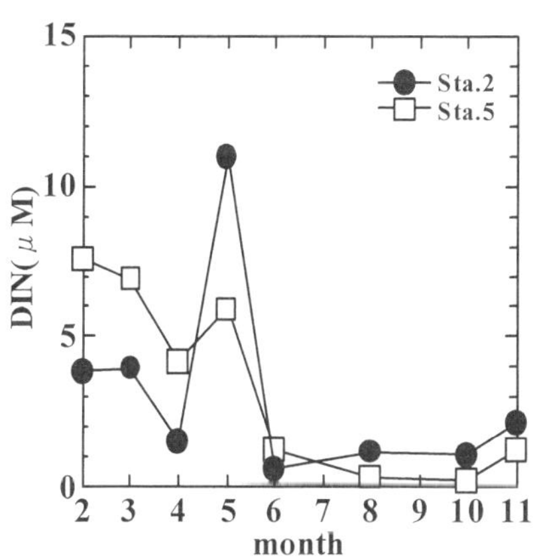

Fig.1 Temporal variation of DIN (20m below from sea surface) at nearshore(Sta.2) and offshore region(Sta.5).

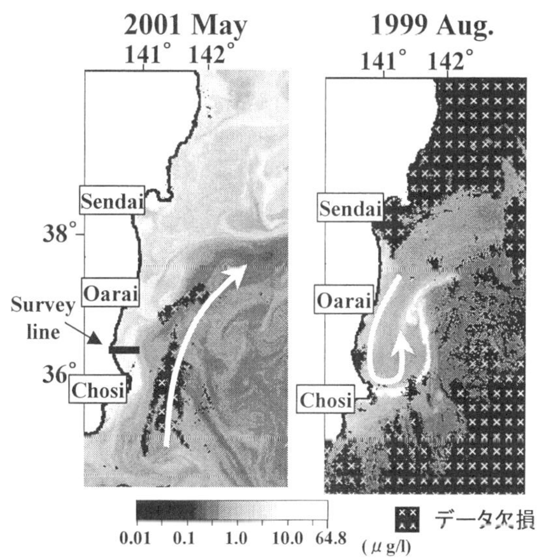

Fig.3 Sea surface Chlorophyll.a concentration from satellite image data (Sea WiFS) during the period of Kuroshio axis approaching to the Kashima coast.

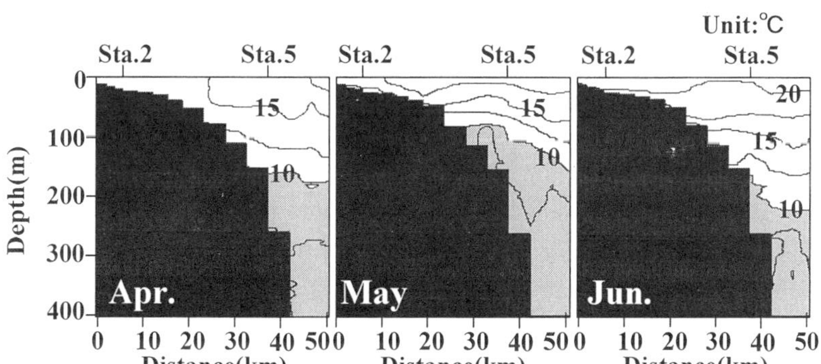

Fig.2 Spatial distribution of temperature along the survey line

CHARACTERISTICS OF ENVIRONMENT AND TIDAL CURRENT IN THE ARIAKE SEA

Kiyoshi Takikawa[1], Chiharu Aoyama[2], Kenji Tanaka[3] and Takaomi Hokamura[4]

[1]Member of JSCE, Dr. Eng., Professor, Dept. of Civil Eng., Kumamoto University
(39-1 Kurokami 2-chome Kumamoto-shi, Kumamoto 8608555, Japan)
Phone and Fax.: ++8196-342-3548
E-mail: takikawa@gpo.kumamoto-u.ac.jp

[2]Member of JSCE, Dr. Fish., Director of Japan's Independent Institute for Comprehensive Research, Inc.
(6-1-1205 Kaigan 1-chome Minato-ku, Tokyo 1050022, Japan)
Phone: ++813-5777-0208
Fax.: ++813-3438-3077
E-mail: aoyamac@hotmail.com

[3]Member of JSCE, Mr. Sci., Research Associate, Dept. of Civil Eng., Kumamoto University
Phone and Fax.: ++8196-342-3601
E-mail:ktanaka@gpo.kumamoto-u.ac.jp

[4]Member of JSCE, Engineering Official, Dept. of Civil Eng., Kumamoto University
Email: hokamura@gpo.kumamoto-u.ac.jp

ABSTRACT

In recent years, Ariake and Yatsushiro sea has encountered serious environmental problems, such as seaweed damage by anomalously multiplied planktons. A lot of aggravating factors can be included on this issue: for example, the decrease of tideland area, large amounts of polluted water into sea by the development in the coastal and land area surrounding these seas, amounts of other polluted solid materials and into sea caused by severe flood or variation of the river shape, variation of tidal stream or influx of warm seawater by typhoon or long term variation of ocean current. However, it is still not clear in detail because of its complexity. In order to make causal analysis and construct guideline for recovery, it is indispensable to make comprehensive attempt including chemical-physical environment and biochemical processes.

In this paper, at first, characteristics of environmental parameters over Ariake sea, one of the most important factors, was investigated by analysis past data, for example, comparison of red tide and inorganic nitrogen in sea water. As a result, it was found that there was high correlation among red tide, inorganic nitrogen and a quantity of sediment of plankton. A correlation between a quantity of seaweed crop and precipitation or atmospheric temperature could also be seen in this analysis.

Secondly, characteristics of tidal current in Ariake sea, one of the most important factors, was investigated by numerical experiment (using a two-dimensional flow simulation). Influence of shutoff bank in Isahaya Bay, control of water gate in the bank, Kumamoto port, and drag by seaweed crop nets were included. In this numerical experiment, following results were obtained: 1) a vibration cycle became shorter by about 20 minutes, after the shutoff in Isahaya Bay, 2) the local circulation inside the shutoff bank evolved due to control of water gate, 3) after Kumamoto Port built, the drift of a current around the port changed by 20-30cm s^{-1}, and 4) the drag induced by seaweed crop nets made fairly decrease of drift of a current in center and inner part of Ariake sea.

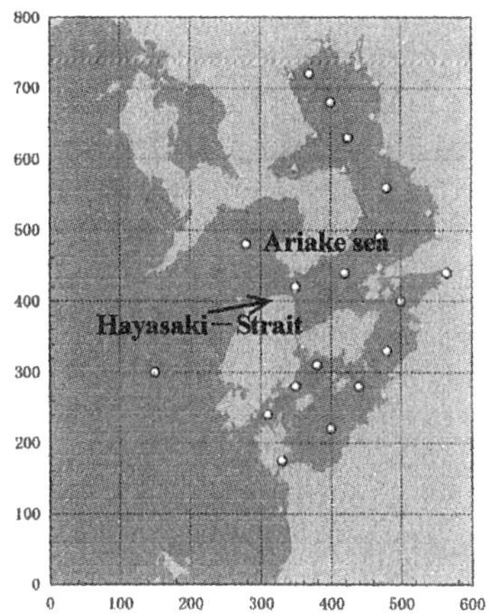

Figure 1. Computational area in this study

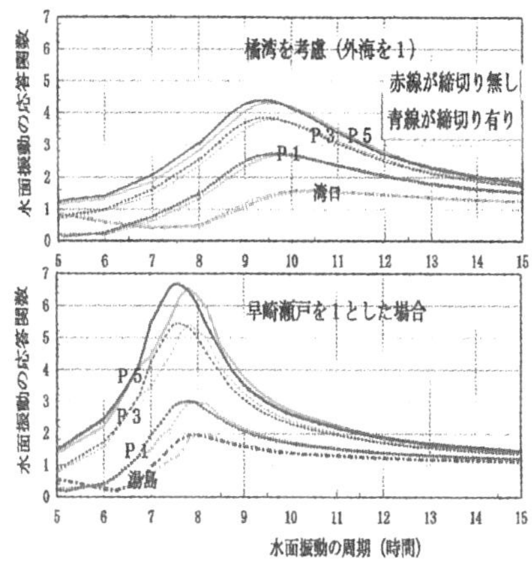

Figure 2. Influence of shutoff bank in Isahaya bay to gulf oscillation in Ariake sea.

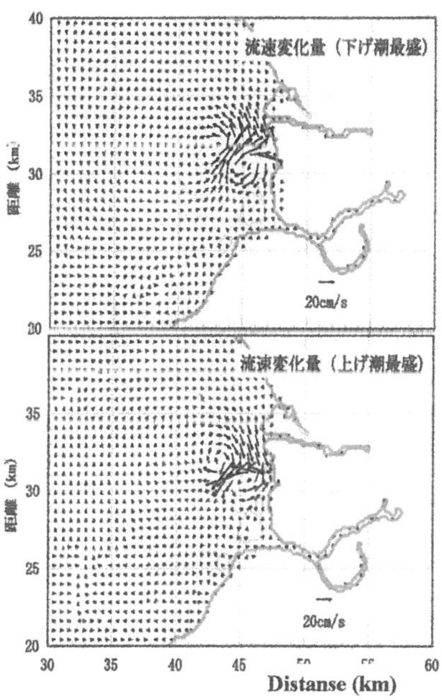

Figere 3. Comparison of the drift of a current before and after Kumamoto port built.

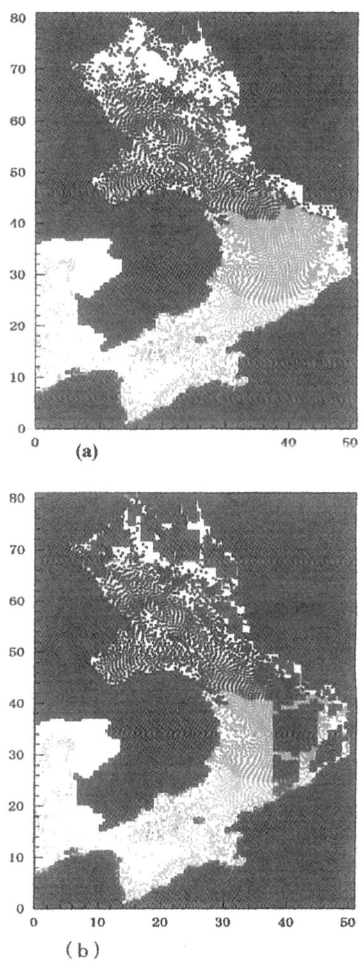

Figure 4. Comparison of current field whether seaweed net existed (b) or not (a), after computation 3 periods of tide by particle tracing method.

Short-period fluctuations of surface circulations in Sagami Bay induced by the Kuroshio warm water intrusion through Oshima West Channel

HIROFUMI HINATA *(National Institute for Land and Infrastructure Management, 3-1-1 Nagase, Yokosuka 2390826,Japan, TEL.: +81-46-844-5023, FAX: +81-46-844-1145,* hinata-h92y2@ysk.nilim.go.jp, *Corresponding author)*

TETSUO YANAGI *(Research Institute of Applied Mechanics, Kyusyu University, 6-1 Kasuga-Koen, Kasuga 8168580, Japan, TEL.: +81-92-583-7932, FAX: +81-92-583-7492,* tyanagi@riam.kyushu-u.ac.jp*)*

Abstract

Near-surface current measurements by two High-Frequency oceanic surface radars (HF radars) of the Frequency Modulated Interrupted Continuous Wave (FMICW) type were carried out during the period from December 7, 2000 to March 6, 2001 in order to investigate the characteristics of short-period fluctuations of surface circulations in Sagami Bay, Japan induced by the Kuroshio warm water (KWW) intrusion (Fig.1). When the KWW intrudes into the bay through Ooshima west channel (OWC), cyclonic and anti-cyclonic circulations are generated in the central part of the bay and in the lee of Oshima Island (Fig.2), respectively as suggested by previous studies based on mooring and drifting buoy observations (e.g. Taira and Teramoto, 1986; Kawabe and Yoneno, 1987; Iwata and Matsuyama, 1989). Radar-derived velocity field clearly shows that these two circulations fluctuate with periods of about 8-11 days. When the KWW strongly intrudes into the bay from OWC, cyclonic and anti-cyclonic circulations become large and the centers of the circulations move southwestward and northeastward, respectively. In addition, warm water intrudes into the mouth of Tokyo Bay and coastal front appears at the bay mouth. On the other hand, KWW intrusion weakens, the circulations become very small (Fig.3).

Statistical analyses, including standard Empirical Orthogonal Function (EOF) analysis and AVHRR sea surface composite image of the Kuroshio indicate that the dominant fluctuation of the circulations are caused by temporal change in strength of the KWW intrusion due to small-scale (10-15km) offshore-inshore movement of the Kuroshio front. And this movement is due to the Ekman transport induced by a southwestward-northeastward wind (Fig.4).

Key words

Sagami Bay, Kuroshio, HF radar, EOF analysis

References

Kawabe, M. and M. Yoneno : Water and flow variations in Sagami Bay under the influence of the Kuroshio Path. *J. Oceanogr. Soc. Japan*, 43, 283-294, 1987.

Iwata, S. and M. Matsuyama : Surface circulation in Sagami Bay: the response to variations of the Kuroshio axis. *J. Oceanogr. Soc. Japan*, 45, 310-320, 1989.

Taira, K. and T. Teramoto: Path and volume transport of the Kuroshio current in Sagami Bay and their relationship to cold water masses near Izu Peninsula *: J. Oceanogr. Soc. Japan*, 42, 212-223, 1986.

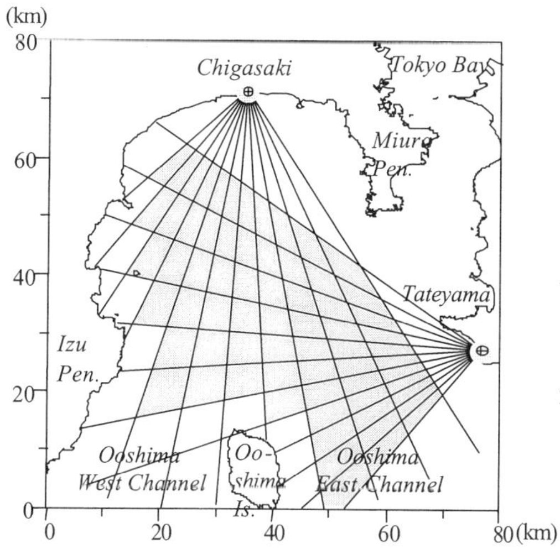

Fig. 1. Observational region.

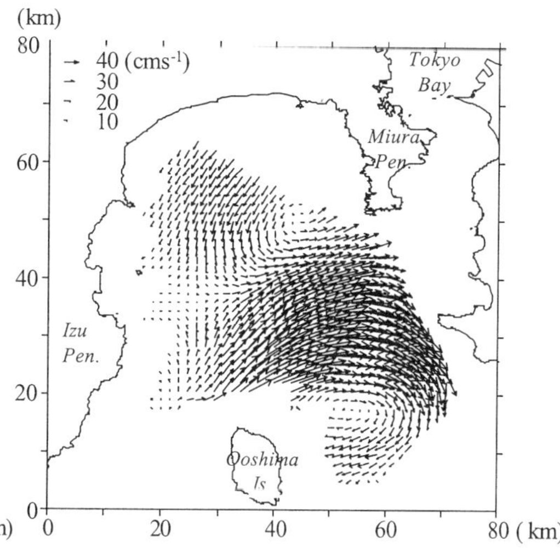

Fig. 2. Average radar-derived vectors for the period from Dec.15, 2000 to Jan. 16, 2001.

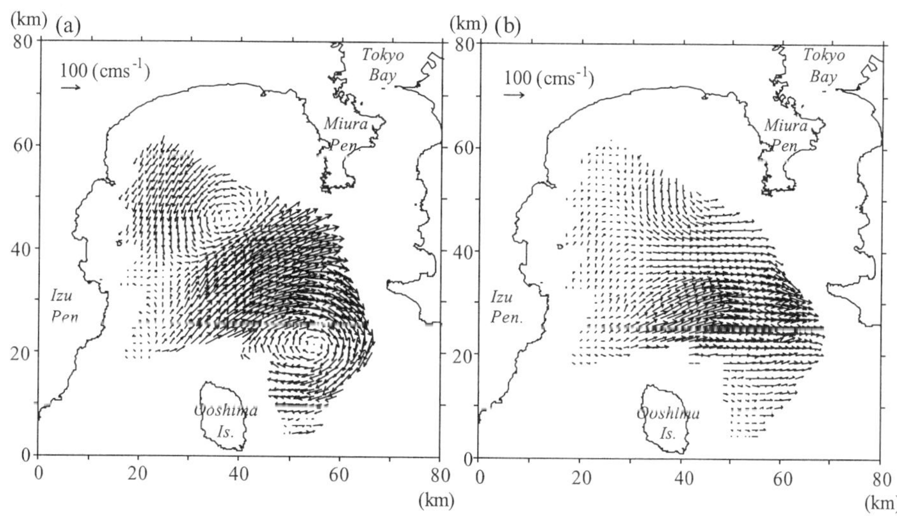

Fig. 3. Surface current maps using EOFs1, 2 plus the mean field. (a) 00:00, Dec. 28, 2000 and (b) 12:00, Jan. 03, 2001.

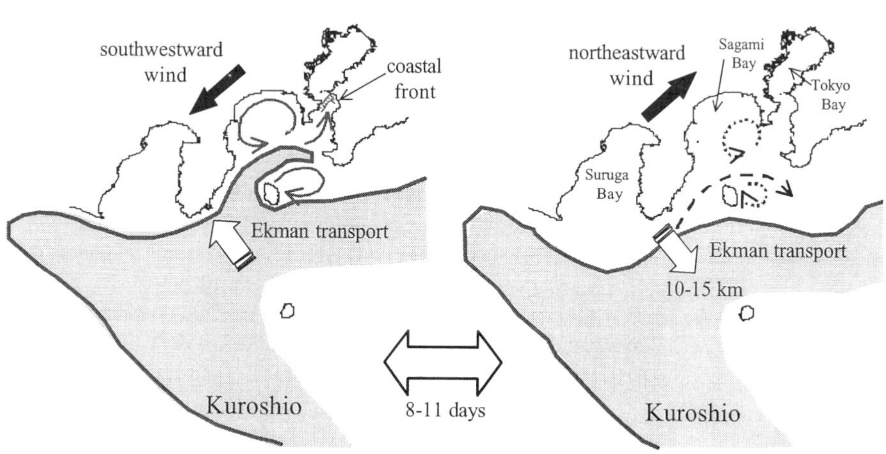

Fig. 4. Schematic image of short-period fluctuations of surface circulations and small-scale offshore-onshore movement of the Kuroshio.

CHEMICAL CONTAMINANTS IN THE HUGLI ESTUARY: ITS IMPLICATION ON MONITORING

Santosh Kumar Sarkar[1*], Pravakar Mishra[2] and Hideshige Takada[3]

1. Department of Marine Science, University of Calcutta, 35, B.C.Road, Calcutta 700019, India Phone : +91 33 430 5232 ; Fax : +91 33 476 4419 ; E-mail : sarkar22@yahoo.com
2. Ocean Research Institute, University of Tokyo, 1-15-1, Minamidai, Nakno, Tokyo 164-8639, Japan Phone : 81-3-5351-6505; Fax : 81-3-5351-6506 ; E-mail : babuli@ori.u-tokyo.ac.jp
3. Department of Environmental Sciences and Natural Resources, Tokyo University of Agriculture and Technology, Fuchu, Tokyo 183-8509, Japan , Phone : 81-42-367-5825; Fax : 81-42-360-8264; E-mail: shige@cc.tuat.ac.jp

* Corresponding author

The Hugli estuary is situated in the western part of Sunderban gangetic delta and is a classical example of a tide-dominated sink for contaminants coming from industrial, aquaculture wastes, domestic sewage, agricultural pesticides, soil erosion, marine traffic and runoff from upstream mining sites. The metropolitan city Kolkata itself discharges about 398 million tons per year of municipal sewage and domestic effluents to this river. The living resources have been degraded recently due to increase in population pressure, pollution and natural resource consumption to the extent of overexploitation. This paper summarizes some of baseline studies on heavy metals, organochlorine pesticides in sediments and water in the estuary and their impact on the biota and outlines some of the future plans required to undertake for better management of the river system. Physicochemical variables of water (temperature, salinity, dissolved oxygen, pH, turbidity, alkalinity, hardness, ammonia nitrogen, nitrate nitrogen, conductivity, total dissolved solids, biological oxygen demand and chemical oxygen demand) as well as sediments (organic carbon, pH and grain size) are also considered and correlated with the level of contaminants.

The distribution of heavy metal concentration in surface waters and sediments fluctuates widely in both inter-metallic and seasonal. These variations are affected by a number of physical and chemical processes including vertical mixing, atmospheric fallout as well as wide variations in different environmental parameters. The descending order of the total concentration of metals in sediment and water is as follows: Fe>Zn>Mn>Cu>Cd>Cr>Pb>Hg with few exceptions while Co and Ni values are below the detection levels in majority of the cases. High level of Fe in the sediment is attributed to the presence of floating old rusty and stranded bergs. Higher concentrations in Fe and Zn show that the inflow water is enriched with these two metals confirms the findings of Braganca and Sanziry (1980) for the coastal regions of Bay of Bengal. The highest concentration of Fe, Hg and Pb is observed in the monsoon months (September-October) a period characterized by low salinity and relatively low pH of water. Such seasonal variation in water and sediment perhaps due to anthropogenic activities and land runoff during the monsoon floods. Phytoplankton activity facilitates the seasonal variations of Fe, as this metal is an essential one for the phytoplankton growth. Levels of dissolved metals registered a seasonal pattern, with an increase during late monsoon months. The concentration of four important metabolites of HCHs in the Hugli river basin reveals a heterogenic nature of distribution. The ΣHCH ranges from 0.003 to 0.33 with a mean value of 0.134 ± 0.073 whereas the ΣDDT ranges from 0.003 to 0.119 with a mean value of 0.058 ± 0.032. A high concentration of β-HCH was observed sharing a major portion of HCH followed by α - and χ- HCH that were present in almost uniform concentration. An overall pattern of accumulation of the pesticides in sediments is in the order of: Σ HCH > endosulfan sulphate > ΣDDT > α-endosulfan. Regarding the HCH isomers, β-HCH was the highest followed by δ-HCH (0.90 to 48.4%). Conversely, the level of α- HCH was below the detection limit in all the samples. It is seen that the β– HCH is always higher than χ-HCH confirming the characteristic feature of more stability and resistance to microbial degradation (Rajendran and Subramanian, 1997). The occurrence of DDT isomers are predominant in the following order: pp′ DDT > pp′ DDE > pp′ TDE > op′ DDE > op′ DDT . Among the

DDT metabolites, pp'- DDT and pp'- DDE accounted for the greatest abundance with approximately 70 – 90% of the total DDT in sediments as the sum of these two forms. These distributions indicate an active degradation of DDT in the sediments and /or inputs of already degraded DDT to the area.

Effects of contaminants in sediment and water are also reflected in the biota inhabiting in the area. A high degree of organ specificity is evident in the bivalve mollusks (*Anadara granosa* and *Meretrix meretrix*) where gill and mantle exhibits higher metal and organochlorine accumulation due to the ion exchange property of the mucous layer covering these organs (Sarkar et al., 1999). Concentration of heavy metals in different organs in the mullet *Liza parsia* reveals that the level is always low in muscle and highest in gill and liver. This is probably due to their physiological roles in fish metabolism. Gastropods and finfishes are differentially selective for a range of metals, reflecting the availability of the elements in the environment and the diet and digestive physiology of these organisms. The study demonstrates that the bivalve mollusk is the major macro invertebrate taxon which can efficiently accumulate biologically available contaminants and can be successfully used in pollution monitoring program (Sarkar et al., 1994).

It is observed that the Hugli river system transports significant amount of pollutants to the coastal waters of Bay of Bengal. The deterioration of water and sediment quality is closely related to insufficiency in environment protection facilities, malfunctioning of wastewater treatment facilities, poor compliance of environmental law, lack of environmental awareness and lack of environmental planning and coordination. To achieve targets of regional sustainability, implementation of realistic conservation and management strategies should be given high priority. There is an urgent need for creating a platform for public participation and communications on issues related to the common environmental interest among the coastal people.

References :

Bhattacharya, B. and Sarkar, S.K. 1996. Total mercury content in marine organisms of the Hooghly estuary, West Bengal, India. *Chemosphere*, Pergamon, 33 (1): 147 – 158.

Braganca, A. and Sanziry, S. 1980. Concentrations of few trace metals in coastal and offshore regions of the Bay of Bengal. *Indian Journal of Marine Science*, 9, pp. 283-286.

Rajendran, R.B. and Subramanian, A.N. 1997. Pesticide residues in water from Kaveri, South India. *Chemistry and Ecology*, 13: 57 – 70.

Sarkar, S.K., Bhattacharya, B. and Debnath, S. 1994. The suitability of tropical marine bivalves as biomonitors of heavy metals in deltaic Sunderbans, northeast India. *Chemosphere*, Pergamon, 29 (4): 759-770.

SEASONAL VARIATIONS IN TEMPERATURE, SALINITY AND DISSOLVED OXYGEN IN THE ENCLOSED AREA AT THE HEAD OF OSAKA BAY, JAPAN

Masayasu Irie, Keiji Nakatsuji, Shuzo Nishida and Kusukatsu Yuasa
Department of Civil Engineering, Graduate School of Engineering, Osaka University,
2-1 Yamadaoka, Suita City, Osaka, 565-0871 Japan
Ph: +81-6-6879-7605; Fax: +81-6879-7607
Email: irie@civil.eng.osaka-u.ac.jp

Abstract:

OBJECTIVE OF THE PRESENT RESEARCH

The coastal area, which becomes the object of this investigation, is enclosed by many artificially-built islands and breakwaters in the head of Osaka Bay as shown in Figs.1 and 2. Every summer algal blooms have been often observed at the sea surface, besides anoxic water has been formed near sea bottom and upwelled dependently of wind's direction and strength. Complains are voiced by the residents who live in high multi-floored apartments built in the reclaimed islands. Under such circumstance, there remains only one small area still in almost natural condition with a sand beach. The final target of this research is to propose the policies from the standpoints of restoring the lost sand environments and actively creating new environments. As the first step, the authors carried out field surveys for better understanding seasonal variation of water qualities. Some researchers have been reported the existence of the anoxic water at the head of Osaka Bay, but it is not well known that the mechanism of the anoxic water in the enclosed coastal areas such as a port and a water area behind the reclaimed island. In the present study, we discuss the spatial and temporal distributions of anoxic water and the physical and biochemical dynamics controlling the water based on long-term field surveys.

OUTLINE OF FIELD SURVEYS

The field surveys were carried out in the Port of Amgasaki-Nishinomiya-Ashiya (ANA-Port), which is shown in Fig.2. The long breakwater divides coastal area into ANA-Port and the head of Osaka Bay. Temperature, salinity, chlorophyll-a, suspended solids (SS) and dissolved oxygen (DO) were measured for clarifying time changes of their vertical profiles at 15 observation points every week from May to November, 2002. In addition, water qualities were measured at 12 points and flow patterns were also measured by ADCP every month.

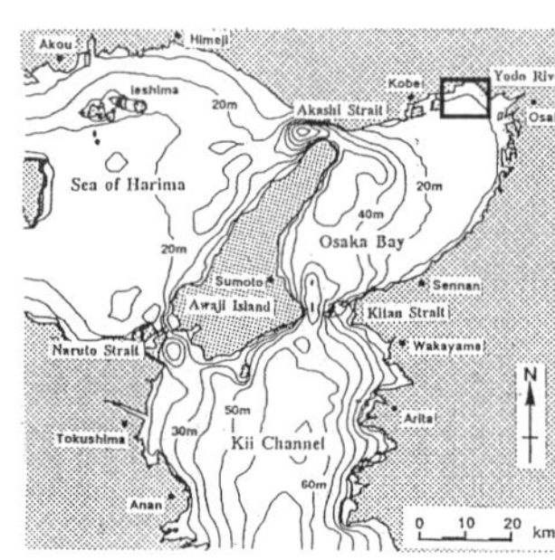

Fig.1 Location of the Port of ANA (See details of **Fig.2**)

OBSERVATION RESULTS

(1) Anoxic water was always observed near the sea bottom in the whole area of the ANA-Port from June to September. Figure 2 shows the horizontal distribution of DO observed at the bottom layer on July 17. The values of DO are very small less than 3 mg/l, which corresponds to the critical standard whether the living things can live or die. The DO value near the sea bottom became almost zero at midsummer. If the thickness of the anoxic water layer is

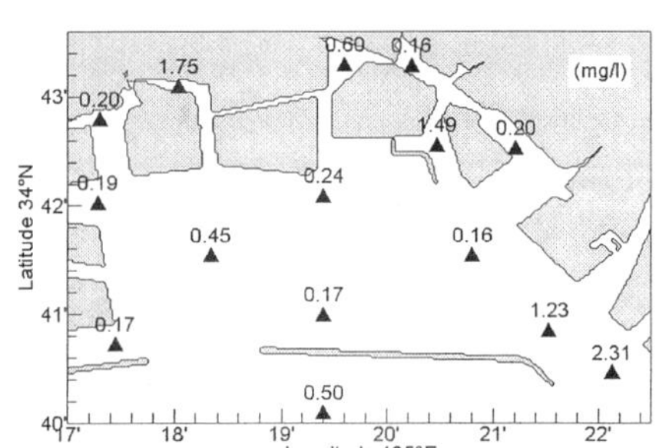

Fig.2 Horizontal distribution of DO at the bottom layer

defined as that of DO less than 3mg/l, it attained around 4 meters and more in the central area of the ANA-Port which is 12 meters deep in July. (See Fig. 3)

(2) Blue tide, characteristic phenomena observed in Tokyo Bay, can be observed several times in Osaka Bay, too. Blue tide is named since the surface water changes milky-blue. It is induced by upwelling phenomena of anoxic bottom water, which is different from red tide. The number of the appearance of blue tide was two during this survey period. It occurred in the sand beaches, which included the natural beach of the subject of this research. It was because the upwelling of anoxic water easily occurred in gently slope beach as compared with upright concrete breakwaters.

(3) Figure 4 shows the vertical distributions of temperature, salinity, Chlorophyll-a and DO. The layer in which DO rapidly decreases corresponds to that of discontinuity in temperature and salinity. The water column consists of three layers. In the upper layer phytoplankton grows abundantly, the lower layer is in which water becomes anoxic because of consumption of oxygen in the sea-bottom, and the rest of them is the middle layer. These three layered structure sometime changes two layered one as shown in the observation point A4 in Fig. 4. It is because the great consumption of oxygen in the bottom layer makes the bottom layer thick and it stretch to the layer of discontinuity. The thickness of anoxic water, therefore would be placed limit on by the layer of discontinuity. It is ensured by that the correlation between the thickness and the vertical variation of temperature and density is very poor. The water depth of the layer of discontinuity would be important to restrict the dynamics of the anoxic water.

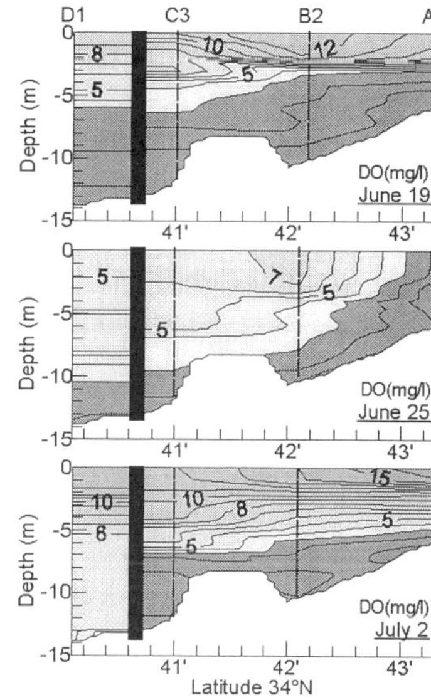

Fig.3 Vertical distributions of DO from out of the port to behind the reclaimed land

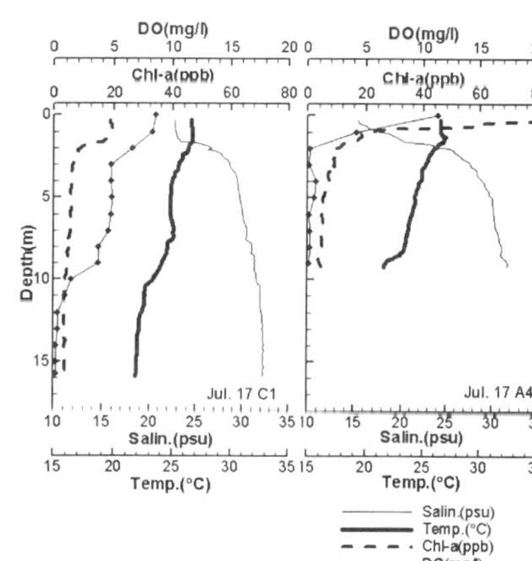

Fig.4 Vertical distributions of two observation points

DISCUSSION AND SUMMARY

In the head of Osaka Bay, it has been well known that the north wind causes upwelling the water at the bottom layer. It makes a contribution to the disappearance of the anoxic water, and to produce the water upwelling in the ANA-Port. The direction of the prevailing wind in summer is southwest, and the anoxic water remains stably over the sea bottom. However, when the northerly wind blows induced by meteorological disturbance, the lower anoxic water is drifted toward the upper layer of the north shore, which results in blue tide appearance for supplementing the upper water transported to south. If the duration time of meteorological disturbances is short, namely one to three days, no vertical mixing occurs between the upper and lower water. When the wind blowing was strong, the anoxic water once disappeared; however it was observed to return to the original shape on the next observation after one week. The north wind, which is effective to decrease the anoxic water over the whole area of the head of the bay, does not have a significant improvement of the anoxic water in the port.

THE CHARACTERISTIC OF FLUID MUD AND COASTAL WATER QUALITY IN THE HIROSHIMA BAY

K. Tada[1], T. Hibino[2], K.C. Tran[3] and H. Matsumoto[4]

[1]Master student, Department of Social and Environmental Engineering, Graduate School of Engineering, Hiroshima University, 4-1, Kagamiyama 1 chome, Higashi-Hiroshima 739-8527, Japan. Tel/fax: +81 824 24 7818, e-mail: tada@hiroshima-u.ac.jp

[2]Associate Professor, Department of Social and Environmental Engineering, Graduate School of Engineering, Hiroshima University, 4-1, Kagamiyama 1 chome, Higashi-Hiroshima 739-8527, Japan. Tel/fax: +81 824 24 7816, e-mail: hibinot@hiroshima-u.ac.jp

[3]Associate Professor, School of Policy Studies, Kwansei Gakuin University, 2-1, Gakuen, Sanda, Hyogo 669-1337, Japan. Tel: +81 795 65 7600, Fax: +81 795 65 7605, e-mail: mv9k-trnt@asahi-net.or.jp

[4]Hiroshima port airport technical investigation office, Chugoku district maintenance office, Ministry of Land, Infrastructure and Transport, 2-10, mikawacyo, nakaku, Hiroshima 730-0029, Japan. Tel: +81 82 545 7017, Fax: +81 82 545 7019

Introduction Hiroshima Bay, surrounded by Kurahashi Island, Yashiro Island and other islands, is the bay connected to Seto Inland Sea. Recently there have been environmental problems of various degrees of severity, namely sea-level rise, eutrophication and 'red tide' occurrences, etc. The incidence of eutrophication due to the overload of nutrients has caused poor health of the ecosystem and economic loss in fishery. Therefore, the study of processes and mechanisms related to eutrophication such as aerobic and anaerobic decomposition of organic matter, and nutrient cycles including phosphorus and nitrogen cycles is important.

By the inflow load from Ota River and Pacific Ocean, or by death and excretion of zooplankton, death of phytoplankton, nutrition salt, and an organic matter sediment, fluid mud layer (high turbidity layer) is formed near a submarine surface. The purpose of this research is to study the characteristic of fluid mud in Hiroshima Bay with many unknown points in submarine substance circulation. The distribution and influence of fluid mud on the ocean space environment must be united and considered.

The generating mechanism of anaerobic decomposition gas Under the conditions in which sulfuric acid ion exists abundantly, anaerobic disassembly of an organic matter takes place by making sulfuric acid ion into an electronic receptor, and hydrogen sulfide is generated as the result.

$$SO_4^{2-} + \text{organic matter} \xrightarrow{\text{Sulfuric acid reduction bacillus}} S^{2-} + H_2O + CO_2$$

$$S^{2-} + 2H^+ \rightarrow H_2S$$

The second reaction is pH dependent: when $pH \geq 8$, the reaction equilibrium shifts towards the left-hand side generating more HS^- or S^{2-}. When $pH < 8$, the reaction equilibrium shifts towards the right-hand side generating H_2S in gaseous state, which is then emitted underwater. In mud, insoluble Fe^{3+} exists in the form of iron compound ($Fe(OH)_3$, $Fe_2(PO_4)_2$) and if it is in a reduction state, it will become Fe^{2+} and will begin to dissolve in water. The soluble Fe^{2+} reacts immediately with hydrogen sulfide, and FeS is generated, which is responsible for the black colour in the mud. FeS exists in a precipitate form.

$$Fe^{III}(OH)_3 \rightarrow Fe^{II}(OH)_2 \rightarrow Fe^{2+} + 2OH^-$$

$$(Fe_2^{III}(PO_4)_3 \rightarrow Fe^{II}(PO_4)_2)$$

$$SO_4^{2-} + \text{organic matter} \xrightarrow{\text{Sulfuric acid reduction bacillus}} S^{2-} + H_2O + CO_2$$

$$Fe^{2+} + S^{2-} \rightarrow FeS \downarrow$$

$$(Fe_3(PO_4)_2 + 3S^{2-} \rightarrow 3FeS \downarrow + PO_4^{3-})$$

However, if too much H_2S is generated, the sulfide ion of H_2S and HS^- and S^{2-} can be detected in both water and sediment, releasing a fouling smell.

Mechanism of the phosphorus cycle Circulation of phosphorus is chemically simple. It is hydrolyzed quite easily, and organic phosphoric acid turns into inorganic phosphoric acid, which is again taken in by phytoplankton in seawater (alkalinity).

Mechanism of the nitrogen cycle Since nitrogen exists in various forms (oxidation states), which do not interchange easily, the nitrogen circulation in the ocean is complicated. The nitrogen circulation is performed by the inorganic-ized metabolism activities of bacteria and excretion of an animal.

Nitration : Oxidation ($NH^{4+} \rightarrow NO^{2-} \rightarrow NO^{3-}$)

Denitrification : Reduction ($NO^{3-} \rightarrow NO^{2-} \rightarrow NO \rightarrow N_2$, O, N_2)

Nitrogen fixation : N_2 changes into an organic nitrogen compound.

Decomposition by bacteria : NH^{4+} Nitration (DON, PON→DIN or in other words dissolved and precipitate organic nitrogen transform to dissolve inorganic nitrogen)

In the above processes, nitrogen changes its valency forms according to the oxidization-reduction reactions. The change in valency of nitrogen is much larger compared to other elements in living process.

The seasonal variation of the bottom mud and water quality in the central part of Kure Bay (2001.10~2002.9)

The study conducted over one year period (October 2001 to September 2002) shows a variation in various chemical forms of nitrogen. NO_2-N concentration was 0.05 mg/l at water above mud in October 2001 (Fig.1(a)). This concentration decreases to near zero in September 2002 (Fig.1(a)). The NH_4-N concentration tends to increase as going deep in the mud (Fig.1(b)). Moreover, the NH_4-N concentration was 3.6 mg/l near fluid mud layer in September (Fig.1(b)). PO_4-P present at concentration of more than 0.1 mg/l at fluid mud layer and near the mud upper layer in July, September, and October (Fig.1(c)). August, since a perpendicular change was not seen and bottom mud has given impact greatly to the water quality above mud for nutrition salt, remarkable perpendicular mixture can be considered through water above mud and bottom mud (Fig.1(b),(c)).

pH value also shows seasonal variation of 8.1 at water above mud and 7.9 at mud gap water in May and October (Fig.1(d)). In July, August, and September, pH is 8 at water above mud and 7.7 at mud gap water, which is smaller than that in May and October (Fig.1(d)).

Chlorophyll-a shows a large concentration, 2 ~ 4 μ g/g at fluid mud layer in August, September, and October, although change is seldom observed in the perpendicular direction in May and July (Fig.1(e)). Pheophytin is present at concentration of 80 ~ 110 μ g/g at fluid mud layer in August, September, and October (Fig.1(f)).

Oxidation-reduction potential (ORP) shows an oxidization state tendency in July, and shows a reduction state tendency in October (Fig.1(g)). Chemical oxygen demand (COD) is about 44 mg/g in August, September and October; while in May and July, it is about 38 mg/g (Fig.1(h)).

Sulfide has the highest concentration near bottom (2-10cm) of the seabed rather than in the fluid mud layer (Fig.1 (i)). It is thought that the stinking smell of mud called "sulfur smell" is emitted from these layers.

CONCLUSION The generation process of fluid mud is hypothesized that firstly, detritus and an organic matter sink to the seabed. In bottom mud, N and SO_4^{2-} are abundantly present. H_2S occurs and mud changes black. When oxidization reduction potential falls, NH_4-N and PO_4-P begin to dissolve from the seabed. DO and COD of water above mud are consumed resulting to oxygen-deficient water mass occurs in the seabed.

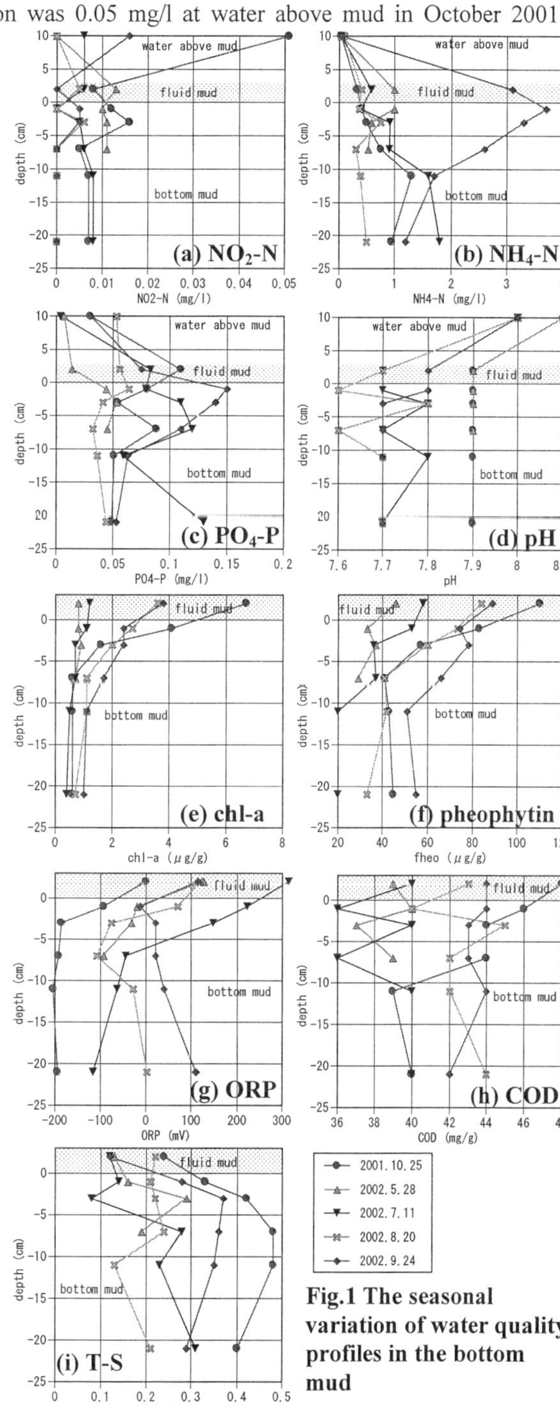

Fig.1 The seasonal variation of water quality profiles in the bottom mud

ENVIRONMENTAL IMPACT OF POSCO CONSTRUCTION, KOREA

Hyoseob Kim[1], Byung Chul Oh[2] and Hoesoo Jeong[3]

Youngil Bay has a rectangular shape on the east coast of Korea, see Figure 1. Hyungsan River supplies pollutants and water to the center of the bay. The tidal range around the bay is as small as 4 ㎝, and the circulation of the seawater in the bay is not sufficient to keep the seawater clean since 1975. Waves are relatively strong at the entrance of the bay but weak along the coastline inside the bay. The area is relatively shallow ; water depth is under 80 m throughout the bay. The extent of contamination is especially higher at some areas in the bay than at other area because of supply of organic materials from the industrial area nearby. The COD at some areas in the bay reaches over 5 mg/L, which is considered to be a very serious situation. The river discharge from Hyungsan River is about 20 t/s on average, and reaches up to about 3600 t/s at extreme flood cases. Possible treatments of the contamination in the bay are under study. Flushing out the bay with the river flood discharge by adding groins around the river mouth is a countermeasure. Artificial circulation with pumping will be another countermeasure. An existing depth-average 2-dimensional horizontal, finite-difference numerical model system, KU-2DHFC was used for the simulation of flow and dispersion of the two cases. Exchange rate with the outer sea was improved for each development scheme.

1 Dept. of Civil & Env. Engineering, Kookmin Univ., 861-1, Jeongnung-Dong, Seongbook-Goo, Seoul, 136-702, Korea, hkim@kookmin.ac.kr
2 Korea Ocean Research & Development Institute, Ansan, Korea
3 Korea Ocean Research & Development Institute, Ansan, Korea

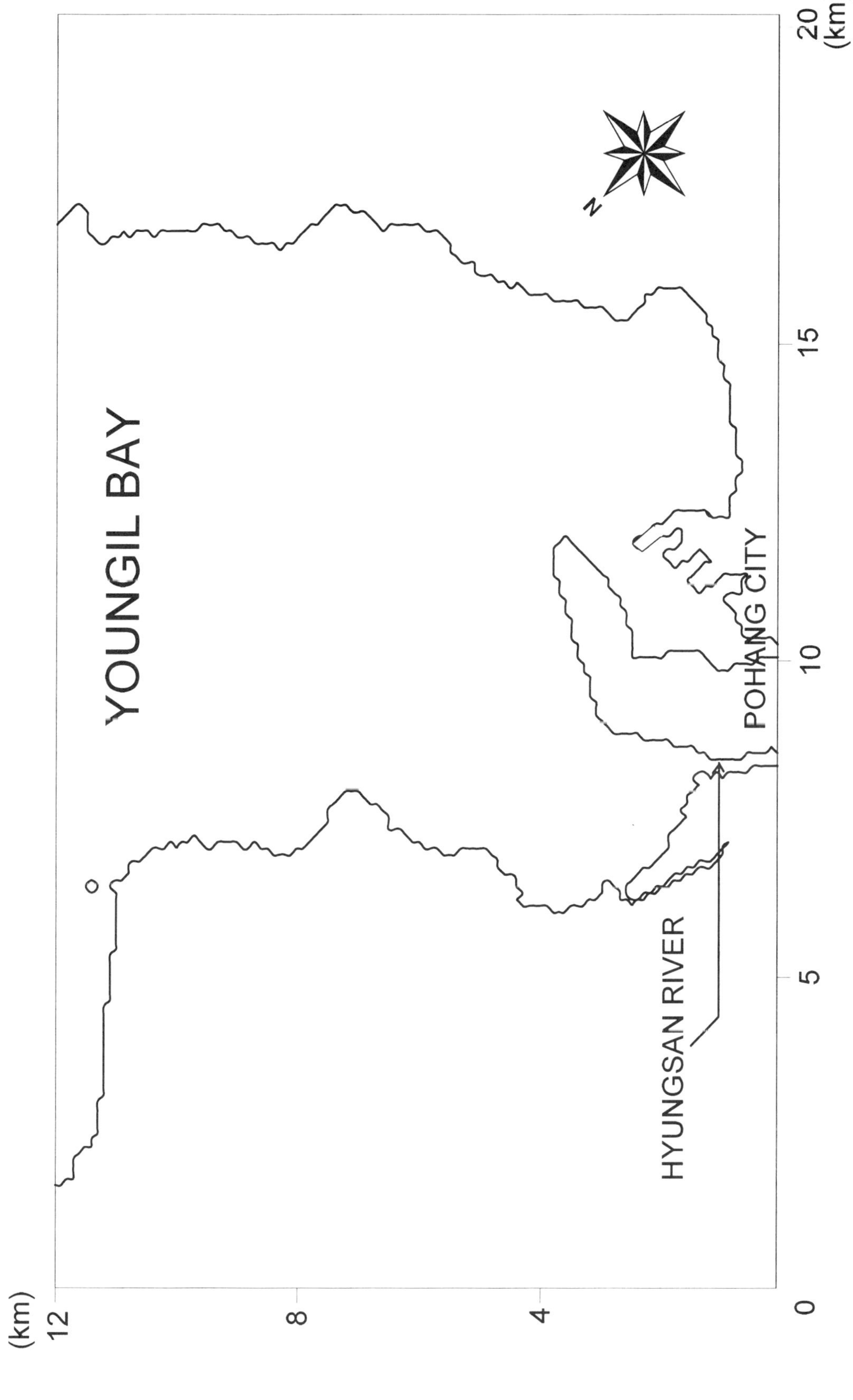
YOUNGIL BAY
POHANG CITY
HYUNGSAN RIVER
N
(km)
12
8
4
0
5
10
15
20
(km)

FIELD OBSERVATION OF WATER ENVIRONMENT IN ARIAKE BAY

Yukio Koibuchi[1], Masahiko Isobe[2]

[1] *Department of Civil & Environmental Engineering, Tokyo Denki University, Ishizaka, Hatoyama, Hiki-Gun, Saitama 350-0394, Japan, Phone: 0492-96-2911, Fax: 0492-96-6501, Email: koi@g.dendai.ac.jp*

[2] *Graduate School of Frontier Sciences, University of Tokyo, 7-3-1 Hongo, Bunkyo-ku, Tokyo 113-8656, Japan, Phone: 03-5841-6112, Fax: 03-5841-8503, Email: isobe@k.u-tokyo.ac.jp*

1. Introduction

In 1980's, red tide began to occur in Ariake Bay. In recent years, occurrence frequency and scale of the red tide is increasing (Takagi et al., 2002). Once red tide occurs, oxygen depleted water is produced near the seabed through aerobic decomposition of excess organic matters settled from the upper layer. Anoxic environment causes mortality of benthic organisms which have the function of purifying the water, resulting in the negative spiral of the water environment.

However, continuous and long-term observation of water quality has not been done in Ariake Bay. Available information on factors controlling the temporal variations of phytoplankton and oxygen depleted water is limited.

The present study aims (1) to describe temporal variations in chlorophyll-*a* and dissolved oxygen on the sea bed based on the continuous measurement of water quality in Ariake Bay, (2) to describe inorganic nutrients during blooms and non-blooms based on frequent observations, and (3) to discuss a possible relationship of formation and duration of anoxic water with local weather.

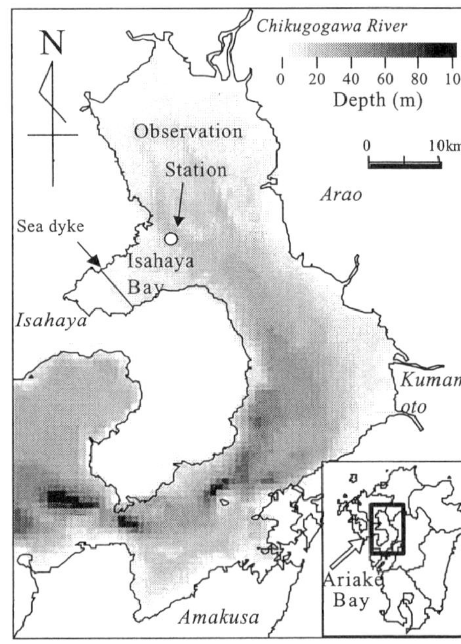

Fig. 1: Map of Ariake Bay, Japan, showing the location of the measurement in the bay.

2. Method of field observation

Field observation was performed at an offshore location with water depth of 17m in the center of the inner part of Ariake Bay (Fig. 1) from July 2001. The observation properties are chlorophyll-*a*, water temperature, salinity, turbidity, and dissolved oxygen (DO). These were measured and recorded every 10 minutes at four elevations by built-in memory.

Seawater was sampled weekly in the surface and bottom and its nutrients and phytoplankton contents were analyzed. (Isobe and Koibuchi, 2002).

3. Results and discussion

Water was warming steadily from July through late August (Fig. 2). The thermal stratification developed in the surface water. This stratification disappeared soon by mixing after the continuous north wind on August 20.

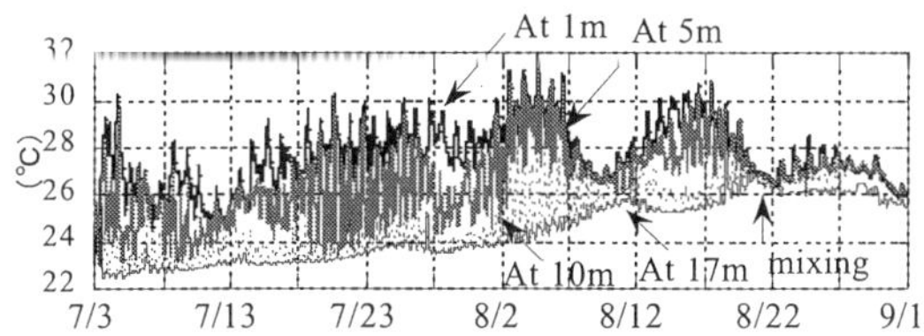

Fig. 2: Time series of temperature.

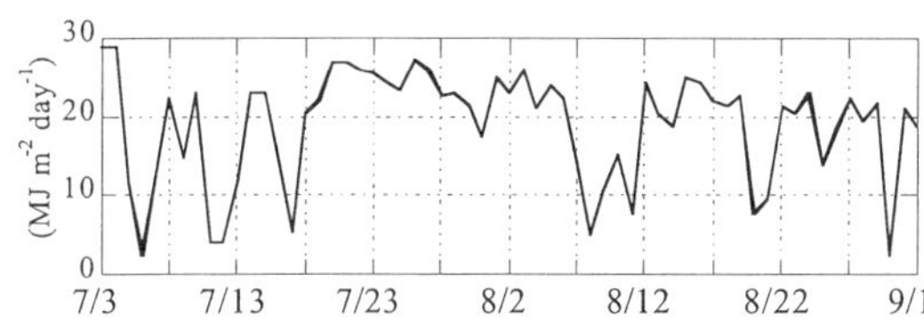

Fig. 3: Time series of daily solar radiation. Measurement of solar radiation was made at Saga (Japan Meteorological Agency).

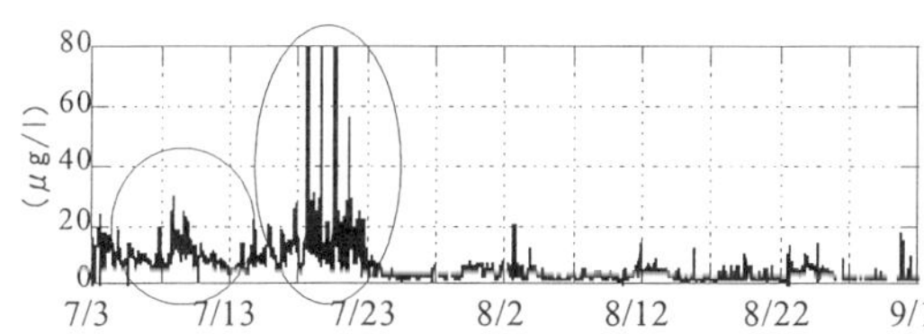

Fig. 4: Time series of chlorophyll-*a* at the surface.

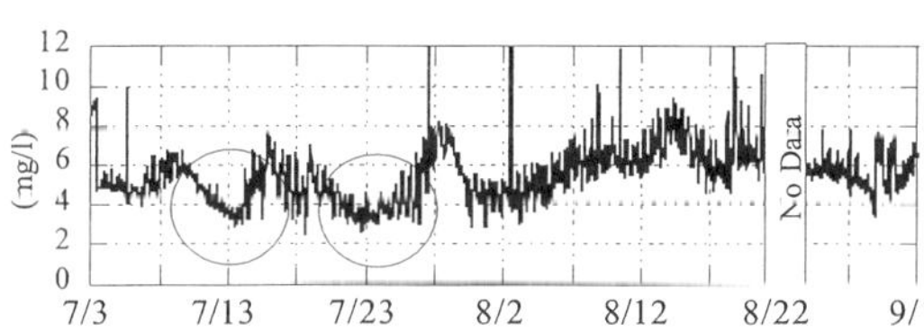

Fig. 5: Time series of dissolved oxygen at the bottom.

Chlorophyll-*a* increased to twice during this period. The first bloom began on July 7 (Fig. 4). During this bloom, chlorophyll-*a* increased steadily for 3 days, and began to decrease on July 10. The second bloom began on July 18. Then chlorophyll-*a* concentration fluctuated sharply, and reached a maximum on July 21. The blooms during the periods of July 8 to 11 and July 18 to 23 were supported by calm weather and high solar radiation (Fig. 3). Stratification in the upper 10-m column (Fig. 2) in these periods seemed to accelerate the bloom.

Dissolved oxygen (DO) decreased to half (from July 9 to 13, and from July 20 to 24), in which the concentrations higher than 6 mg/l decreased to that lower than 3 mg/l. These periods corresponded closely to increase of chlorophyll-*a*. Therefore, low DO concentrations were caused by the large amount of oxygen consumption after the blooming in the surface layer. Dissolved oxygen was lowest in the center of Isahaya Bay.

Then, the factor to restrict growth of phytoplankton after July 23 under sufficient solar radiation is discussed. From the change of nitrogen and phosphorus concentrations in the period, dissolved inorganic nitrogen in the surface layer remarkably decreased after July 20. This period agreed closely with period of the maximum chlorophyll-*a*. Therefore, it is concluded that dissolved inorganic nitrogen was exhausted by the temporal phytoplankton blooming (on July 20 to 23) and this exhaustion inhibit the growth of phytoplankton after July 23.

4. Conclusion

Solar radiation and nutrient concentration are very important factors for red tide occurrence and anoxic environment in Ariake Bay. The change of nutrient supply and transportation by human activity may bring about a serious effect on the water quality in the bay.

References

Isobe, M. and Koibuchi, Y. (2002). Present water environment in Ariake bay and possibility of its numerical simulation, *Proc. 51st Nat. Cong. of Theoretical & Applied Mechanics*, pp.11-14.

Takigawa, K., Tanaka, K., Hokamura, T., Yoshimura, S. and Furuichi, M. (2002). Factiorial analysis of environmental change in the middle area of Ariake- Sea, *Proc. Coastal Eng.* JSCE, **49**, pp.1066-1070.

DEVELOPMENT OF SEAWATER EXCHANGE SYSTEM USING THERMAL ENERGY AND METAL HYDRIDE ACTUATOR

Mitsuhiro Sakikawa[1],Youichi Atsumi[2],Atsushi Kubouchi[3],Kazuhiro Matsumura[4],
Tsutomu Endo[5],Shizuo Yoshida[6]

Purpose of study:

Development and use of new energy sources such as renewable energy and recycled energy have become much more common in recent years due to concerns about the global environment and depletion of traditional energy sources. Hydrogen is a clean energy source that can replace fossil fuels, which are some of the causes of global warming. Metal hydride (MH) has been studied as a method for storage of hydrogen. Application of the energy storage capabilities of MH in various fields is eagerly awaited. The MH actuator is being developed as a method for directly obtaining a mechanical force using hydrogen pressure generated by a temperature difference applied to a MH.

At the same time, seawater exchange is being considered as a measure against deterioration of water quality and freezing, which are becoming problems in harbors and fishing ports. Because commercial power sources will generally be used to drive the pumps for these proposed systems, the running costs are expected to be quite high. It is therefore desirable to develop a seawater exchange system with low running cost to put these systems into practical use.

We have devised a seawater-pumping device using a MH actuator (Fig. 1). This device incurs almost no running cost because it pumps seawater by raising and lowering a plunger powered by a MH actuator. The actuator is powered using the temperature difference between sources in nature. The high-temperature heat source could be subterranean heat or a hot spring, and the low-temperature heat source, seawater or subterranean water. In this study, a seawater-pumping device was produced experimentally using a MH actuator. Its performance and characteristics were observed in a laboratory test, and the feasibility of its practical application was examined.

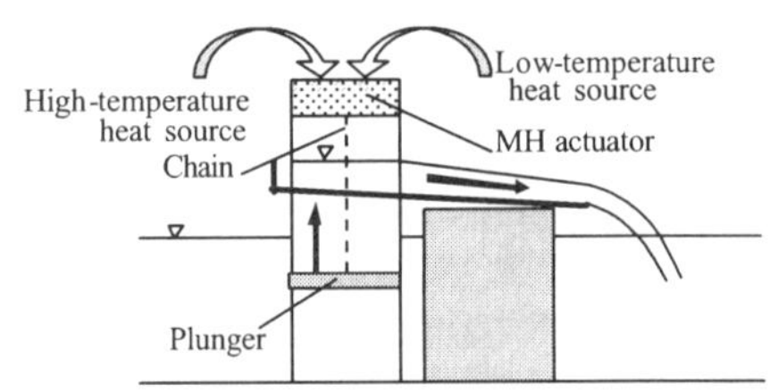

Fig. 1 Structure of seawater pumping device

Contents of study:

A seawater-pumping device using a MH actuator was constructed experimentally and a laboratory test for its practical application was conducted in a tank. The device consists of a MH container charged with LmNi-type metal hydride, a gas cylinder connected to the container, and a seawater transport pump driven by a chain attached to the cylinder. The MH container was alternately exposed to high-temperature water and low-temperature water, simulating the use of water from a hot spring and ambient sea water. This caused the MH to release and absorb, respectively, hydrogen gas, which pressurized and depressurized the cylinder, cycling the pump through the chain. In the experiment, the hydrogen pressure in the MH container, the temperature of the MH container (water temperature in the hot water supply system), the plunger position of the seawater transport pump, the level and volume of pumped water and other values were measured continuously during operation of the device.

Major conclusions:

(1) Figure 2 shows the relationship between the pressure difference $\triangle p$ and the volume pumped by the MH actuator. The result revealed that the pumping volume increased almost proportionally with increase in pressure difference and decrease in pump head h_0.

(2) The motion of water transported in the pump can be expressed by Eq. (1) if the clearance between the plunger and piston shaft can be ignored. In the equation, M: mass of the plunger; m: mass of the chain; d_1: inner diameter of the plunger; d_2: inner diameter of the cylinder; ρ : density of water; y: position coordinate of the water surface in the pump; h': distance from

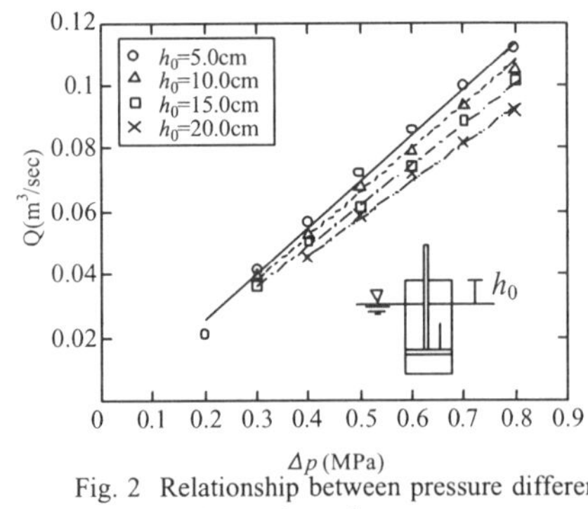

Fig. 2 Relationship between pressure difference and pumping volume

the plunger to the water surface at bottom of stroke; f: force due to mechanical friction in the system; and F: drag force due to the pressure difference across the plunger (= ky'). The frictional force f was set to be 65 N from the result of a preliminary experiment, and 850 kg/s was used for k to correspond with the maximum value of y obtained in the experiment. Figure 3 shows the changes of water surface in the piston shaft over time under the conditions of h' = 46 cm and h_0 = 30 cm. The measured values and the values calculated by Eq. (1) agreed well with each other. Because similar results were obtained for other h' values, changes of water surface in the piston shaft over time could be predicted for any given h' value.

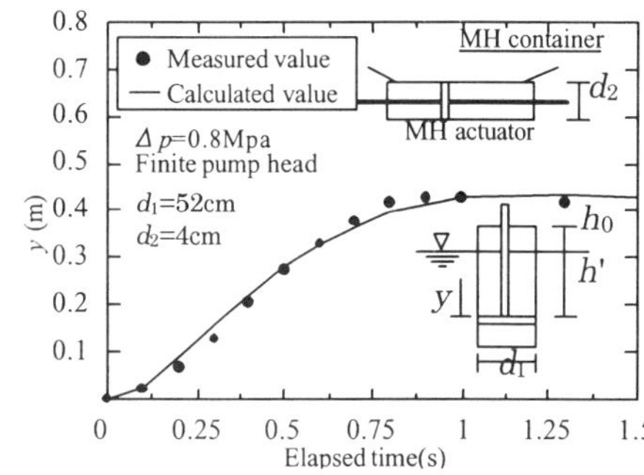

Fig. 3 Changes of water surface in piston shaft over time

(3) In the condition without pump head h_0, the movement of water transported in the pump can be expressed by Eq. (2). L is the total length of the piston shaft. Figure 4 shows the relationship between the measured values of changes of water surface in the piston shaft over time and the value calculated by Eq. (2) under the condition of h'= 106 cm without pump head. Under this condition, the plunger rises to the upper limit at once under the same power input from the MH actuator as that in Figure 3; the seawater transport volume increases dramatically.

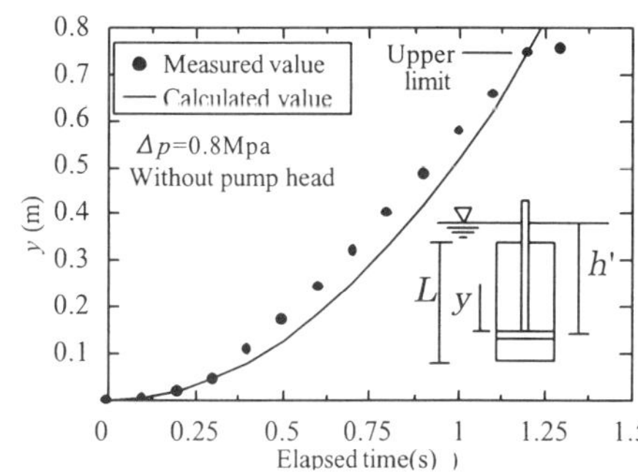

Fig. 4 Changes in plunger level in piston shaft over time

(4) Supposing that the current system can be operated continuously under a condition without pump head and the time required for one cycle is 10 seconds, the daily seawater transport volume of this device would be 1,287.4 m^3/day. It was found that exchange of the entire water mass of a port with a capacity of 50,000 m^3 is possible in approximately 39 days even if the current test device is used as it is, and that the feasibility of its practical application is high.

$$\left\{2M + m + \rho \pi \frac{d_1^2}{4}(h'+y)\right\}\frac{d^2 y}{dt^2} = -\rho \pi \frac{d_1^2}{4} gy + \Delta p \pi \frac{d_2^2}{4} - f - F \quad \cdots\cdots \quad (1)$$

$$\left\{2M + m + \rho \pi \frac{d_1^2}{4} L\right\}\frac{d^2 y}{dt^2} = \Delta p \pi \frac{d_2^2}{4} - f - F \quad \cdots\cdots \quad (2)$$

---

1 Researcher of Port and Harbor Engineering Division, Civil Engineering Research Institute of Hokkaido, Hiragishi 1-3-1-34, Toyohira-ku, Sapporo, 062-8602, Japan, tel: +81(11)841-1684 (direct), fax: +81(11)842-9169, e-mail: sakikawa@ceri.go.jp

2 Senior Researcher of Port and Harbor Engineering Division, Civil Engineering Research Institute of Hokkaido, Hiragishi 1-3-1-34, Toyohira-ku, Sapporo, 062-8602, Japan, tel: +81(11)841-1684 (direct), fax: +81(11)842-9169, e-mail: 99258@ceri.go.jp

3 Director of Port and Harbor Engineering Division, Civil Engineering Research Institute of Hokkaido, Hiragishi 1-3-1-34, Toyohira-ku, Sapporo, 062-8602, Japan, tel: +81(11)841-1684 (direct), fax: +81(11)842-9169, e-mail: 81119@ceri.go.jp

4 Chief Researcher, Research and Development Department, Hokkaido Industrial Technology Center, 379, kikyou-cho, Hakodate, 041-0801, Japan) , tel: +81(138)34-2600, fax: +81(138)34-2602, e-mail: matsumura@techakodate.or.jp

5 Numerical Analysis Section, Nippon Data Service Co Ltd, N16E19, Higashi-ku, Sapporo, 065-0016, Japan, tel: +81(11)780-1117, fax: +81(11)780-1123, e-mail: t-endou@ndsinc.co.jp

6 Associate Professor, Physical Fluid Dynamics Lab., Hokkaido University, N13W8 Kita-ku, Sapporo, 060-8628, Japan, tel: +81(11)706-6723(direct), fax: +81(11)706-6723, e-mail: Yoshidas@eng.hokudai.ac.jp

[1]Correponding author.

VERIFICATION OF HYDRAULIC MODEL TEST FOR NORTH DRAINAGE GATE ON ISAHAYA-BAY SEA RECLAMATION PROJECT AT SEAWATER INTAKE

Hirohide Kiri[1], Hajime Tanji[2] and Tetsuo Nakaya[3]

1 Senior Researcher, Laboratory of River and Coast, National Institute for Rural Engineering, 2-1-6 Kan-nondai, Tsukuba, Ibaraki 305-8609, Japan. kiri@nkk.affrc.go.jp
2 Chief, Laboratory of River and Coast, National Institute for Rural Engineering, 2-1-6, Kan-nondai, Tsukuba, Ibaraki 305-8609, Japan.
3 Researcher, Laboratory of River and Coast, National Institute for Rural Engineering, 2-1-6 Kan-nondai, T sukuba, Ibaraki 305-8609, Japan.

1. Introduction

A hydraulic model test is one of the important tools for design of hydraulic facilities. However, floods are often assumed as the condition of hydraulic model tests, because a lot of hydraulic model tests aim at the verification of hydraulic feature of the facilities. Therefore, the example that evaluated the reproducibility of result of the hydraulic model test after construction is few.

On the other hand, an introduction of seawater through drainage gates aiming at investigation of relation between deterioration of water quality in Ariake Sea and the Isahaya sea area reclamation project was carried out in 2002. The forecasting of flow velocity distribution of influent was important because some facility might be broken with the bottom mud that lifted up by the seawater introduction.

In this paper, we clarified the flow velocity distribution at the seawater introduction by the hydraulic model test. Moreover, we verified the reproducibility of the hydraulic model test compared with the result of field observation.

2. Outline of the hydraulic model test

The outline of hydraulic model used on this test is shown in **Figure 1**. The hydraulic model test used the non-distortion model of scale 1/50 which extracted the surrounding of the north drainage gate. The gate openings were assumed to be two kinds (0.6m and 0.9m) due to the verification result of the forced vibration at the gate by the influent (**Figure 2**).

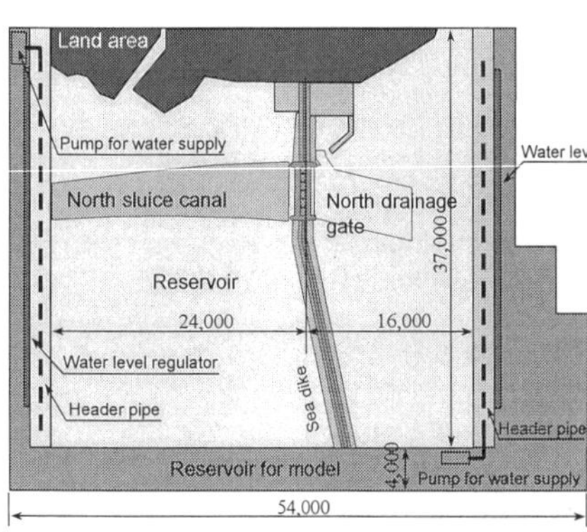

Figure 1 Outline of hydraulic model

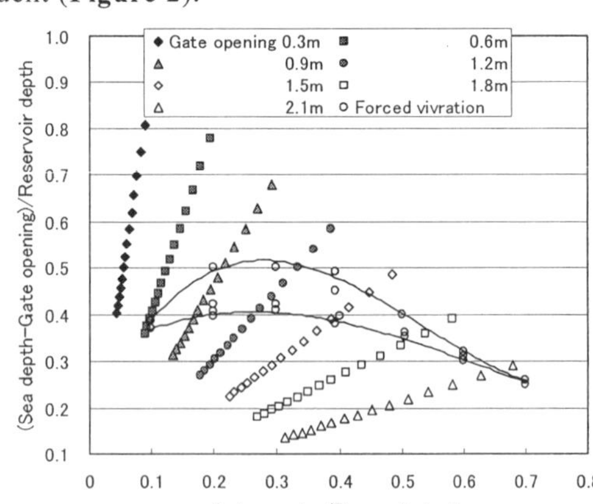

Figure 2 Verification of forced vibration

The water level in the reservoir is managed within the range from -1.0m to -1.2m. Moreover, water level in the outside of the reservoir changes by tide. The relation between tide level and reservoir water level was decided with the results of numerical simulations.

3. Feature of reduction of flow velocity

Figure 3 shows the relation between the distance from the gate and a dimensionless flow velocity that is averaged at each section. In any case of 0.6 and 0.9m in gate opening, it is understood that flow velocity decrease in proportion to the -1/3rd power of the distance from the gate. As a result, it is able to say that the tendency of the reduction of flow velocity through the north drainage gate was similar to that of maximum flow velocity of two-dimensional laminar jet flow.

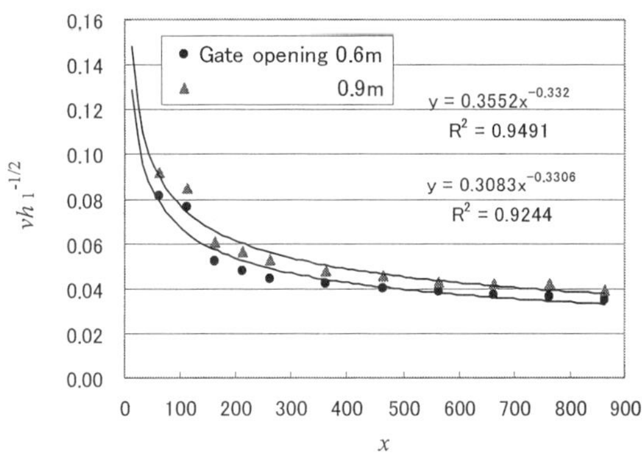

Figure 3 Relation between distance and velocity

4. Comparison between hydraulic model test and field observation

The reproducibility of the hydraulic model test was verified by comparing the results of the hydraulic model test with the field observation results that observed by mooring buoy. There is an influence such as the winds in the field observation results, and some differences in the condition of the bottom shape exist between the prototype and the model. However, it is thought that an appropriate comparison can have been done about flow velocity by using observation results on the day of steady current direction. As a result of these comparisons, it is thought that we can forecast the actual flow velocity by the result of the hydraulic model test, though the bottom shape influences current direction.

Substances Transported by the Ishikari River and Seasonal Water Quality Changes in the Sea Near the Estuary

Toshihiko YAMASHITA[1], Tsukasa UMEBAYASHI[2], Tsuyoshi SUGANUMA[3], Daisaku SAITOU[4] and Shin-ichi YAMAZAKI[5]

[1]*Graduate School of Engineering, Hokkaido University*
Kita-13jo, Nishi-8, Sapporo 060-8628,JAPAN
Tel +81-11-706-6184
Fax+81-11-706-6184
E-mail y-toshi@eng.hokudai.ac.jp
[2]*Graduate School of Engineering, Hokkaido University*
[3]*Shizuoka Prefectural Government*
[4]*Hokkaido Bureau, Land, Infrastructure and Transportation Ministry*
[5]*Civil Engineering Research Institute of Hokkaido*

In considering environmental management of the coastal sea, it is important to understand the influence of various materials discharged from a river. Although the sea and the river basin often are studied separately, it is necessary to regard both areas as a single unit and to clarify the cycle of water, sediment, nutrients, etc. over a wide area.

Our research targeted the Ishikari River basin and coastal sea area in Hokkaido, Japan, (Fig.1). We made long-term observations throughout the year, to clarify characteristics of substances supplied from the river, their behavior in the coastal sea, and water quality changes. From the results, we considered the influence of discharged materials on the coastal sea environment.

For the river water, changes in year-round concentrations of suspended solid (SS), nitrogen, and phosphorus were measured and analyzed. Data were gathered from 86 surveys, which included the third-largest summer flood in history (approximately 7000 m^3/s). Most of the nitrogen substances were found to be dissoluble, and the nitrogen concentration was high at the early stage of snowmelt season. Most of the phosphorus substances were found to be particulates. The concentration of SS was shown to be proportional to 1.5th power of discharge, and

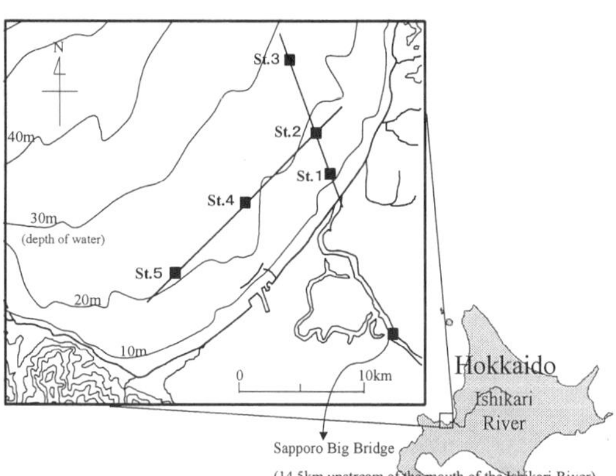

Fig.1 Survey areas

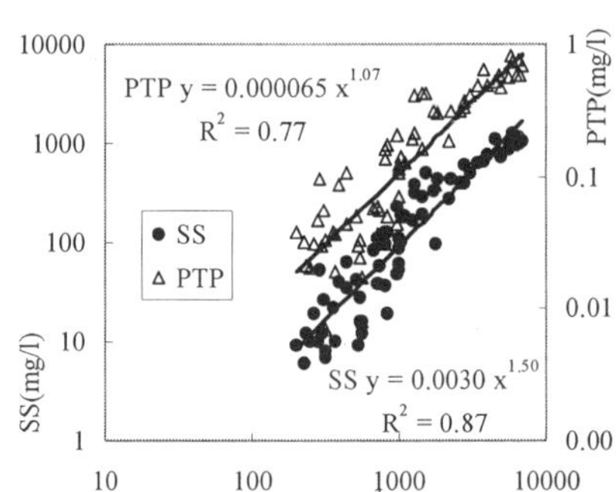

Fig.2 Relationship of discharge and concentrations of SS and PTP

that of particulate total phosphorus (PTP) was shown to be proportional to discharge (Fig.2). Approximately 80% of water quality components discharged in a year flow out of the river during snowmelt season and summer flood.

The concentration of nutrients in the sea within 10 km from the estuary increased during snowmelt flood and summer flood, and was followed by an increase in the concentration of chl-a (Fig.3). The proliferation of phytoplankton is attributed to the great volumes of flood-supplied nutrients. The upwelling of deep, nutrient-rich seawater is thought to greatly contribute to the increase in the chl-a concentration between January and March.

During snowmelt season and summer flood, areas of decreased salt concentrations in the coastal sea roughly coincided with areas of increased chl-a concentrations (Fig.4). On-site flow observation showed that during flood, the Coriolis force induces river water to flow northward, mainly along the coastline. After the river water gradually mixes with seawater, the mixed water flows back southwestward due to the effects of drift and compensation currents. This mixed water, which has been resident in the Ishikari Bay for a relatively long time and includes nutrient salts from the river, is considered to have caused propagation of phytoplankton.

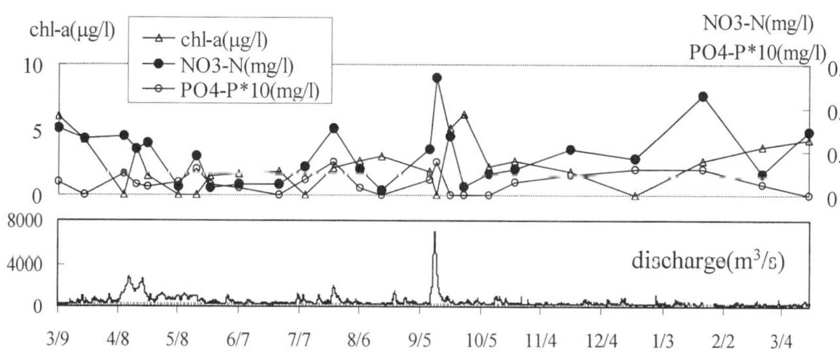

Fig.3 Concentrations of chl-a, NO^3-N, and PO^4-P at St. 2

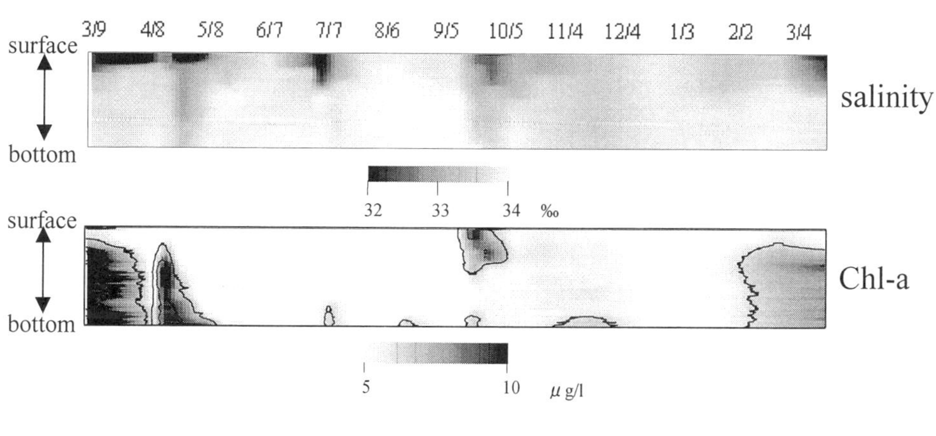

Fig.4 Concentrations of salt and chl-a at St. 5

FLOOD EFFECT ON NUTRIENT DISTRIBUTION IN OPEN COASTAL WATERS OFF THE ABUKUMA RIVER

YONIK MEILAWATI YUSTIANI[1], AKIRA MANO[2]

[1] Graduate student, Civil Engineering Dept., Graduate School of Engineering, Tohoku University, Aoba 06, Sendai 980-8579, Japan. Phone: +81-22-217-7525, Fax: +81-22-217-7525.
E-mail: yonik@potential1.pote.civil.tohoku.ac.jp (*address for correspondence*)

[2] Professor, Disaster Control Research Center, Graduate School of Engineering, Tohoku University, Aoba 06, Sendai 980-8579, Japan. Phone/ Fax: +81-22-217-7512. E-mail: mano@civil.tohoku.ac.jp

Extended Abstract

Nutrients have always existed in natural water, but not at the present excessive concentrations. Excess amounts of phosphorus and nitrogen cause rapid growth of phytoplankton, creating dense population, or blooms. These blooms can lead into many environmental problems such as sunlight penetration reducing, and oxygen depletion. Therefore, for water quality management, including in the area of coastal waters, quantitative evaluation of nutrient condition is required.

Many researchers investigated the transport process of nutrient in water, but most of them were dealing with freshwater body, mostly lake (Asaeda and Van Bon, 1997; Bonnet and Wessen, 2001). Several study were also carried out to study the coastal eutrophication, but almost all directed to bay, lagoon, or semi-closed bay (Arhonditsis, et.al, 2000). Coastal area, which is located facing the open sea, is usually functioned for many purposes such as marine culture site, fishing ground, etc. Yet, the study about nutrient distribution in this area is still scarce.

The Abukuma River discharges its water into the Pacific Ocean at Iwanuma, Miyagi Perfecture. There are several researches conducted in Abukuma River, investigating its water quality. Black mud was found in the coastal sediment near Abukuma River Mouth (Sato, et. al, 1998). Excessive nutrient can be the cause of the depletion of dissolved oxygen concentration and anaerobic condition in this area, which lead into the black mud formation. During flood, suspended load is closely related with the concentrations of nitrogen and phosphorus over a wide area of the river course (Nagabayashi, et. al, 2000). Therefore, further study to understand the nutrient distribution in coastal waters, considering the river flood effect, is important.

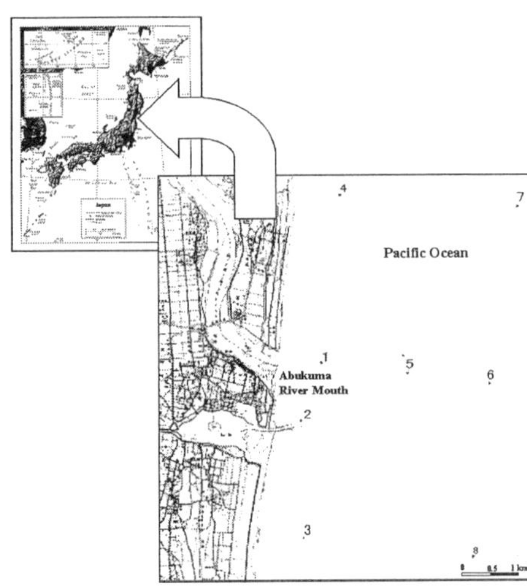

Fig 1. Sampling stations in Abukuma River Mouth.

Field measurements were conducted in coastal waters of Abukuma river mouth on November 2, 2001, representing the usual condition, and on July 20, 2002, representing after flood condition, which caused by typhoon on July 6. Fig. 1 describes the sampling points location. Station No. 1 is located around 1 km from the river mouth, and the farthest stations (No. 6, 7, and 8) are located about 5 km from shoreline. Station no. 5 is located in the middle between station no. 1 and 6. Water samples were taken from 3 different depths every station, representing surface, middle, and bottom layer. Station No. 2 is located near to the port gate.

Abukuma River discharge extremely increased during flood caused by typhoon (Fig. 2). Average discharge was about 100m^3/s, and flood peak discharge reached about 5500m^3/s occurred on July 11, 2002. Small flood was also occurred before sampling. The discharge reached approximately 1000m^3/s on July 16 and 17. Discharge occurred in usual condition was about the average, 100m^3/s.

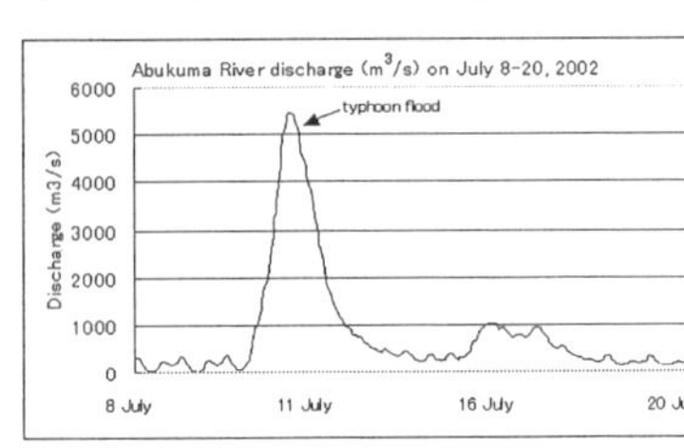

Fig 2. Abukuma River discharge (m^3/s) on July 8-20, 2002.

Fig. 3 shows the concentration of suspended solid (SS) in surface, middle, and bottom layers. SS concentrations after flood condition were still remarkably high compare to those in usual condition, although the measurement was conducted about 9 days after the typhoon. The highest value was measured in bottom layer, reaching almost 0.5gr/L, indicating settlement process of particulate into the coastal bed.

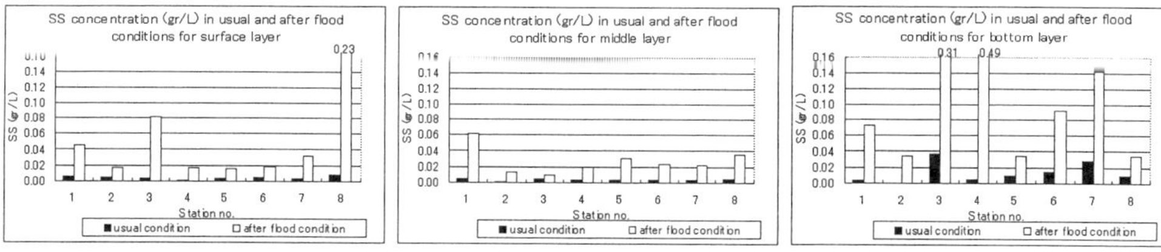

(a) (b) (c)

Fig. 3. Suspended solid concentration measured in usual and after flood conditions, for surface (a), middle (b), and bottom (c) layers.

As shown in Fig. 4, nitrogen did not always have higher concentration after the flood. It appeared that nutrient had been washed away from the river mouth. Surface layer had the highest concentration compared to middle and bottom layer. Nitrification process was likely to be carried out in surface layer. High concentration of nitrogen in bottom layer was probably caused by decomposition process of detritus.

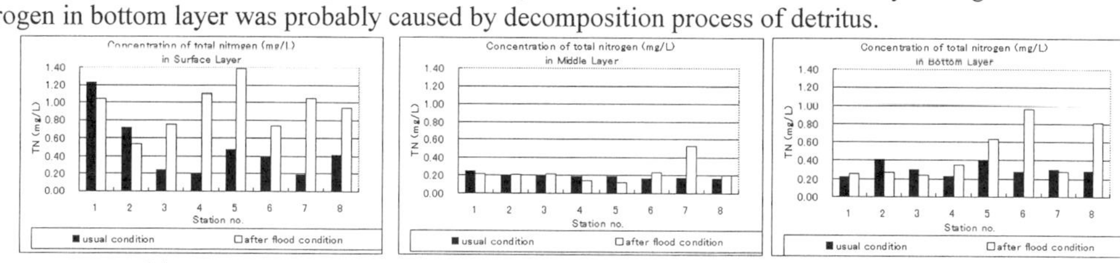

(a) (b) (c)

Fig. 4. Total nitrogen measured in usual and after flood conditions, for surface (a), middle (b), and bottom (c) layers.

Similar to nitrogen, concentration of phosphorus also appeared higher in offshore location (Fig. 5). Compare to those in usual condition, several stations near the coast had lower phosphorus concentration after the flood. River waters contains high concentration of phosphorus in first stage of the flood but then decreased gradually. Comparison between layer shows that phosphorus is easily attached to particles and sunk into the bottom layer of coastal waters.

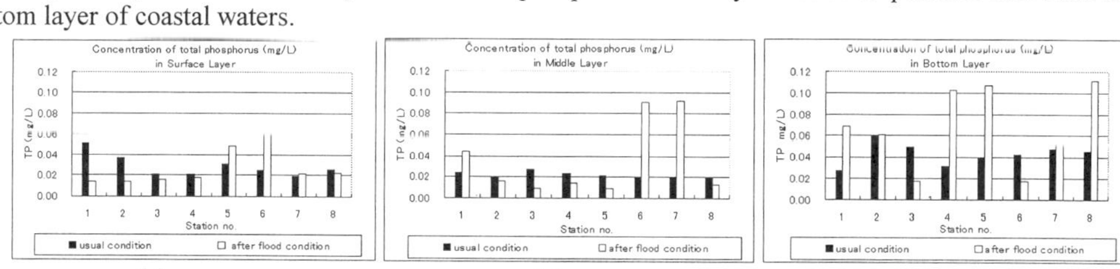

(a) (b) (c)

Fig. 5. Total phosphorus measured in usual and after flood conditions, for surface (a), middle (b), and bottom (c) layers.

Unlike usual condition, which has nutrient distribution decreasing seaward, distribution of nitrogen after flood did not apparently has any relation with river mouth distance. The influence of river discharge fluctuation seems dominantly affected the distribution of SS, and nutrients.

References

Arhonditsis, G., Tsirtsis, G, Angelidis, M.O., Karydis, M., 2000. *Quantification of the effect of nonpoint nutrient sources to coastal marine eutrophication: applications to a semi-enclosed gulf in the Mediterranean Sea*, Ecological Modellling 129, pp. 209-227.

Asaeda, T., Van Bon, T., 1997. *Modelling the effects on algal blooming in eutrophication shallow lakes.* Ecological Modelling 104, pp. 261-287.

Nagabayasi, H., Kobayashi, N., Maki, Y., Mano, A., 2000. *The mechanism of nutrient load and material transportation of flooding river.* Proceedings of the 12th Congress of the Asia and Pacific Regional Division of the International Association for Hydraulic Engineering and Research, Bangkok, Thailand.

Sato, Mano, A., Sawamoto, 1998. *Characteristic of sediment in coastal area of Abukuma River Mouth*, Collection of Coastal Engineering Theses, Civil Engineering Volume 45. pp. 611-615 (In Japanese).

A PREDICTION SYSTEM OF COASTAL CIRCULATION AND SUSPENDED SEDIMENT TRANSPORT ON GUI

Tae Sung Jung[1], See Whan Kang[2], Tae Sik Kim[3] and Sung Gon Kim[4]

Over the past decade dredging has been increased to deepen navigation channel in port and harbor of Korea. Fine sediment resuspended by the dredging activity is transported into coastal waters in the form of plume. The transport results in increases in SS (suspended sediment concentration) above background levels and sedimentation on the seabed. To assess the environmental impact we have to know the behavior of the plume exactly. Numerical models have been developed to understand and predict the behavior of suspended materials in coastal waters.

To estimate the dispersion of the suspended fine sediments occurred by dredging, a SS prediction system for Kunsan coastal waters has been developed The real-time prediction system on GUI (Graphic User Interface), KUSSMOS (Kunsan SS Modeling System), uses a finite element model for calculating depth-integrated tidal currents and a random-walk particle tracking model for calculating SS concentration distribution. The system was applied to forecast the tidal currents and SS concentration in coastal water of Korea. The predicted tidal currents by the tide model showed good agreements with the observed currents. The transport model was verified for the analytic solutions showing good agreements. Improvement of the model accuracy will be continued through the comparison with field data.

The system consists of pre-processing, modeling of tidal currents and SS dispersion, and post-processing. Model parameters such as date, SS source, and settling velocity are entered into the model on Window. The tide model can predict the tide and tidal currents only by inputting the wanted date. he simulated currents can be displayed on PC screen like Figure 1. Water depth is displayed by using color-filled contours. The computed velocities are shown as vectors meaning the speed and direction. The system has many functions such as zooming, static and dynamic visualization, information bar showing water depth, velocity, and SS concentration of the point selected by mouse, and so on. The distribution of sediment particle and the SS concentration are displayed on Window like Figure 2. The system has been developed with C++ language and OpenGL graphic library.

This system will be used to predict the tidal currents and SS concentration. By using the system we can get the information of environmental impacts quickly and conveniently. The system will help in preventing the water quality problem generated by dredging.

1 Professor, Department of Civil & Environmental Engineering, Hannam University, Ojung-dong 133, Daeduk-gu, Daejeon 306-791, Korea. jungts@mail.hannam.ac.kr

2 Principal Researcher, Coastal and Harbor Engineering Research Laboratory, KORDI, Ansan P.O.Box 29, 425-600, Korea. swkang@kordi.re.kr

3 Graduate Student, Department of Civil & Environmental Engineering, Hannam University, Ojung-dong 133, Daeduk-gu, Daejeon 306-791, Korea. kts@mail.hannam.ac.kr

4 Engineer, Hangil I&T, Seoul Venture Town 1003, Yuksam-dong, Kangnam-gu, Seoul 135-978, Korea. ksg@aroad.co.kr

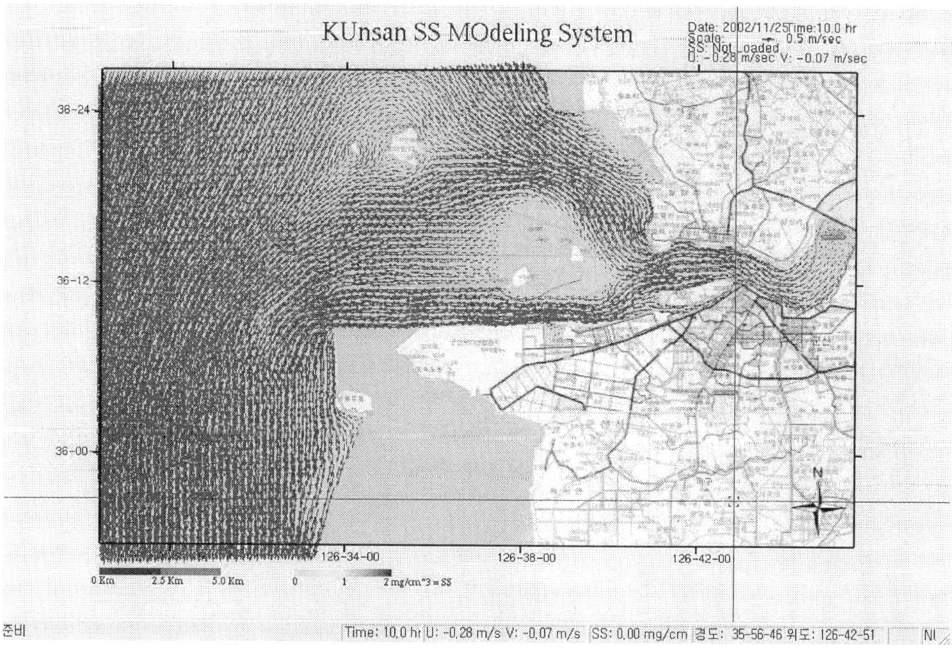

Figure 1. Example of plotting of velocity vectors

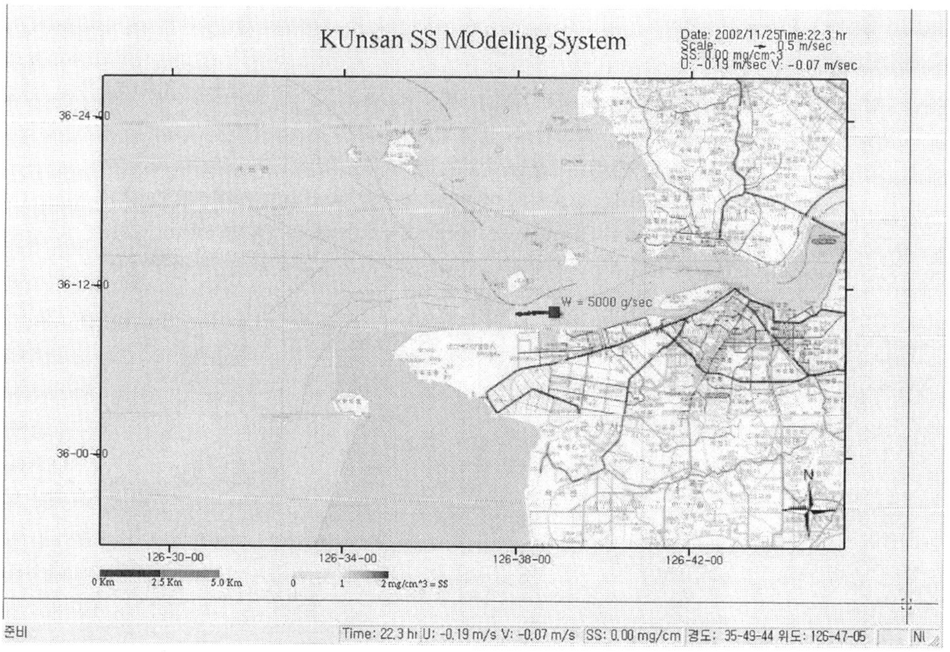

Figure 2. Example of plotting particle distribution and SS concentration

NUTRIENT DYNAMICS IN ARTIFICIAL PEBBLE AND ROCK BEACH

Masato NIKI [1], Tetsuo SAKAI [2] and Hiroyuki NAKAHARA[3]

[1] Graduate Student, School of Civil Engrg., Kyoto Univ., Sakyo-ku, Kyoto, 606-8501, Japan
tel.: +81-75-753-5099, fax.: +81-75-761-0646, e-mail : niki@coast.kuciv.kyoto-u.ac.jp
[2] Prof., Dept. of Civil Engrg. Kyoto Univ.
[3] Prof., Division of Applied Life Sciences, Graduate School of Agriculture, Kyoto Univ.

Objective

When an artificial beach is designed, its safety and landscape are recognized as the important factors from the viewpoint of citizen use. Then, the beach becomes semi-enclosed in some cases, and the seawater exchange rate is low. The micro-topography such as tide-pool, crack and notch in natural beach dose not appear. It is expected that the water quality is bad and the appearing lives are poor. The purpose of this study is to discussed substance cycle especially of nutrients in an artificial beach by using observed data and a numerical model.

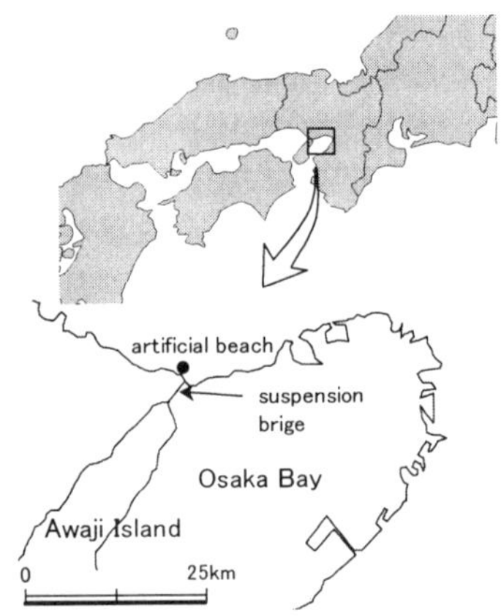

Fig.1 Osaka bay

Observation

Near Akashi-Kaikyo-Ohashi suspension bridge (Fig.1), an artificial pebble and rock beach was constructed 5 years ago. The beach is located west of Kobe, facing the Akashi strait in the Seto Inland Sea. Then in every season an observation has been conducted, in which the water quality and species of appearing algae and animal were observed. The plan view of the artificial beach is shown in Fig.2. The area is about 6ha, and the length is about 400m. A caisson breakwater having two entrances protects the front line. The beach consists of two parts, a deep basin (2.0m depth) and a shallow pool (0.5m depth). The bed in the deep basin is covered with flat stones of about 2m diameter, while the bed in the shallow pool is covered with pebble of about 2cm diameter. In the shallow pool many large rocks are distributed randomly. The observation of water quality has been done at two points in the

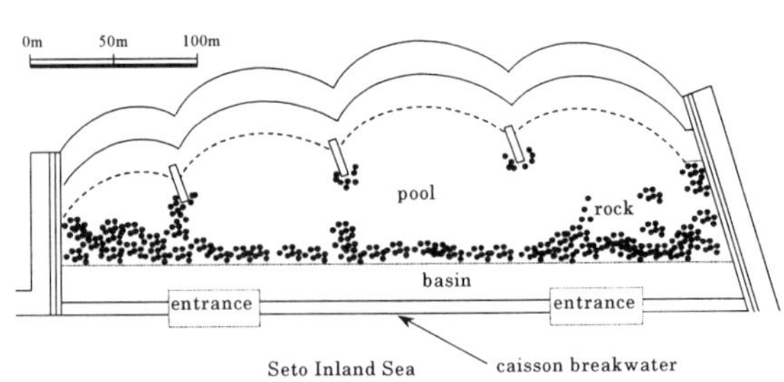

Fig.2 Plan view of artificial pebble and rock beach

shallow pool and one point in the deep basin.

Analysis

At first, the data of the water quality was analyzed. The result shows that the total phosphorus (TP) concentration at the shallow pool increases in summer. The increase is due to the concentration of organic phosphorus (OP) (Fig.3). An additional observation of chlorophyll-a was done. The result suggests the phytoplankton activities. In spring, when the total phosphorus concentration is low, the short life seaweed such as *Ectocarpus siliculosus* appears in the shallow pool. In order to explain this phenomenon, an ecosystem model was developed. It is necessary to explain the competition between the phytoplankton and the short life seaweed. For the water area, the model of Nakada (1993) was applied, and for the sediment area, the model of Nishimura (1998) was applied.

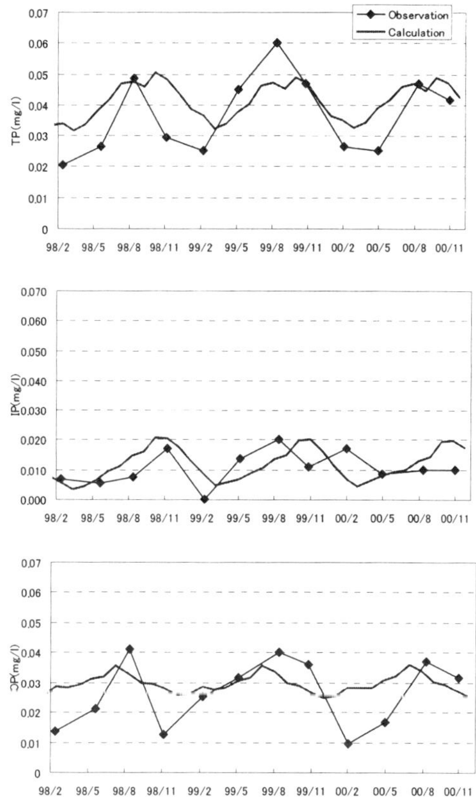

Fig.3 The change of the phosphorus in shallow pool for the first 3 years

Result

Fig.3 also calculated result of the ecosystem model. The calculated result roughly produces the increase of the total and organic phosphorus in summer. From the calculated results, the following things are found. The increase of the organic phosphorus concentration in summer is due to the multiplied phytoplankton. The seaweed is multiplied at beginning of spring, and the phytoplankton is multiplied in summer. This tendency makes a difference between the substance cycles in two seasons. In summer, the phytoplankton occupied 26% of the organic phosphorus concentration. While in winter, the phytoplankton occupied only 12% of the organic phosphorus concentration. In summer, the factors related to the phytoplankton (for example, the zooplankton) are active. In winter, the factors related to the seaweed (for example, detritus in sediment area) are active.

References

Nakada (1993): Environmental fluid pollution, Morikita pub. Co., Ltd., pp.194-220 (in Japanese)

Nishimura (1998): Numerical Analysis for the Effect of Artificial Tidal Flat on Water purification Capacity, Jour. Japanese Assoc. Coastal Zone Studies, Vol.10, pp.137-149 (in Japanese)

MIXING PROCESSES OF DEEP-SEA WATER DISCHARGED INTO STEADY TIDAL CURRENT AND THEIR SEASONAL VARIATION

Masanobu Hasebe*, Takumi Ohyama*

1. Purpose

As the utilization of deep-sea water is extended, the volume of discharged wastewater may become greater than before. Under such circumstance, it is more important to precisely evaluate the environmental impact by the discharged wastewater. In the case of submerged outfall diffusers, complicated interactions between the discharged water jet and ambient tidal flow takes place, which may influence not only the near-field mixing processes but also the diffusion away from the outfall. In addition, the diffusion pattern may significantly vary with the seasonal fluctuations of temperature and salinity in coastal area. Using a three-dimensional hydrodynamic model, the present paper discusses the effects of ambient flow and seasonal variation of coastal environment on the mixing and diffusion processes of discharged deep-sea water.

2. Numerical Model and Condition of Case Studies

The numerical model adopted here is a fully three-dimensional one and is constituted by the continuity and momentum equations, the heat, salinity and dye transportation equations and the k-ε turbulent equations. Hydrostatic approximation is not applied so as to reproduce complex three-dimensional mixing phenomena in the vicinity of the outfall. A finite volume method and SIMPLE scheme for steady state are utilized for numerical formulations. First-order upwind scheme is employed in the discretization of the convection terms.

Figure. 1 shows the configuration of coastal area considered in the computation. A submerged drainpipe with a diameter of 2.2 m is fixed on a constant slope of 1/25 at 20 m deep and is directed to the offshore with an angle elevation at 10 degree. We assumed that the discharged deep-sea water is warmed up at 25 degrees C. after utilizing as cooling water of power plant facilities, and that the amount of discharged water is one million ton per day, corresponding to the discharge velocity of 3.0 m/s. In addition, we assumed that wastewater contains dye so as to give dilution factor as C_0/C, here C_0: initial concentration of dye, C: concentration of dye. Initial distributions of the temperature and salinity are based on field observations, in which the ambient water is stratified in summer but not in winter. Steady tidal current of 0.5 m/s is considered in both seasons.

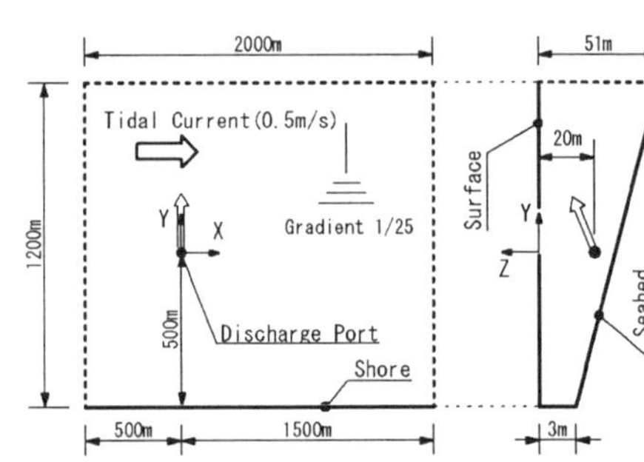

Fig. 1 Coastal area considered in computations

3. Numerical Results and Discussions

The computed vertical trajectories of wastewater plumes in winter and summer are indicated in Fig. 2. The dots represent the point at which dilution factor was minimum in each y-z sectional plane. In the case of winter, the water plume rises to the water surface because its temperature is higher than the ambient water. In contrast, the plume is trapped at a certain submerged level of neutral buoyancy in the case of summer where the ambient

* Institute of Technology, Shimizu Corp. 4-17, Etchujima 3-chome, Koto-ku, Tokyo 135-8530, Japan
Tel: +81-3-3820-5528 Fax: +81-3-3820-5955 E-mail: hasebe@sit.shimz.co.jp, ohyama@sit.shimz.co.jp

water is stratified.

Figure 3 shows the velocity fields in the y-z section at 50m downstream locations (x = 50 m). The gray area indicated in the figures represents the wastewater plume where the dilution factor is fewer than 50. It has already been reported that a pair of vortices is originated around a jet in cross-flow and persists far downstream (e.g., Kelso et al., JFM., Vol. 306, pp. 111-144, 1996). As shown in Fig. 3, such vortices clearly appear in our numerical results of both cases, but their relative positions are different between summer and winter. In the winter case where the plume rises to the surface, the two vortices are arranged horizontally at some distance away from the outfall (Fig. 3b), whereas their relative position is kept to be almost vertical in summer (Fig. 3a). In the former case, furthermore, the two vortices move apart from each other, which causes plume branching. This phenomenon is indicated in Fig. 4, in which the horizontal trajectories of the plumes are plotted. In the winter case, the plume is divided at around 50 m downstream, and the major plume is carried toward the shore side.

The formation of the vortices also influences the dilution of wastewater. Figure 5 shows the dilution factor along the trajectories. At 1500m downstream, the dilution factor attains to 90 in summer but only 50 in winter. As shown in Fig. 3, the ambient water flows horizontally into the plume between the vortices in summer. Such smooth entrainment leads to higher dilution compared to the case of winter, where the seabed disturbs the entrainment of the ambient water.

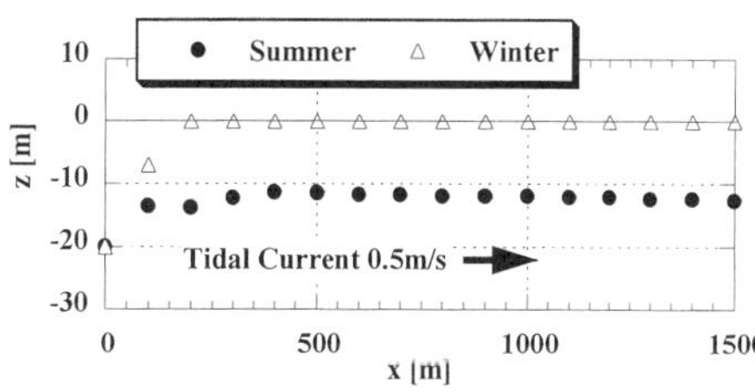

Fig. 2 Vertical trajectories of plumes

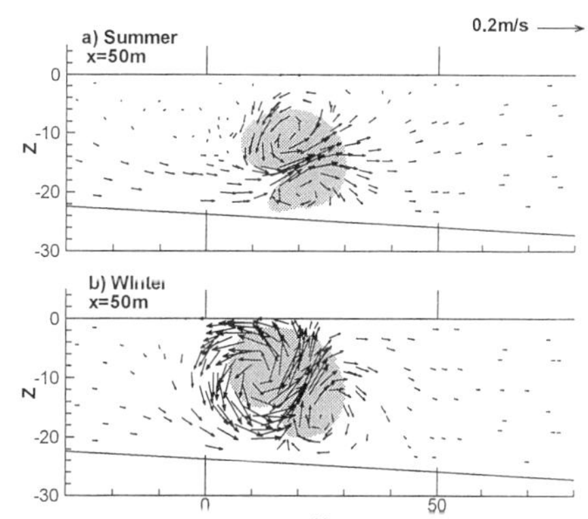

Fig. 3 Velocity fields on y-z section at 50 m downstream

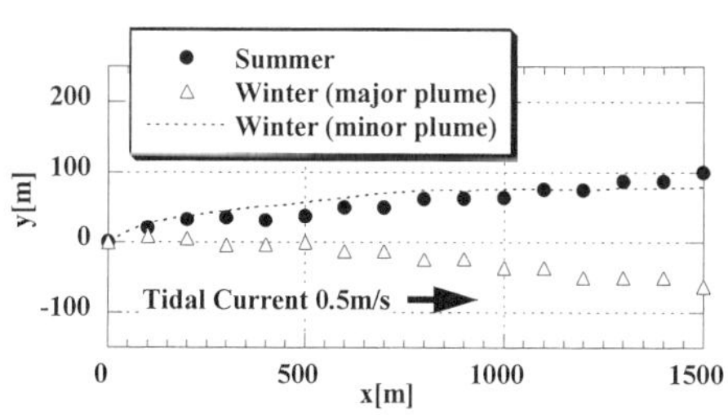

Fig. 4 Horizontal trajectories of the plumes

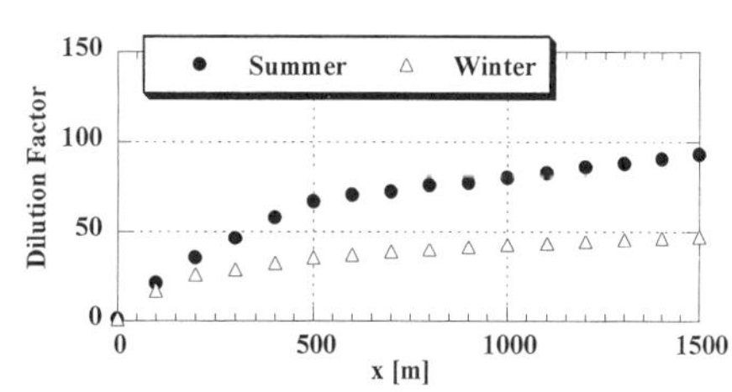

Fig. 5 Dilution factor along the trajectories

4. Conclusions

A three-dimensional hydrodynamic analysis was conducted to investigate the influence of background flow on the mixing and diffusion processes of discharged deep-sea water. Two different seasonal conditions were considered for the coastal environment. Both for cases of summer and winter, a pair of vortices originally organized at the outfall affects not only near-field mixing but also the far-field behavior of the plume. In the summer case where the coastal water is stratified, the ambient water is smoothly entrained between the vertically arranged vortices. For the winter case, in constant, the vortices are horizontally stretched at downstream, which results in the plume branching and the lower dilution of wastewater.

Evaluation of Thermal and Hydraulic Characteristics in a Tidal Flat

By Aiko KAKIZUKA *, Wataru KIOKA **and Kazuhiko TASAKA***

**Corresponding Author; Professor, Dept. of Civil Engineering, Nagoya Institute of Technology, Gokiso-cho, Showa-ku, Nagoya, 466-8555 Japan. (Email: kioka@ace.nitech.ac.jp; Fax: +81-52-735-5503)

*Graduate Student, Dept. of Civil Engineering, Nagoya Institute of Technology

***Under Graduate Student, Dept. of Civil Engineering, Nagoya Institute of Technology

PURPOSE OF THE STUDY

Tidal flat is dynamic systems, responding to periodic tidal inundation and climate change. Thermal and hydraulic structures in a tidal flat are controlled by alternating inundation and evaporation. Tidal flat provides an essential refuge and habitat for a great diversity of plants and animal species, and it is essential to manage the tidal flats to ensure the survival of this critical link in the ecosystem. It is, therefore, necessary to understand thoroughly the physical and biological processes in the ecosystem. In particular, temperature, subsurface flow and salinity processes influence plants and animal life, and they are an important factor determining the background state of health and sustainability of the ecosystem.

In this study, first, we estimate experimentally the bulk transfer coefficients for sensible heat and evaporation over a tidal flat surface, which are the most important parameter for determining thermal characteristics in case of using the bulk aerodynamic transfer formulas. A numerical model of temperature, subsurface flow and salinity response in a tidal flat is examined, and finally, thermal and hydraulic structures are discussed through the numerical calculations.

OBSERVATIONS AND NUMERICAL MODEL

Field observations were conducted in Shiokawa tidal flat located in Mikawa Bay, as indicated in Fig. 1. We observed thermal characteristics and weather conditions on 26 July and 4 October, 2002. Air velocity, air temperature, relative humidity and the tidal flat surface and subsurface temperature (at depths of 0, 0.2, 5, 10, 20, 30cm) were measured. In particular, the gradients of air velocity, air temperature and specific humidity were measured at 2 levels, 3m and 0.26m or 3m and 0.5m above the tidal flat surface, and the tidal flat surface temperatures were measured at the same time. Sea water temperature and velocity were also measured while the tidal flat was inundated by spring tide. The sampling interval of temperature was 10 minutes, the others were 20 minutes.

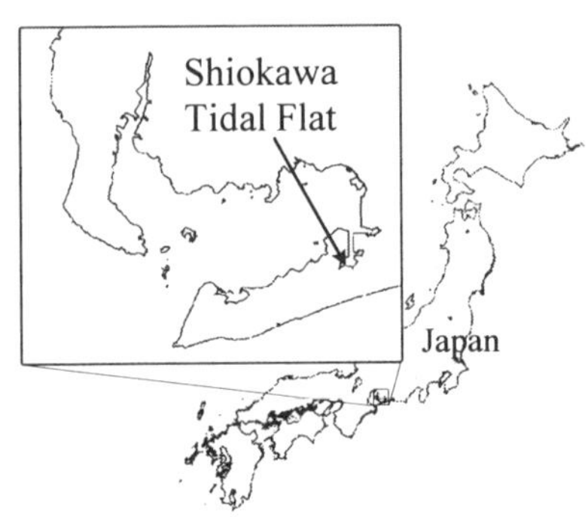

Fig. 1 The location of field observation

Monin-Obukhov's similarity theory is adopted to estimate the bulk transfer coefficients of momentum, sensible heat and water vapor. The coefficients vary depending on the atmospheric stability. The data in very stable and unstable conditions were excluded, and the coefficients in near neutral conditions were estimated.

The tidal flat equations, the field equations for flow, salinity transport and heat transport in the tidal flat flow domain for both unsaturated and saturated regions are used to evaluate numerically the thermal and hydraulic process. The dependent variables are pressure head, temperature and salinity concentration. The tidal flat equations are nonlinear and solved numerically in one dimension, allowing fluid density to vary with salinity, pressure and temperature. This model calculates the surface heat flux and the water thermal structure. The surface temperature of the tidal flat is given by the balance of the heat transfers. When tidal flat surface is exposed, sensible heat transfer and latent heat transfer are calculated using the bulk aerodynamic transfer formulas with the bulk transfer coefficients.

RESULTS AND DISSCUSSIONS

Fig. 2 shows the observed bulk transfer coefficients $C_{H.1}$, $C_{E.1}$ versus wind velocity for a reference height of 1 m. The average values of $C_{H.1} = 1.7\times10^{-3}$, $C_{E.1} = 2.3\times10^{-3}$ are obtained from this figure. From the figure, it may be seen that both coefficients $C_{H.1}$, $C_{E.1}$ increase with increasing wind speed for U_1<3.4 m s^{-1}, and these are nearly constant for U_1>3.4 m s^{-1}.

Fig. 3 shows temporal variations of temperature distribution calculated using the bulk transfer coefficients while the tidal flat surface is exposed. The tidal flat surface temperature is given water bottom temperature while the tidal flat is inundated. The computed temperature is lower near the surface and this fluctuation is smaller as compared with the observed distribution.

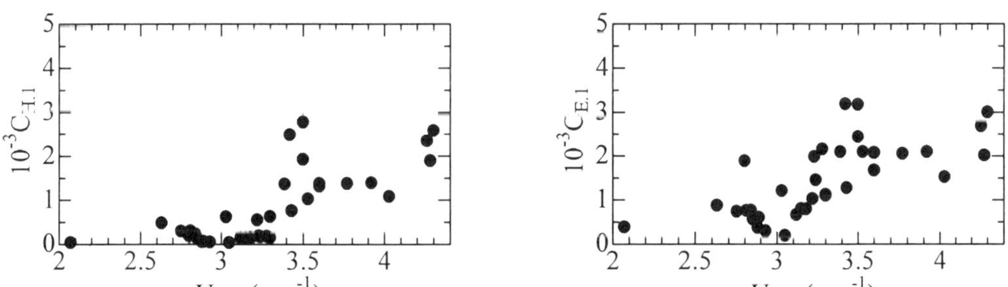

Fig. 2. The observed values of the bulk transfer coefficients for sensible heat (left), evaporation (right)

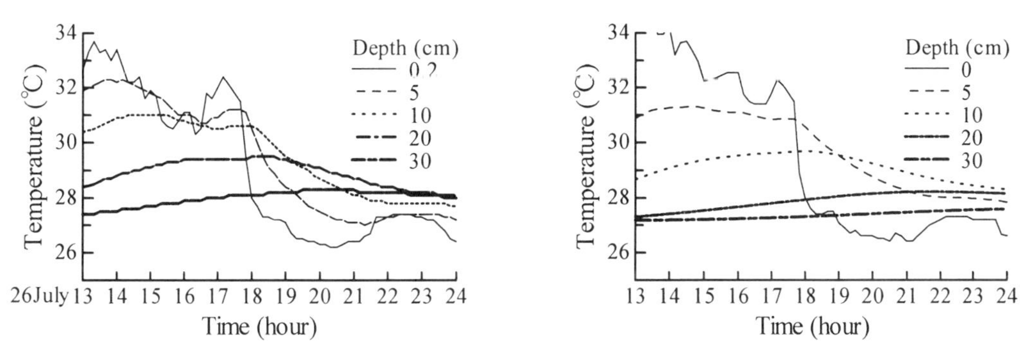

Fig. 3. Temperature distribution in tidal flat, (left) observed, (right) calculated

Numerical Prediction of Physical Environmental Change After Construction Of Offshore Airport Planned in Tokyo Bay

Jun SASAKI[1], Takayuki HAGIWARA[2] and Masahiko ISOBE[3]

Introduction

In response to a recent increase in plane's passengers in Tokyo metropolitan district, extension of Haneda Airport and furthermore construction of a new offshore airport in Tokyo Bay, the third airport in the capital sphere, is planned to solve future shortage of slots for planes. On the other hand, there is growing concern in conservation of environment in the bay. Recent increase in the knowledge of the importance of shallows and tidal flats for the conservation of water quality and marine ecosystems, reclamation of these areas has started to be avoided considering its impacts on the bay environments. Development in the offshore areas in the bay, however, has not necessarily become a social environmental issue because its environmental impacts are still unclear.

In the present study, we investigated the impacts of construction of a new offshore airport on physical environments in the bay. Nowadays, the concept of nowcasting is proposed, and as a result, a number of numerical models for nowcast/forecast system have been developed, which reproduce a continuous monitoring dataset in the field (*e.g.* Sasaki et al., 1999). Nonetheless, a general method friendly to the public has not been established yet, which can describe the essence of environmental impacts. Accordingly, we proposed a methodology to describe change in physical environment.

Methodology

Physical environment in bays is characterized by tidal, density, and wind induced flows, which have multiple spatial and temporal scales. Thus, prediction should be made considering not only seasonal changes but also the effects of short-term events due to meteorological disturbance. We, therefore, developed a prediction technique that includes the effect of short-term events as well as seasonal variation.

Prediction results of physical environment are often represented by velocity vector maps for residual current. It is, however, not easy to represent the effect of short-term events or to interpret them for even researchers and engineers. As a remedy for the problem, we proposed a sedimentation index applying the Euler-Lagrange method, tracking of the movement of particles (*e.g.* Yanagi, 1999). The most important phenomenon considered after the construction of an offshore airport would be appearance of stagnant regions and its resultant pollution of sediment and hypoxia due to accumulation of organic material. The organic material is produced in the upper layer in the bay due to photosynthesis of phytoplankton and land loads discharging through the rivers. We introduced particles with the sinking velocity of the order of 1m/s that represents the particulate organic material. The particles were given in the surface water uniformly in space and time. The sedimentation index S_d is defined as a sum of the particles deposited on the bottom at each horizontal grid, normalized by the total number of input at each horizontal grid. Furthermore, difference in the sedimentation index between the present and the future runs is defined as follows,

$$\delta S_d = S_{d\text{future}} - S_{d\text{present}}$$

Preliminary considerations were made to confirm the generality of the present method, such as the variation of S_d among years due to difference in meteorological and oceanographic conditions. Fig. 1 shows the sedimentation index S_d in the year of 1994, and δS_d between the year of 1994 and 1996. The magnitude of overall difference in them was within the value of 10 of the sedimentation index, and thus, δS_d larger than the value of 10 could be considered to be the resultant change in physical environment caused by construction of an airport.

Results and Conclusions

Firstly, we considered the effect of the extension of the Tokyo International Airport (Haneda Airport). Fig. 2 shows spatial distribution of δS_d for the case. It is found that particles were accumulated more around the southwestward side and less around the northeastward side of the extension than before. This is mainly caused by the interruption of the residual current having directed from southwest toward northeast around the extension. The effect is straightforward and easy to understand. In addition, it is possible to compare the effect of each alternative quantitatively by applying the present method.

[1]Associate Professor, Yokohama National University, Fax: +81-45-348-4565, E-mail: sasaki@cvg.ynu.ac.jp
[2]Graduate student, The University of Tokyo, E-mail: hagi@planner.t.u-tokyo.ac.jp
[3]Professor, The University of Tokyo, Fax: +81-3-5841-8503, E-mail: isobe@k.u-tokyo.ac.jp

Secondly, we applied the method to several cases of the construction of offshore airports planned in Tokyo Bay. Change in total water exchange rate through the mouth of the bay was predicted to be around a few percents for each case. Significant influence was addressed in the change in variation of accumulation areas of the particles. For instance, Fig. 3 shows spatial variation of δS_d after the construction of a large off-Kawasaki-Yokohama airport. The result shows that accumulation is accelerated around the southwestward of the airport as well as the area between the extension of Haneda airport and the off-Kawasaki-Yokohama airport. Furthermore, the annual mean residual current field in the bay is divided into two ring currents; one is in the inner part of the bay and the other is in the central part blocked by the airport. Inside each ring current, accumulation is slightly accelerated and the marginal area of each current, decelerated. Thus, construction of an offshore airport would influence not only a local environment around the airport but also a bay scale environment through a significant change in residual current field.

References

Sasaki, J and M. Isobe (1999): Development of a long-term predictive model of water quality in Tokyo Bay, Estuarine and Coastal Modeling, Vol. 6, ASCE, pp. 567-580.

Yanagi, T (1999): Coastal Oceanography, Terra Scientific Publishing Company, Kluwer Academic Publishers, 162p.

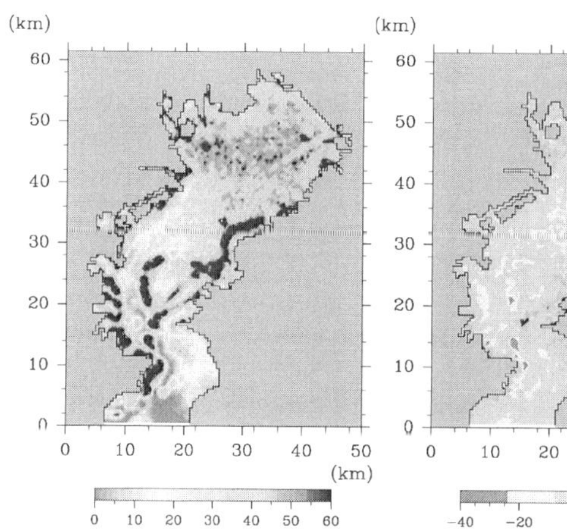

Fig. 1: S_d during 1994 and δS_d between 1994 and 1996

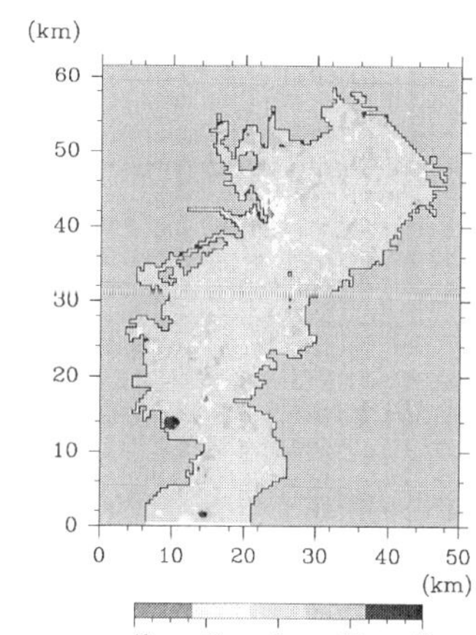

Fig. 2: δS_d after extension of Haneda Airport

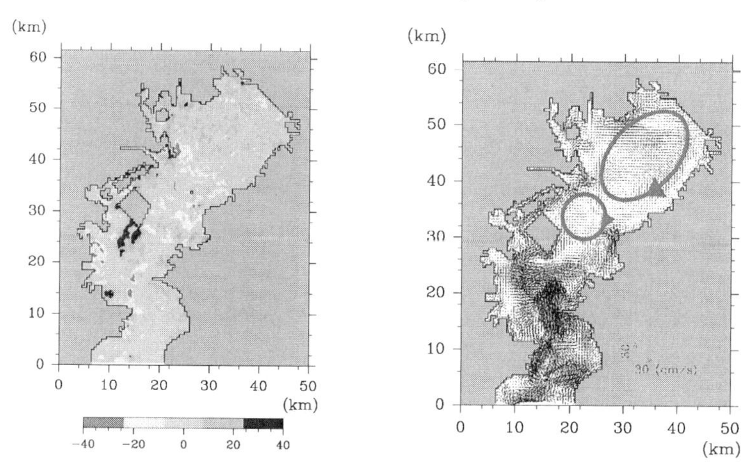

Fig. 3 δS_d and residual current field after construction of an off-Kawasaki-Yokohama airport

EVALUATION OF THE IMPACTS OF HARBOUR ENGINEERING IN ANIBARE BAY, REPUBLIC OF NAURU

Russell J. Maharaj, SOPAC Secretariat, PMB, GPO, Suva, Fiji Islands; E-Mail: rossi@sopac.org.fj ; Fax: (679) 3370 040; Tel: (679) 3381 377

Abstract: This paper presents the results of a coastal engineering, and environmental evaluation, of design parameters, for a fishing boat harbour facility at Anibare Bay, Republic of Nauru. The Government of the Republic of Nauru (RON) requested the South Pacific Applied Geoscience Commission (SOPAC) Secretariat, Fiji Islands, to evaluated the development, and comment on the impacts on the harbour facility on sedimentation and erosion of the adjacent coastline, and also, advise on maintenance of the constructed facility at the development site. An in-country site visit was made and a follow-up visit was done, so as to gain further information and insight on response of the beach-coastal system to the engineered facility. It also provides additional and new time-series observation data .

This project is executed on behalf of the Nauru Fisheries and Marine Resources Authority (NFMA), Government of the Republic of Nauru (RON). Following the visit, and after briefing of officials of Nauru Fisheries and Marine Resources Authority, SOPAC Secretariat indicated that it would provide appropriate guidelines, for shoreline management in the form of technical reports, which will contain recommendations for appropriate shorefront development, in particular guidelines for preparing EA and EIA for the coastal development. Key engineering and environmental issues will also be discussed and highlighted in the text as well as guidelines for preparation of a full EIA. This project was funded and executed under SOPAC Task NR 99.045.

Anibare District is located in the Eastern part of Nauru. The development site is located in Anibare Channel, on the southern half of the Anibare Bay. The developed coastline is just about 700 m North of the Mennen Hotel facility. RON received bi-lateral funding from the Government of Japan (GOJ), for the development of a port and harbour facility at Anibare Bay. The development site chosen is a former cut and excavated channel in Anibare Bay, which was also used for fishing activities on the East coast of the Island. The purpose of this harbour facility is to provide RON with additional docking, launch and berthing facilities to support their offshore tuna fishing fleet.

Coastal infrastructure and development, like this one, on a small island state, like RON, can impact adversely on coastal processes. In addition, deleterious impacts can also result along the natural and built shorelines at the site, and in the vicinity of the development. Consequently, it is important to monitor development activities and their impacts, both positive and negative, on the coast. In this way, negative and deleterious impacts of the development can be captured immediately and remedied, where possible, in a timely and appropriate fashion. Further, information gained from such monitoring can be useful for future coastal development and engineering design of similar and related facilities on the island. Since it is the intention of RON to develop another harbour facility on the western side of the island, information gained from this monitoring study can be crucial to optimum planning and design of that new facility The present study was set-up for this very purpose, with the objective of advising RON with timely and appropriate advice on the impacts of the harbour facility and strategies for addressing these impacts. Since RON is a small island, with limited land area, the loss of any land or coastal property represents a significant one. In addition, developing economies, like RON, can be seriously affected by damage to or loss of civil infrastructure and residential facilities from "silent" natural hazards, like coastal erosion. Therefore, addressing these types of coastal development problems is of paramount importance to Nauru's coastal communities and for the livelihood and well-being of residents and the nation as a whole. In the context of sustainable natural resource management, coastal development and it's impacts on coastal erosion is also an important process that need to be addressed, if future generations are to derive benefits from coastal resources, like beaches and reefs.

The harbour development plans were consulted, for assessment of dimensions and quantities of the various components of the Anibare Harbour facility (Tetra, 1999). Environmental (including geological and geotechnical) information and data were collected during site visits to the eroding coast. This included wave and littoral information, beach sediment characteristics, erosion characteristics, and documentation of damage to any critical facilities and infrastructure. Beach sediments were described according to American Society for Testing Material guidelines, which are in accordance with the Unified Soil Engineering Classification System. Rock classification and descriptions are based on ASTM guidelines and conforms to the international criteria set out by the International Society for Rock Mechanics and the Engineering Group of the Geological Society of London guidelines. Beach profiles were measure with a Sokkia Automatic level, using a Sokkia survey staff, a Brunton compass and a hand-held, Garmin Global Positioning System (GPS). Erosion scarps were measured with these surveying instruments. Steel lamp poles along the main coastal road were used as benchmarks for surveying beach profiles. Beach profiles elevations were corrected to the Nauru's Chart Datum, by precise levelling loops, using the Sokkia Automatic Level, from surveyed benchmarks at the development site. Several benchmarks, which were established for construction of the harbour, were used in this assessment. Wind speeds and direction was measured with a digital, hand-held anemometer.

Positions in the field were determined with a Garmin hand-held GPS. Information on waves was also obtained from Tetra (1998) and vertical, colour, 1992 stereopairs/aerial photographs. The dolomite limestone rock strength was measured in-situ, with an ELE Schmidt L-Type Hammer, while concrete strength was measured with an ELE Concrete Testing Hammer. Rip-rap/armourstone description and revetment evaluation are based on current international design criteria. Design analysis has been produced after Numerical Analysis with International Institute for Hydraulic Engineering, Delft University of Technology (IHE-Delft), Coastal and River Engineering Support System (CRESS) and the U. S. Army Corps of Engineers, Automatic Coastal Engineering System (ACES). Computations of rip-rap dimensions, armourstone stability and wave run-up are based on numerical equations of van de Meer and are modifications and improvements of the typical Hudson formula. Computations are therefore for random wave attack, more typical in the natural environments.

In summary, the coast is part of an emergent, Holocene reef-carbonate system, with the beach being comprised entirely of carbonate sediments developed on phosphate-rich, cavernous, dolomite limestone bedrock. The coastline is partly rocky with classic karst limestone pinnacles found throughout the bay. The reef is a coral dominated system and is narrow and well-flushed, with many closely-spaced reef channels. The coastline at Anibare Bay is an active and dynamic one. The relatively coarse admixture of abraded sand and gravel and highly abraded karts pinnacles testify to this. The beach at the development site is moderately steep and has experienced erosion in the recent past and show sign of recent and current erosion, with fresh erosion scarps. The harbour development appears to have increased this erosion at the localized level, at the adjacent, undeveloped coastal segments. The topographic elevation of the coastal land areas are relatively low with respect to CDL and MSL and under EHWST or during windy and low pressure systems, when large (3 m +) waves approach shore, from the East, the beach, coastal road and adjacent areas can be easily overtopped. The relatively narrow and almost flat backreef and reef flat, together with the numerous closely-spaced reef channels make it almost impossible to dissipitate significant wave energy and prevent overtopping during these conditions. In addition, the almost featureless backreef cannot trap sediments entrained in longshore currents and therefore, sediments removed from the local areas can be completely lost from the coastal system in Anibare Bay.

The harbour development acts as a large groin, which breaks the continuity and smoothness of the concave Anibare Bay. As a result it interrupts Southerly longshore currents and will cause erosion of downdrift areas (to the South). By its very nature, the harbour also acts as a headland, protruding into the bay. As a result of this "headland-like" morphology, wave diffract around it, and cause much agitation and disturbance of beach sediments immediately North and South of the harbour. Therefore, the facility can cause erosion on both the North and South aspects of the harbour. With respect to the concrete breakwaters built to protect the harbour and mooring basin, these structures are already overtopped by 3 m high spring tide waves. While the design and construction firm (Tetra, 1999) indicated that a 50-year, of 5.34 m design wave was used to design the harbour, the fact that a 3 m high offshore wave, after undergoing decay over the reef crest, can overtop the main breakwater, raises some concern as to what acceptable risks were allowed/selected for this facility.

To that end, numerical analysis was performed for the facility. Numerical computations were done for only the design (5.34 m) and average wave heights (3 m) chosen for the harbour facility and under EHWST conditions, as these represent extreme and common wave height respectively. The reef crest was modeled as a dynamically stable, submerged breakwater, situated below MSL and 1.5 m below CDL. In addition it has a relatively straight offshore slope of 25°-40° and the backreef is unusually flat and smooth for a reef environment, with a constant seaward grade of less than 1° (see Section 3.10). Wave transformation was done using numerical equations proposed for a dynamically stable submerged breakwater.

For a 5.34 m, 1-in-50-year wave height proposed by Tetra (2000), wave period of 10 sec, a reef crest of about 12 m wide, a forereef slope of 1: 1.2, in fore reef water depth of about 10 m (seaward of the reef crest), under EHWST (freeboard level of -4.14 m; with the reef crest at –1.5 m below CDL) and under Easterly (the modal) wave approach, the transformed wave computed was 3.7 m. This wave transformation corresponds to 30 % decay in wave height across the reef crest. Interestingly enough, this is consistent with a wave that will run-up and overtop the coastal road under EHWST, and also that, which has been observed by residents, at the site, for more than a decade. For such a wave height, the size and density of boulders required to maintain stable structural conditions under the design/ and transformed wave, in the backreef, and on parts of the facility (e.g. on the groin and spending beach rip-rap), may be larger or more dense than those specified. For the local dolomite limestone used at the harbour site, which is dense, but porous, and with an estimated unit weight of 2650 kg/m³, and under a 3.7 m transformed wave, during EHWST, the nominal stone diameter required would be about 1 m. This is about 3-5 times the diameter specified by Tetra (2000). Tetra's (2000) diameter is 500-1000 kg/pc or about 18-35 % of that computed by the author (assuming a rock density of 2650 kg/m³). If a 3.7 m high transformed wave impacts on the vertical seaward face of the main breakwater, under an EHWST, overtopping of the structure will be about 0.457 m³/sec. At mean water level (1.57 m above CDL) overtopping will be 0.13 m³/sec.

In addition, the navigational channel will not cause significant wave decay, as would the adjacent intact reef crest. This is because it was dredged to –2.5 m below CDL (1 m deeper than the level of the existing reef crest), with a 30 m wide funnel-like entrance that narrows to 20 m. The freeboard height above EHWST level is therefore –5.14 m, estimated from Tetra (2000) designs, while these award channel slope is 1: 16. If a 5.34 m high wave break over the navigational access channel, the transformed wave will be about 4 m high or about 0.6 m higher than the transformed wave over the intact reef crest (3.4 m). This will then run-up on the spending beach rip-rap, and enter the mooring area, causing choppy conditions to develop within the harbour. Despite the fact that there is a spending beach of rip-rap, some reflection and refraction will occur on the landward side of the access channel.

It is therefore important and necessary to cater for routine and regular maintenance of the spending beach rip-rap so as to ensure that any rip-rap dislodgement, erosion or damage is repaired. Some numerical analysis was also performed for average wave climate for the same reef architecture and harbour design. For an average offshore non-broken wave height of 3 m (within a 1-year return interval computed by Tetra, 2000), with a 6 sec period, the transformed wave on EHWST, over the same reef morphology will be much smaller, at 2.4 m or 20 % decay. For such a transformed wave, the required rip-rap for the groin, under EHWST, should be at least 0.65 m diameter or 741 kg (assuming a rock density of 2650 kg/m³). The rip-rap required for stability at the spending beach should be 0.3 m or 60 kg, also assuming a rock density of 2650 kg/m³. Overtopping of the main breakwater by a 2.4 m transformed wave will be about 0.07 m³/sec, smaller, but nevertheless, noticeable.

Management of built shorelines, like those in Anibare Bay, is a dynamic process based on assessments and re-assessments and therefore, cannot be pursued by targeting specific activities, continuously, through time. Shoreline management strategies must be developed to reflect the current and future/forecasted needs. Forecast should also be for the short-term, medium-term and long-term, and therefore, strategies must be developed which reflect these changing needs through time.

KEYWORDS: Anibare District, Republic of Nauru, port and harbour engineering, coastal development, coastal protection, erosion assessment, coastal monitoring.

ENVIRONMENTAL CONDITIONS AND STRATEGIES FOR SUSTAINABLE MANAGEMENT OF CHILIKA LAKE, INDIA

Pravakar Mishra[1*], Pratap Kumar Mohanty[2] and Takashige Sugimoto[1]

1. Ocean Research Institute, University of Tokyo, Tokyo, Japan,
2. Department of Marine Sciences, Berhampur University, Orissa, India

* Corresponding author: pravakarmishra@yahoo.com,

Abstract:

Chilka lake (19°28'-19°54'N; 85°05'-85°38' E) along the eastern seaboard of India is one of the largest brackish water bodies in tropical Asia supports more than 150,000 people's livelihood. Its geomorphic, physiographic, hydrographic and faunistic features conform to the characteristics attributed to a 'typical' lagoon. Since 1981, the Government of India has declared it as a Ramsar site and this year the lake received the "Ramsar Award" for successful restoration program monitored by different federal agencies.

The lake is formed by accumulation of coastal sediments in the beaches of the barrier berm spit over a period of time. The water-spread area of the lake varies between 1165 to 906 km2 during the monsoon and summer seasons, respectively. Until late 2000, a narrow intricate channel of 35km long and 150m wide was connected to the main body to the Bay of Bengal. The tidal fluctuation inside the lagoon is small (around 0.2m). Though tides of about 0.9 -2.4 m occur along this coast, it is not strong enough to override the long channel and initiate an active circulation inside the lake. Huge amount of coastal sediment moves along this coast during the monsoon season (April to September) results in blockage of the channel mouth or if carried through the channel, leads to siltation. All these factors lead to spatial stratification of the lake. The lake is broadly classified into four/five major ecological sectors on the basis of its hydrographic distribution. Starting from early seventies, the lake had undergone rapid transformation such as shrinkage of total lake area, high siltation, eutrophication, biological sedimentation resulted from enormous weed growth and depletion of fishery resources. All these problems attributed to closer of its opening to the Bay of Bengal by natural processes, poor circulation pattern inside the lake and huge influx of fresh water from tributaries of a major river system. As a part of the restoration program and to enhance the lake fishery conditions, several developmental programs are undertaken in the last couple of years and are in pipeline. A new mouth was cut across along the outer channel to the Bay of Bengal on the 23rd of September 2000 in order to improve the water quality and circulation pattern and overall fishery condition of the lake.

Events in the lake are some of the best examples of interaction between natural processes, human impacts & interventions, and socioeconomic conditions of coastal population and government policies. In the two succeeding years of the initiation of the restoration projects, the authorities claim that the fishery of the lake has rebound unexpectedly and it is a complete success that the lake has been restored to its original condition. However, unfortunately, it is observed that the entire lake has been changed from brackish to a more saline condition initiating sudden disruption in the ecosystem and prompting damages to its pristine ecology and provides a classic example of the effect due to human alternations and interference in the ecosystem. Some of the ongoing ecological problems have created controversial debates among environmentalists, administrators and fisherman communities to stop the developmental work in the lagoon. It is also observed that the lake is over-exploited in the last two years in the name of restoration of fishery of the lake and wrong policies adopted to boost the economy of the area without proper planning is the principal reason in deterioration of lake's environment. This paper highlights some of the major changes in the environmental conditions of the lake in the last two decades based on our long-term field survey and Indian Remote Sensing (IRS) satellite data. A comprehensive assessment of the integrative physical, chemical and biological factors crucial to the functioning of the lake's ecosystem is reviewed and discussed. Some of the fundamental questions related to the pollution and fisheries of the lake are examined. At last, we suggest that involvement of local communities and development of community based hatcheries or fishing zones are highly essential for a sustainable management of fisheries of the lake.

MECHANISM OF RAPID CHANGE FROM NATURAL TO ARTIFICIAL COAST IN JAPAN – THE EXAMPLE OF NODE COAST IN KUJYUKURI COASTAL PLAIN

Takaaki Uda[1], Satoquo Seino[2] and Toshiro San-nami[3]

[1] Dr. Eng., Executive Director, Public Works Research Center, 1-6-4 Taito, Taito, Tokyo 110-0016 Japan. Tel +81-3-3835-3609, Fax +81-3-3832-7397, mail uda@pwrc.or.jp

[2] Univ. of Tokyo, 3-8-1 Komaba, Meguro, 153-8902 Japan

[3] Coastal Engineering Laboratory Co., Ltd, 1-22-208 Wakaba, Shinjyuku, Tokyo 160-0011 Japan (Tel +81-3-3359-7821)

1. INTRODUCTION

In the analysis of the shoreline changes on a coast, where continuous longshore sand transport is interrupted by structures and the downcoast is eroded, comparison of the shoreline changes based on the aerial photographs is often used and useful results have been obtained on several coasts. However, the analysis of the shoreline change does not necessarily lead to sufficient understanding of the events occurring on the coasts and rather direct comparison of aerial photographs is useful. The disappearance of the natural coastal sand dune and sandy beach in Japan has been caused not only by the recession of the shoreline due to decrease of longshore sand supply, but also by the excess advance of the coastal forest as a part of the land use change behind the coastline. This study aims at investigating this issue by using the method of direct comparison of aerial photographs taking the Node coast as the example. Finally it is concluded that in order to prevent further devastation of the coastal environment, overall comprehensive management in the widespread coastal zone including the coastal forest is needed.

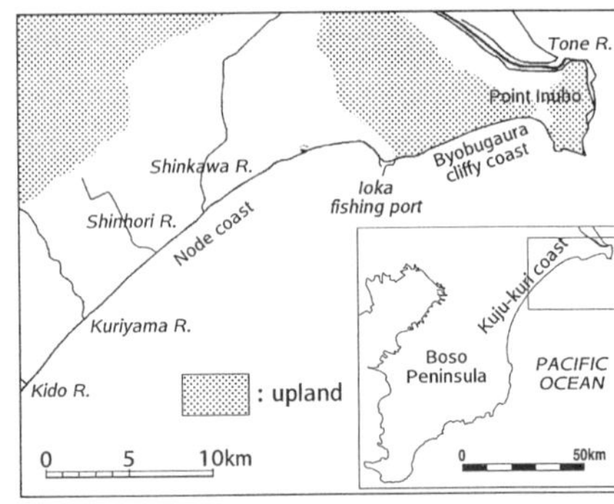

Figure 1. Location of Node coast in Kujyukuri coastal plain and Byobugaura cliffy coast.

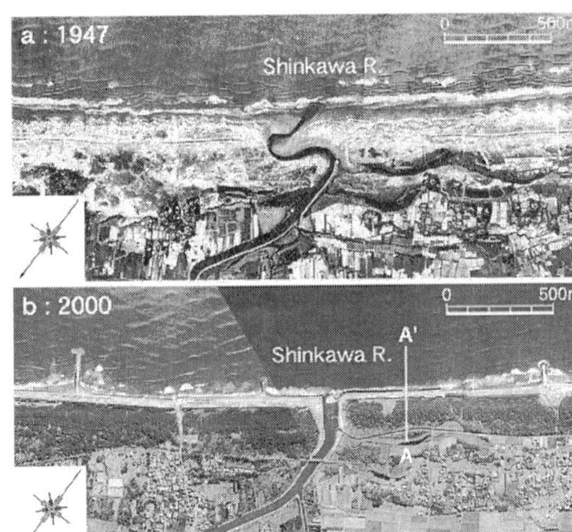

Figure 2. Aerial photographs of Node coast

2. CHARACTERISTICS OF NODE COAST

The Node coast is located at the northeast part of the Kujyukuri coast as shown in **Fig. 1**. East of this coast there is a sea cliff named Byobugaura extending over around 10km length and of the height of 30m through 50m. This sea cliff has been retreated about 0.75m/yr and sand supplied from this cliff fed the Kujyukuri coastal plain of 60km length and 10km width. On the Node coast, southwestward longshore sand transport dominates, carrying sand supplied from the sea cliff to the central Kujyukuri coast through the Iioka coast located on the down drift side of the Iioka fishing port located at the west end of this cliff.

In order to investigate the overall changes of the Kujyukuri coast, including the Node coast, the shoreline changes were investigated through the comparison of aerial photographs. On the Node coast the shoreline had retreated around 40m since 1970, and the shoreline recession was severe in the past, but in recent years the shoreline seems to approach the stable condition, showing there are no environmental and shore protection issues. In reality, however, severe changes in coastal environment continue at present, which can not be understood only from the shoreline changes.

Although the longshore sand transport is flowing from northeast to southwest as a whole in this area, wave dissipating breakwaters had been built along the foot of the sea cliff to prevent erosion during several decades at Byobugaura. As a result, around 90% of all the entire stretch of the sea cliff has been protected against wave erosion in recent years. At Iioka fishing port, the breakwaters of the fishing port have been built during 38years since 1964. In the age without the breakwaters of Iioka fishing port, which continued during around 6,000 years since the Jomon transgression of the sea, sand has been transported through the present location of Iioka fishing port toward the central part of the Kujyukuri coast, forming the coastal plain. However, in order to enhance the safety and efficiency of the fishery, long breakwaters were extended and as a result the movement of sand supplied from the sea cliff toward Kujyukuri coast became difficult. Sand supply from the sea cliff was originally decreased by the protection works along the sea cliff, and in addition to this, longshore sand movement was blocked by the long breakwaters. This is the cause of the beach erosion on the northern Kujyukuri coast.

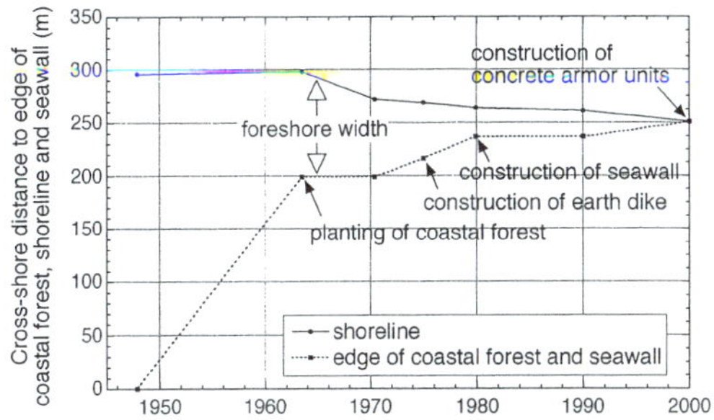

Figure 3. Temporal change in cross-shore distance from the reference point to the edge of coastal forest and shoreline position at the Node coast.

3. DIRECT COMPARISON OF AERIAL PHOTOGRAPHS

Long-term shoreline changes of the Node coast between 1947 and 2000 were analyzed through the direct comparison of aerial photographs. **Figure 2(a)** shows the coastal condition in 1947. In the central part of the picture flows the Shinkawa River. The river stream largely meandered near the river mouth and it opened to south. In 1947 there were wide natural sand dunes of around 300m width around the river mouth. Until 2000, seawall was built along the overall shoreline as shown in **Fig. 2(b)** and natural sandy beach disappeared. Headlands were built as a measure against beach erosion. Although a sandy beach of triangular shape was formed at the corner of the headland, sufficiently wide sandy beach was not formed because of the decrease in total sand amount of the beach in this area.

4. CAUSES OF NARROWING OF FORESHORE WIDTH

On the Node coast, a wide sandy beach including the sand dune area extended in 1947, but it changed to the artificial coast protected by seawall and concrete armor units installed along the shoreline during a period of around 50 years. **Figure 3** shows the time change in the cross-shore distance from the reference point to the shoreline, seaward edge of the coastal forest and seawall along the survey line A-A' set at 300m southwest of the Shinkawa River mouth as shown in **Fig. 2(b)**. At this location, the foreshore width was around 300m in 1947, but around two thirds of the foreshore disappeared until 1963 due to the development of coastal forest. Simultaneously the shoreline recession had started since this time and it retreated 26m until 1970. The shoreline recession continued thereafter and shoreline recession from 1947 to 2000 reached 45m. However, shoreline recession contributes only 15% of all the narrowing of the foreshore width. Furthermore, it should be noted that earth dike to protect coastal forest against wind blown sand was built in front of the coastal forest despite the widening of coastal forest until 1963 in contrast to the shoreline recession. However, the shoreline retreated further because of decrease in longshore sand supply caused as a large-scale imbalance of sediment supply in this area and sandy beach totally disappeared in front of the seawall. As a result, the intensity of wave overtopping increased and materials filled under the seawall was washed away by local scouring. As a measure against wave overtopping, concrete armor units were installed continuously in front of the seawall. It is clearly understood that artificial coastline

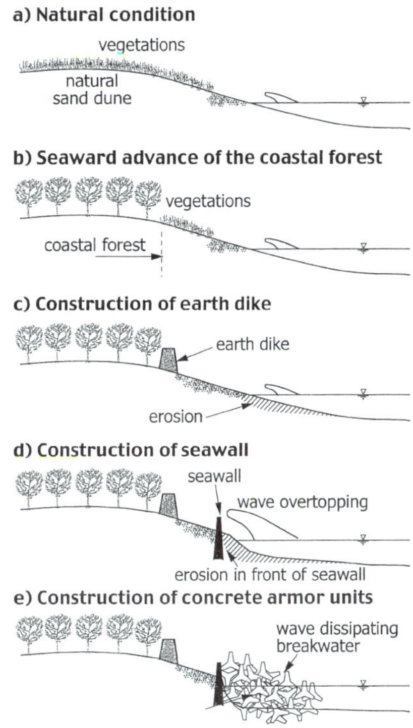

Figure 4. Change process from natural sand dunes to artificial coast.

has been created as a dilemma by the excess advance of coastal forest as well as the construction of the protection facilities of the coastal forest on land and beach erosion due to the exhaustion of sand supply from the upcoast. **Figure 4** summarizes these processes.

5. CONCLUDING REMARKS

Despite that beach erosion and shoreline recession have been caused as a structural problem, fundamental measures have not been taken, and instead coastal forest and related facilities have rather advanced seaward, increasing disaster potential of these facilities against waves. In Japan coastal forest and related facilities are controlled under the forest law, whereas the coastal zone is controlled under the seacoast law. Under this seacoast law, the coastal protection area is basically determined in a range between a location of 50m landward of the shoreline position in high tide level and a location of 50m seaward of the shoreline position in low tide level in March, and shore protection works can be done only in this narrow band without cooperation with coastal protection. The obstruction in solving beach erosion problem as mentioned above is not in the simple barrier of the scientific research, but in the adjustment method of social trade-off issues, which were caused from various senses of values. Based on 'forest law', their action to protect the coastal forest is merely their proper duty to be carried out. A solution acceptable for many people must be found not only under these complicated restraint conditions. Overall comprehensive plan regarding protection and preservation of coastal environment must be considered in the widespread coastal zone including the coastal forest in order to improve extremely armored coast.

NUMERICAL SIMULATION OF BED TOPOGRAPHY CHANGES INDUCED BY DEEP WATER NAVIGATION CHANNEL PROJECT IN THE YANGTZE ESTUARY

Jingxin ZHANG, Hua LIU* and Yousheng HE
School of Civil Engineering and Mechanics, Shanghai Jiao Tong University,
Shanghai 200240, P.R. China
* Email: hliu@sjtu.edu.cn

Abstract

With the development of industries in the Yangtze estuary, more and more projects, such as port and waterway construction, land reclamation in deeper waters as well as coastal protection works, have been conducting in the estuary. More detailed knowledge of the tide hydrodynamics and sediment transport mechanism in the water is needed for construction and management of the projects because of difficulties in modeling the characteristics of fine cohesive sediment movement and the complicated flows driven by river discharge, tide, storm surge and wind induced wave.

One of the key projects under construction is the deep water navigation channel project in the North Channel, which is part of the integrated regulation plan of the Yangtze estuary for improvement of the effective water depth of the North Channel, the main navigation channel, from the 7.5m up to 12.5m by the engineering and dredging approaches. In response to the changes in flow and sediment transport during the construction of the hydraulic structures, i.e. parallel dikes and spur dikes, the channels of the estuary tend to adjust in order to establish the dynamic equilibrium of their regime. The imposed changes in the flow and sediment transport will eventually results in morphological changes to the estuary, particularly to the North Channel, i.e. in its water depth alignment. The behavior of time-dependent variation of bottom is therefore crucial in understanding the effects of the parallel dikes and spur dikes on the morphological processes in the Yangtze estuary and to propose an optimized option for the design of the spur dikes.

The main purpose of this paper is aimed at investigating the changes of bottom in the North Channel induced by the parallel dikes and spur dikes of the deep water navigation channel project using the mathematical modeling technique, the seasonal morphological variation due to the discharge of fresh water and sediment is ignored. The hydrodynamics module of DELFT3D is used to compute the tidal flows to serve as a basement for the development of a fine cohesive sediment transport module, in which a three dimensional shallow water equations are solved in orthogonal curvilinear coordinates using the finite difference method. The mathematical formulation and implementation of the model are provided. Results of the model calibration and validation based on the measured data from the field study are reported in the paper.

Figure 1 shows the computational domain for the simulation of sediment transport. Comparisons between the predicted bed form and the measured data at some cross sections along the deep water navigation channel project in the North Channel are provided in Figure 2.

References

[1] He, Yousheng and Liu, Hua. 1999. On Complicated flows in large estuaries. In Mechanics and Engineering, editors: Guohao Li and Yousheng.

[2] Liu, Hua et al. 2000. A Study on Mathematical models of Water Environments in the Yangtze Estuary. *Journal of*

Hydrodynamics, No.2. (in Chinese)

[3] Liu, Hua and He, Yousheng. 1999. A review on 3-D Mathematical modeling of Hydrodynamic Flow in estuarine and coastal waters. *Journal of Ocean Engineering*, No.2. (in Chinese)

[4] Liu, Hua et al. 1999. A 2-D vertical mathematical model of morphological processes of a trench. *J. of Hydrodynamics*, No.2.

[5] Leo.C.Van Rijn. 1984. Sediment Pickup Functions. *Journal of Hydraulic Engineering*, Vol.110,No.10.

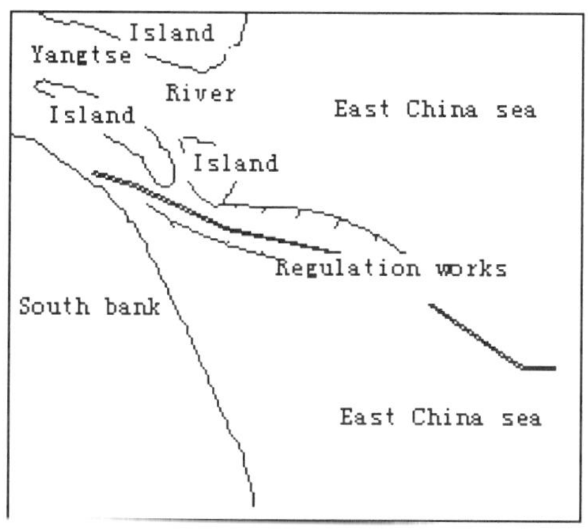

Fig.1 Computational domain for sediment transport model

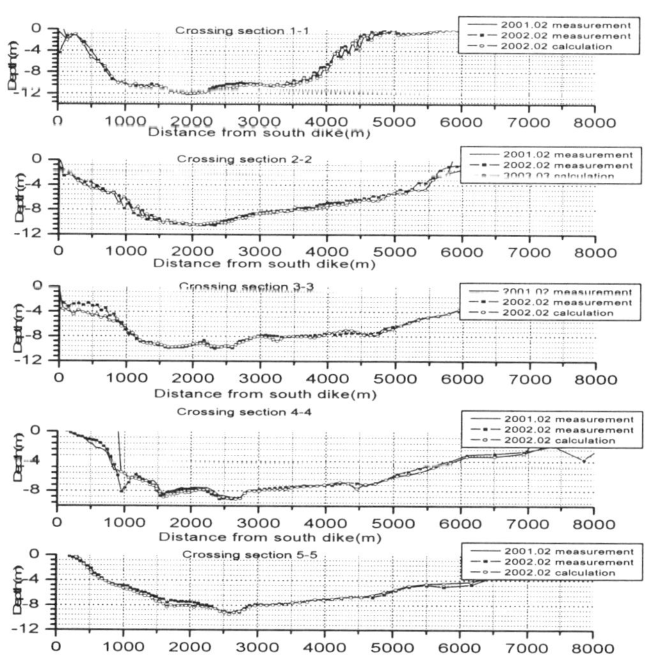

Fig.2 Comparison of the changes of bedform

HISTORICAL TRANSITION OF TIDAL FLAT EFFECTED BY SEWERAGE SYSTEM

H. Tatsumoto, Chiba University, Japan
Y. Ishii, Chiba University, Japan
M. Machida, Chiba University, Japan
K. Murakami, Chiba Institute of Technology, Japan
K. Taki, Chiba Institute of Technology, Japan

Key Words: artificial tideland lake, sewerage system, ecosystem structure, historical transition, urban coastal zone

Introduction

Yatsu tidal flat is located in the inner zone of Tokyo Bay, Japan, and is extremely unique because of its environmental condition that is enclosed with concrete revetment with the reclaim of its circumference. Once, the Yatsu tidal flat was a part of vast tidal flat where developed in the front of the coastline in Tokyo Bay. However, it has been reclaimed for the circumference reclamation from the year 1971, and the circumference has been changed into the closed tidal flat (the total area is 40.1 ha) as shown in Photo 1, surrounded by concrete shore protection. This tidal flat is similar to an artificial tideland lake in regard to shape. Yatsu tidal flat where was registered under Ramsar Convention at June 1993 is very precious tidal flat as a left behind to the urban area, although it has passed 20 years or more since reclaiming and became independent from sea. As the main factors that constitute and influence to tidal flat, topography, sediment, water quality, living things and externally action are considered. Transformations of the ecosystem structure composed of environmental factors in Yatsu tidal flat are investigated in this paper.

Transformation of Water Quality

As shown in Fig.1, it is recognized that the inflow load from its basin area to Yatsu tidal flat is decreased 16.8% for these 12 years from 1983 to 1995. The reason of the decreasing is that the amount of drainage inflow from the household was decreased with improvement of public sewerage system of a basin area. Furthermore, improvement of public sewerage is not only reduction of inflow loads, but also brought about the result of reduction of the amount of fresh water supply. Chloride ion concentration in Yatsu-Funadamari located in the most inner place of Yatsu tidal flat is continuously increased as shown in Fig.2. This phenomenon namely means the water quality in Yatsu tidal flat is changing from brackish water to seawater.

Transition of Ecosystem Structure

Resent years, it is observed that the *Ulva* spp. in this tidal

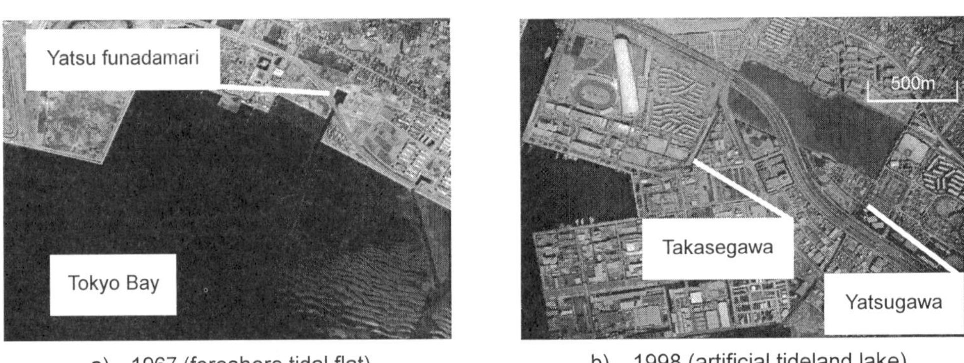

a) 1967 (foreshore tidal flat) b) 1998 (artificial tideland lake)

Photo.1 Aerial Photograph of Yatsu Tidal Flat

flat grew irregularly. The main species of *Ulva* spp. was identified as *U. pertusa* and *U. japonica*.

The growth area has increased year by year. The area was not observed in 1984, which is the start year of the observation. However, it is observed on 6.8 ha in 1995, on 13 ha in 1999, and over 20 ha in 2000. In 2001 summer, *Ulva* spp. was occupied near 2/3 surface area of this tidal flat. *Ulva* spp. is classified as one species of large green algae that like sand or sandy-mud sediment condition rather than mud sediment condition. And then, it is thought that the irregular growth of *Ulva* spp. is brought about by existence of sufficient nutrient salts and progress of seawater and sandy-mud sediment in this tidal flat. It is therefore, able to consider that the reason for the irregular growth of *Ulva* spp. is due to the increasing of (sand/mud) ratio and the quantity of (seawater/brackish water) ratio. Sediment condition of the area where *Ulva* spp. is much grown (ORP, -200～-250mV) is presenting the black reduction state compared with the sediment of the area where *Ulva* spp. is not observed (ORP : -50～-100mV). So, in such reduction area, there cannot be observed any macrobenthos that is supporting a tidal flat ecosystem as main predator. It is considered that *Ulva* spp. decayed and made anaerobic sediment state after its irregular growth. In comparison with 1984 and 1995, the total species number of macrobenthos has increased from 13 to 39, however, the total individual number has clearly decreased, and no living point was also observed. That is, it is suggested that biological production mechanism has been changing from "large quantity of small species" condition to "small quantity of large species" condition, namely, change to brittle bio- production state.

Mechanisms of Historical Transition for Yatsu Tidal Flat

At Yatsu tidal flat, the inflow loads has decreased as improvement of public sewerage system that is started since 1990. As the result, water quality has changed from brackish water to seawater and sediment condition has also changed from mud to sand, and it is caused the irregular growth of *Ulva* spp., which is the indicator of eutrophicated state. That is, as shown in Fig.3, at the first step of transition, a tidal flat just like tideland lake is formed artificially with reclamation and development of its circumference, and the water quality

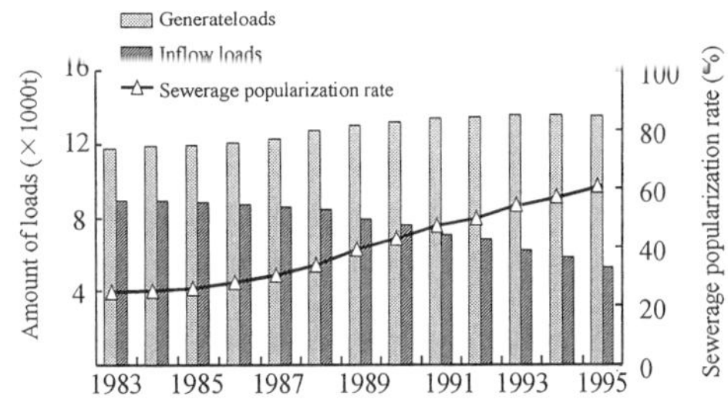

Fig.1 Change of popularization of sewerage and amount of loads

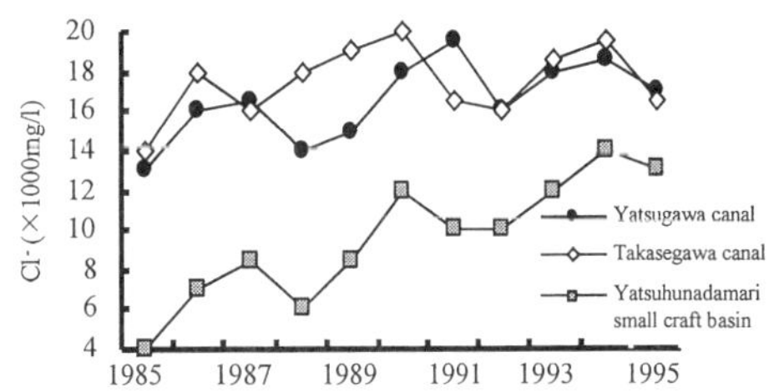

Fig.2 Change of chloride ion concentration

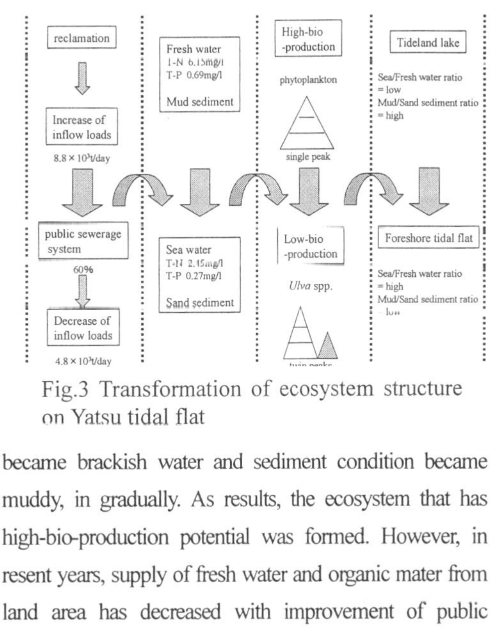

Fig.3 Transformation of ecosystem structure on Yatsu tidal flat

became brackish water and sediment condition became muddy, in gradually. As results, the ecosystem that has high-bio-production potential was formed. However, in resent years, supply of fresh water and organic mater from land area has decreased with improvement of public sewerage system, so the transition from brackish water and mud sediment condition to seawater and sand sediment condition have progressed, and bio-production potential has reduced. Moreover, tide action becomes more influence, such as flow out of sand from Yatsu tidal flat. From these outcomes, it was made clear that Yatsu tidal flat has been changing to the foreshore tidal flat again with remaining the topographical characteristics as tideland lake.

Quantification of oxygen cycling in seagrass beds

Noriko IYODA[1], Jun SASAKI[2] and Masahiko ISOBE[3]

Introduction

Seagrasses are unique marine flowering plants, which grow in shallow, subtidal or intertidal unconsolidated sediments. Seagrass beds have been recognized as a valuable resource critical to the health and function of coastal waters. They have a function of purification of polluted waters through nutrient uptake, supplying foods for aquatic animals and providing shelters to fries, which provide the basis for a highly productive ecosystem (Fonseca *et al.*, 1998). During the past a few decades, however, tremendous losses of this habitat have occurred as a result of development within the coastal zone. The Environmental Agency (2000) reported that about 4% of seagrass and seaweed beds in Japan disappeared between 1978 and 1991. Those losses have been recognized as one of the causes of the degradation of coastal marine ecosystems, such as decrease in coastal habitats and biodiversity, and increase in nuisance algal blooms due to the lack of water purification function.

Nowadays, preservation, restoration and mitigation of seagrass beds has become a matter of concern under the amended law of coast and developers of major projects have been obligated to assess their environmental impacts under the law of environmental impact assessment, which have been enforced from 1999. For these purposes, it is of great necessity to quantify and evaluate the functions of seagrass beds. So far, a large number of indoor experiments and field observations have been conducted. Kawasaki *et al.* (1986) estimated the material cycling in seagrass based on an indoor experiment. Hibino *et al.* (1998) performed monitoring of water quality including oxygen in seagrass beds. However, most of the previous works were limited to the qualitative estimation of the material cycling because of a difficulty in excluding the effect of advection of seawater. Thus the objective of the present study is to develop a new method to estimate oxygen cycling quantitatively in the field using a chamber to exclude the effect of advection of seawater. A simple numerical model is also developed to reproduce the observed oxygen cycling in seagrass beds.

Methodology

A proposed method to estimate *in-situ* oxygen cycling in seagrass beds consists of two parts; field monitoring of water quality using a chamber and water quality sensors, and laboratory experiments including the analysis of sediment cores sampled from the seagrass bed to quantify benthic processes. The estimated flux of each process obtained through the field and laboratory experiments was used for modeling of dissolved oxygen (DO) cycling processes.

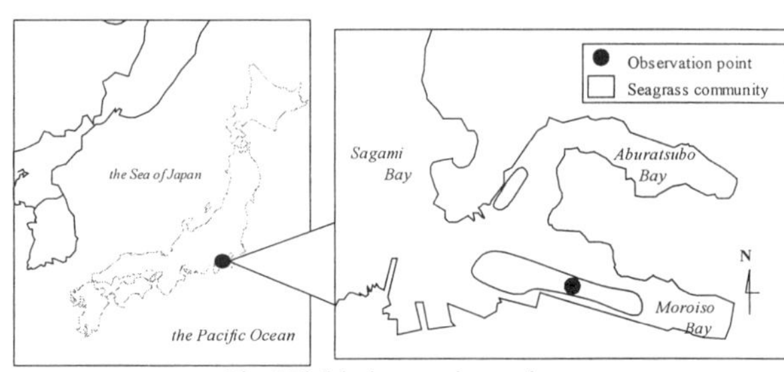

Fig.1Field observation point

(1) Field monitoring and experiment

Field observation was performed in Moroiso Bay, Kanagawa, Japan (see Fig. 1). We designed and installed a chamber on a seagrass bed to cover a little amount of seagrass, *Zostera marina*, and monitored temporal variation of the concentration of DO in the chamber as well as water quality such as temperature, salinity, DO concentration, chlorophyll *a* (Chl-*a*) concentration, turbidity and photosynthetically active radiation (PAR) outside of the chamber. Thus, we could obtain time series of net DO concentration in the seagrass bed excluding the effect of advection. Each monitoring was conducted for a few days and at the end of each run, sediment core samples and leaves of the seagrasses with root in the chamber were collected, which included benthic and epiphytic animals.

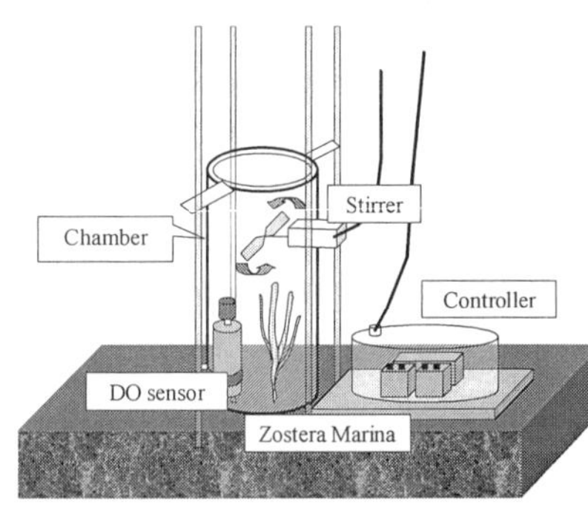

Fig.2 Field monitoring system observation

[1]Technical Research Institute, Penta Ocean Corporation, 1534-1, Yonku-cho, Nishinasuno-machi, Tochigi 329-27, Japan, Fax: +81-287-39-2133, E-mal:Noriko.Iyoda@mail.penta-ocean.co.jp
[2]Associate Professor, Yokohama National University, Fax: +81-45-348-4565, E-mail: sasaki@cvg.ynu.ac.jp
[3]Professor, The University of Tokyo, Fax: +81-3-5841-8503, E-mail: isobe@k.u-tokyo.ac.jp

(2) Estimation of DO flux in each process

We conducted a series of sediment core experiment to quantify DO fluxes of relevant processes such as respiration of benthic macro fauna, photosynthesis of benthic algae, respiration of bacteria, etc. For instance, photosynthesis of benthic algae was estimated through a comparison of the time derivatives of DO concentration under light and dark conditions. Contribution to DO flux due to respiration of epiphytic animals was also considered and quantified through an experiment of DO consumption.

(3) Numerical Simulation

Based on the estimation of DO flux for each process, a simple numerical box model was developed to predict a time series of a temporal change in DO concentration and each DO flux in the chamber. Fig. 3 shows DO cycling in the seagrass bed. The model was forced by PAR, temperature, etc., obtained by field monitoring to reproduce the resultant time series of DO concentration in the chamber. A comparison between observed and simulated variation in DO concentration is shown in Fig. 4. DO concentration increases mainly due to photosynthesis of *Zostera marina* during daytime and decreases due to respiration during nighttime. Though there are some discrepancies between observed and simulated results, overall agreement is considered to be acceptable.

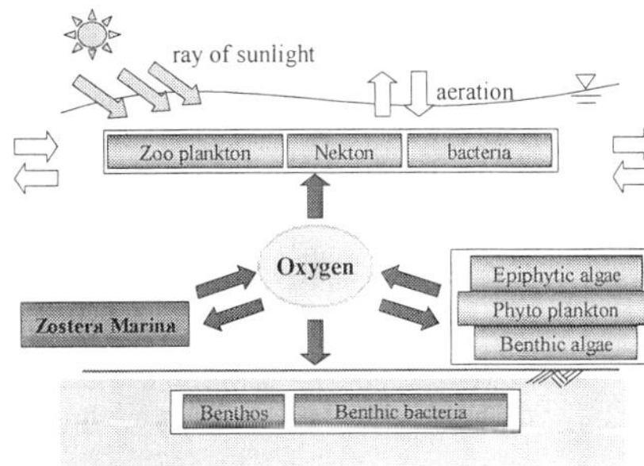

Fig.3 DO cycling in seagrass bed ecosystem

Results and Conclusions

Based on numerical simulations, we quantified the contribution of each process to DO flux. We found that respiration of benthic animals and benthic bacteria is an essential compartment in the ecosystem, which contribute to the negative DO fluxes. Primary production due to seagrasses exceeded respiration due to bacteria, benthic and epiphytic animals even in fall when the seagrass is inactive. This means that seagrasses have a function to prevent from appearing of hypoxic water even in the seasons when the seagrass is inactive and play an important role in conservation of a rich and productive ecosystem in the coastal zone sustaining a large amount of benthic and epiphytic animals.

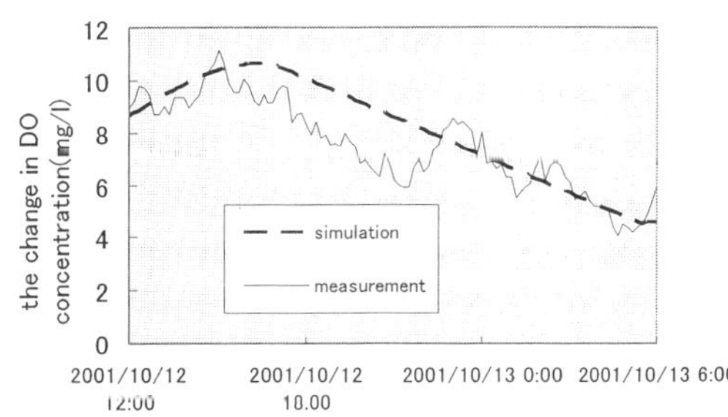

Fig.4 Comparison of observed and simulated DO concentration

References

Bocci, M., G. Coffaro and G. Bendoricchio (1997): Modeling biomass and nutrient dynamics in eelgrass (*Zostera marina L.*): application to the lagoon of Venice (Italy) and Oresund (Denmark), Ecological Modelling, Vol.102, pp.67-80.

Environmental Agency (2000) : Threatened wildlife of Japan : red data book Plants I ,660p. (in Japanese)

Fonseca, M. S., W. J. Kenworthy and G. W. Thayer (1998): Guidelines for the conservation and restoration of seagrasses in the United States and adjacent waters, NOAA's Coastal Ocean Program, Decision Analysis Series No. 12, 222p.

Furukawa, K., A. Kuwae and Y. Hosokawa (1999): Fieldworks on water quality purification mechanisms at sandy beaches and its modelling, Technical Note of The Port and Harbor Research Institute, No.947, 91p. (in Japanese)

Hibino, T., K. Tsuruya and D. Nishimori (1998): Ocean water intrusion and its effects on estuary ecosystem, Proceedings of Coastal Engineering, JSCE, Vol.45, pp.1081-1085. (in Japanese)

Kawasaki, T., S. IIzuka, H. Goto, T. Terawaki, Y. Watanabe and S. Shimo (1986): Effects of temperature on *Zostera Marina L.* seed germination and seed ling development, Technical report, U14, 231p.

Matsunashi, S (1993) : A predictive model for concentrations of nitrogen, phosphorus, and dissolved oxygen in semi-enclosed coastal seas, Proceedings of Coastal Engineering, JSCE, Vol. 40, pp.1076-1080. (in Japanese)

CONSERVATION HISTORY OF HORSESHOE CRAB TACHYPLEUS TRIDENTATUS AND ITS SPAWNING GROUND: A DESIGNATED NATURAL MONUMENT IN KASAOKA BAY IN OKAYAMA PREFECTURE

Satoquo Seino[1], Takaaki Uda[2], Yasufumi Tsuchiya[3] and Keiji Tsuchiya[4]

[1]Graduate School of Arts and Sciences, Univ. of Tokyo (3-8-1 Komaba, Meguro, Tokyo 153-8902 Japan (Tel/fax +81-3-5454-6793, fwid6176@mb.infoweb.ne.jp)

[2]Public Works Research Center, 1-6-4 Taito, Taito, Tokyo 110-0016 Japan (Tel +81-3-3835-3609, Fax +81-3-3832-7397,uda@pwrc.or.jp)

[3]Kasaoka International Exchange Association (6-35 Midori, Kasaoka, Okayama 714-0048 Japan)

[4]Research Group for Conservation of Horseshoe Crab in Japan (209-9 Tomioka, Kasaoka, Okayama 714-0029 Japan)

I. INTRODUCTION

The horseshoe crab Tachypleus tridentatus is one of the endangered species and the ecological key species representing the coastal environment. In the past its habitats spread on many coasts of the Seto Inland Sea and the coasts in Kushu. In recent years, however, the number of this animal has been decreasing rapidly in spite of various measures to protect the environment. There are several causes, but one of them is due that many measures had been taken locally in a superficial manner because of the difficulty of selection of the fundamental measures. In this study the history of conservation of horseshoe crab and its spawning ground assigned to a natural monument in Kasaoka Bay in Okayama Prefecture and measures for protection are investigated. Kasaoka Bay was one of the very famous habitats of horseshoe crab in the western Japan. In this bay large-scale land reclamation had begun since 1968, resulting in the disappearance of the ebb tidal flat, being a habitat of larval horseshoe crab. As a measure, a new habitat area of this animal was assigned to a national monument as well as many other activities. However population of the horseshoe crab came to be endangered. Lack of thinking of environmental capacities and physical factors is a critical point. This history tells us the useful information to reconsider the present and future environmental conservation activities, especially in the eco-coast works, in which many kinds of protective measures are to be taken on the Japanese coasts.

II CONTENTS

In the past Kasaoka Bay was a closed shallow sea with the only opening to the southwest except a very narrow channel connecting to the Sea of Mizushima on the southeast as shown in **Fig.1**. Large part of the bay was reclaimed as shown in **Fig.1** since 1968 and the name of towns in reclaimed land was named after horseshoe crab. The kabuto means horseshoe crab in Japanese. There was wide ebb tidal flat on both shores along Kounoshima channel and many habitats of horseshoe crab were located in this area. The conservation history in Kasaoka Bay was investigated based on old research materials as well as the comparison of past aerial photographs and old maps. For example, **Table 1** summarizes the chronology of protection activities of horseshoe crab in Kasaoka since the Meiji and Taisho eras in Japan. As summarized in this table, conservation history in Kasaoka is back to around 80 years.

Fig.1. Map of past Kasaoka Bay.

Fig.2 and **3** show the distribution of the habitat of horseshoe crab and confirmed spawning grounds in the vicinity of Kasaoka Bay in the past. There were many habitats in the ebb tidal flat in the vicinity of the Oyehama beach in the northern part of Kasaoka Bay. On the other hand, spawning ground of horseshoe crab concentrated on the shores extending along Kounoshima channel as shown in **Fig.3** as well as the concentrated spawning ground at Oyehama beach. Of these spawning grounds, at the spawning ground facing Kamishima channel there were some sandy beaches composed of coarser materials, a condition of which is necessary as the appropriateness for spawning ground of horseshoe crab. Also fresh water can be supplied from the Imadate and Ohshima Rivers and there exists ebb tidal flat necessary for the habitat of young horseshoe crab in the vicinity. All these give sufficient condition for the spawning ground of horseshoe crab. Therefore, a new habitat area of this animal was assigned to a national monument as well as many other activities in place of buried Oyehama beach by land reclamation. **Fig.4** shows the recent engineering measures for the protection such as beach nourishment and construction of detached breakwaters to improve environment condition of this area. In spite of these efforts, population of the horseshoe crab came to be endangered.

Table 1. History of research on horseshoe crab and conservation measures at Kasaoka.

Time	Events
Meiji era	Field observation of spawning ground by famous zoologists
Taisho Era	Research on field ecology.
	Importance of the Oyehama beach as spawning site was confirmed
1927	Investigation by Geographical Div., the Ministry of Home Affairs
1928	Oyehama beach was assigned to a national natural monument
1958	Completion of land reclamation (106 ha) in Tomioka Bay next to Kasaoka Bay
1961	Teacher K. Tsuchiya set up 'research club on horseshoe crab' in junior high school in Kasaoka City.
1962	The Crown Prince and Princess formal visit
1965	Determination of land reclamation (1,800ha) of Kasaoka Bay
	Change of the initially assigned national monument to dry land is inevitable.
1966	Land reclamation office consigned research on ecology to Okayama Univ.
1967	Initiation of land reclamation in Kasaoka Bay
1968	The Cultural Properties Protection Committee of the Ministry of Education submits a report of addition of assigned spawning ground.
1970	Conservationist H. Nishii et al. set up 'club to protect horseshoe crab'.
	'Declaration of emergency in horseshoe crab'
	Request of protection measures and mitigation of spawning ground
1971	Announcement of addition of assigned spawning ground
	Set-up of ' boy scout protecting horseshoe crab in Kasaoka'
1975	Opening of Center for Conservation of Horseshoe Crab
1990	Completion of land reclamation in Kasaoka Bay
	Construction of Horseshoe Crab Museum in Kasaoka
1995	Cancellation of assigned national monument at past Oyehama beach

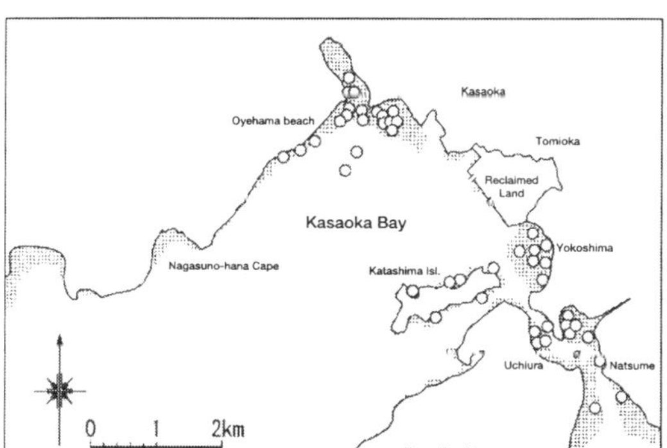

Fig.2. Habitats of horseshoe crab in the vicinity of Kasaoka Bay.

III. CONCLUSIONS

Before the land reclamation, there existed sea water exchange in Kasaoka Bay by the tidal currents through the southwestern opening of the bay and Kounoshima channel as shown in **Fig.1**. This sea water exchange was cut off by land reclamation, resulting in deterioration of the water quality, and Kasaoka Bay became inappropriate area for many varieties of lives to live. Based on these considerations, past measure is considered to be inadequate that horseshoe crabs lived in the planed area of the land reclamation are moved to the other areas and these new places are assigned to a new protection area with appropriate protection facilities . Because recognition of the environment of the habitat of the horseshoe crab was insufficient, and the appropriate

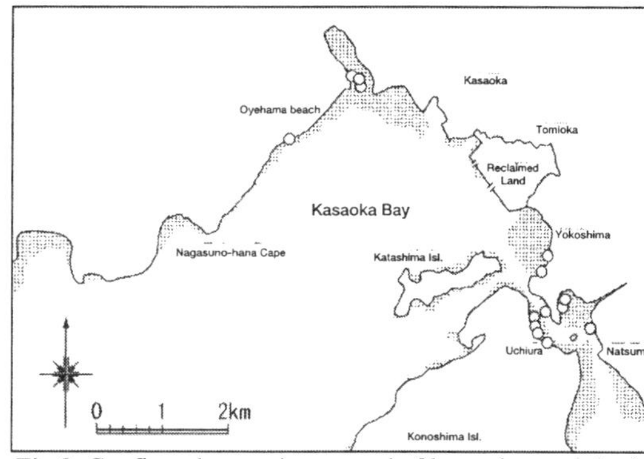

Fig.3. Confirmed spawning ground of horseshoe crab in the vicinity of Kasaoka Bay.

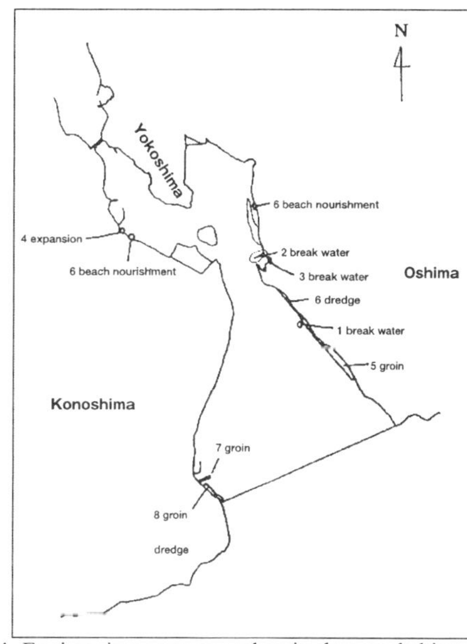

Fig.4. Engineering measures taken in the new habitat area.

environment of the Kasaoka Bay in the past including the new assigned protection area around Kounoshima channel had to inevitably change a great deal by the land reclamation. Large-scale change in bay form brings decisive effects to the wave field and tidal currents with the resultant change in the overall habitats of horseshoe crab. Land reclamation in Kasaoka Bay and its results give clear evidence. It must be considered that these investigations were carried out in 1966, when scientific prediction of theenvironmental change in the coastal water was not difficult, because of insufficient knowledge of science. However, in order to consider future protection of horseshoe crab and environment in Kasaoka, new idea must be taken into account. The environment of the habitats of endangered species should be globally understood and physical conditions that such an environment is sustained must be deeply studied. Protection movement of specific species in the micro-scale manner becomes no meaning, if overall environment does not satisfy the condition necessary for establishment of habitats. We can learn much from the experiences of protection of horseshoe crab in Kasaoka.

A Mangrove Mitigation Project in Singapore

Yuko Tanaka[1], Keiji Arita[2], Eiji Yauchi[3]

[1]Toa Corporation, Development Dept.,
5-banchi, Yonbancho, Chiyoda-ku, Tokyo, 102-8451, Japan
E-mail:yu_tanaka@toa-const.co.jp
[2]Toa Corporation, Overseas Dept., ditto.
[3] Associate Professor, Dept. of Civil Eng., Chiba Institute of Technology,
2-17-1, Tsudanuma, Narashino-city, Chiba, 275-8588, Japan,

Introduction

Mangrove forests are among the most productive wetlands on earth, representing specialized communities of organisms adapted to the saline environment of the tropics. However, they are also rapidly disappearing throughout the world.

In 1996, the government of Singapore began work on a project to create a site for the disposal of incinerated ash and other waste materials. However, as this project would displace a mature stand of mangroves on the island of Pulau Semakau, one of the Pulaun Islands, the Singapore Ministry of the Environment chose to replant a replacement mangrove forest equal to the area (13 ha) of the mangrove forests lost.

This study reports on the outcome to date of the replacement project.

Outline of the Pulau Semakau Project

Located about 8 km to the south of the Singapore main island, Pulau Semakau island is wholly covered with mangrove forest. This project connects two islands through the construction of strong granite bunds to form a central basin between the islands for the disposal of incinerated ash and other waste materials. The incineration ash is scheduled to be accepted for 30 or more years into the future. The revetment is stone-tiled, and a shielding membrane is paved at the basin side.

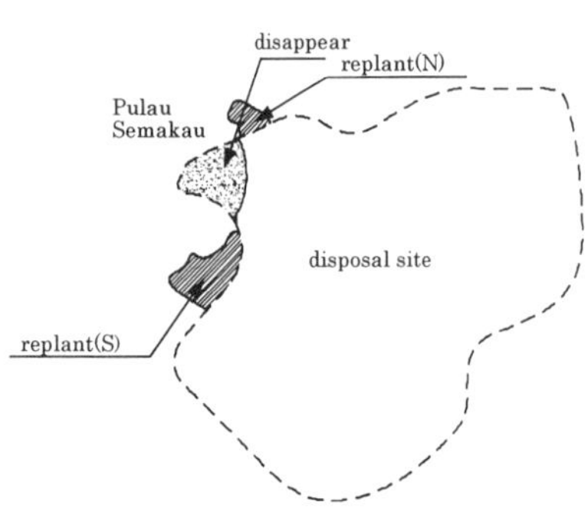

Figure 1. Replant site

The new mangrove forest is intended to replace the 13 ha destroyed by construction, as shown in Figure 1. The replant area is divided into two sites: a north area of 5.8 ha and a south area of 7.2 ha. An existing mangrove forest is to be preserved behind the new forest.

Results

(1)Field survey

The flora of each district mangrove is shown in Table 1. District 1 was a wetland with sparsely distributed shrubs. The dominant species was Rhizophora apiculata in district nos. 1, 2, and 3, and other species were common in those districts. More species were found in district no. 4 than in other districts; and the dominant species were Avicennia spp. and Sonneratia alva. These two species often form a seaward colony in mangrove forests. Given the characteristics of each district, the mangrove forests appear to have expanded from 4, to 3 and 2, to 1.

The values for the area and density of the replant zone – 13 ha and 25,000 boles – are based on these field surveys.

(2)Replant method

It is difficult to replant a grown tree. Therefore, viviparous seeds gathered from existing mangrove forests were used. R. mucronatum was planted in littoral zones, because it is seed often found in such areas. Figures 2 and 3 show the arrangement of R. apiculata and R. mucronatum, selected as dominant species. R. apiculata was planted in from +2.2 to +1.9m C.D.L blocks of both the north and south districts. R. mucronatum was planted at four +1.8 m C.D.L blocks and two wave-affected blocks in the

Table 1. The mangrove flora

District	No. of Main Composition	Size
1	7	3-4.5m
2	5	3-4.5m
3	6	5-10m
4	16	10-15m

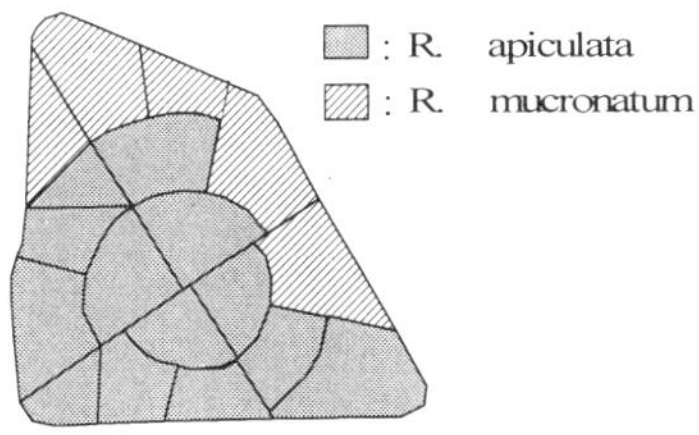

Figure 2. Arrangement of replant mangroves (North)

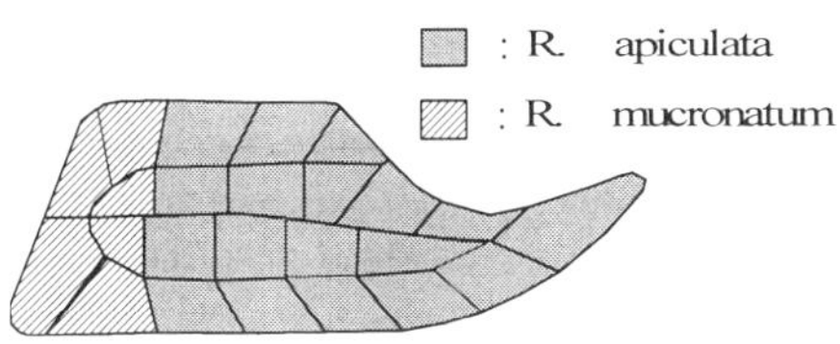

Figure 3. Arrangement of replant mangroves (South)

north district, and in six +1.8 m C.D.L blocks in the south district.

However, some viviparous seeds flew out without taking root, while other seeds withered after planting. Therefore, the nursery was installed on the floor of an existing mangrove forest, and the resulting saplings after two months planted. Floating nurseries and the sufficient space for nursery transport were prepared to supply saplings efficiently, unaffected by tidal effects.

Growth was improved by this treatment, but parts of the seedlings still died. This was discovered to be due to the admixture of acidic soil from a deeper layer during digging. In particular, the growth of R. apiculata is impeded when soil pH drops below 6.3. Upon determination of the cause of this problem, the affected soil was repeatedly exposed to sea water, raising pH. The adverse effects on plant growth disappeared within several months.

In all, some 400,000 seedlings were finally planted. They were tended to prevent the adhesion of acorn barnacles or seaweed, the weight of which might bend and break the saplings

(3)Growth Process

When mangrove growth was inspected 25 to 26 months following planting, R. apiculata saplings that had been 0.25 meters at planting were found to have grown to 0.9 m to 1.3 m in the north district, while R. mucronatum planted at a height of 0.45 m were found to have grown to 1.3 m to 1.4 m. The average growth rate per month was 0.036 m in the former and 0.037 in the latter species. Average growth over

Figure 4. Just after replant

Figure 5. After two years

twenty months for the two species was 1.9 m for R. mucronatum and 1.2 m for R. apiculata.

In the south district, 0.25 m R. apiculata saplings were found to have grown to a height of 0.9 to 1.2 m, while R. mucronatum seedlings that were 0.45 m at planting were found to have grown to 1.2 m to 1.4 m. Average monthly growth rates were 0.036 for the former and 0.037 for the latter. Average growth over twenty months was found to be 1.0 m in R. mucronatum and 1.2 m R. apiculata. Figure 10 shows the relationship between growth and elevation for R. apiculata by planting district. The growth was large as ground level being low in the south district.

Conclusions

The study confirmed the survival of replanted mangroves on at least 80% to 90% of the replanted area. The most important factors for these results are summarized below.

(1) Efficient supply of sapling
(2) Consideration of distribution tendency
(3) Securing a buffer zone

References

1) S.K. Lee and W.H. Tan : The regeneration and establishment of mangroves in Singapore, the Proceedings of Symposium on Science and Management of Mangrove Ecosystem, pp.18-22, 1998.

2) Masayasu Tatani, Yuko Tanaka, Keiji Arita, Eiji Yauchi : Mangrove Replant and the Growth Process as a Mitigation Project, Proceedings of 29th Meeting of The International Association of Hydraulic Research, Theme B, pp.390-395, 2001.

Landscape Analysis by Using Collective Unconscious and Its application to Design of Coastal Structures

Tomaya Shibayama, Yuki Sakai and Yuichirou Morichika

Department of Civil Engineering, Yokohama National University
Tokiwadai, Hodogaya-ku, Yokohama 240-8501 Japan
e-mail:tomo@ynu.ac.jp, phone:+81-45-339-4036, facsimile:+81-45-348-4565

1. Introduction

Coastal structures were designed for protection of infrastructure or protection of human life. However due to recent modification of coastal zone law in Japan in 1999, the basic concept of coastal engineering work has been changed. In designing structures, we, Japanese coastal engineers should consider harmonization of structure with surrounding environment. In the present paper, we will consider landscape in coastal environment and the role of dynamic waves in the landscape design. In order to design better landscape, it is necessary to analyze the classic spectacles, which are maintained over the generations in these 1500 years in Japan. Also it is necessary to clarify existing common recognition by applying the concept of collective unconscious.

The coastal protection methodology gradually changes from the energy reflection type to energy dissipation type. Goda and Takagi (1998) proposed detached breakwater perpendicular to shoreline and named it as longitudinal reef system. This structure is a typical energy dissipation type breakwater. The longitudinal reef system uses energy dissipation process under breaking waves. Figure 1 shows a bird-eye view of the longitudinal reef system.

The difference in wave field among three structures, longitudinal type reef, lateral type reef, and submerged breakwater are considered. The change of water surface is calculated by using unsteady mild slope equation, and surface spectacle is analyzed.

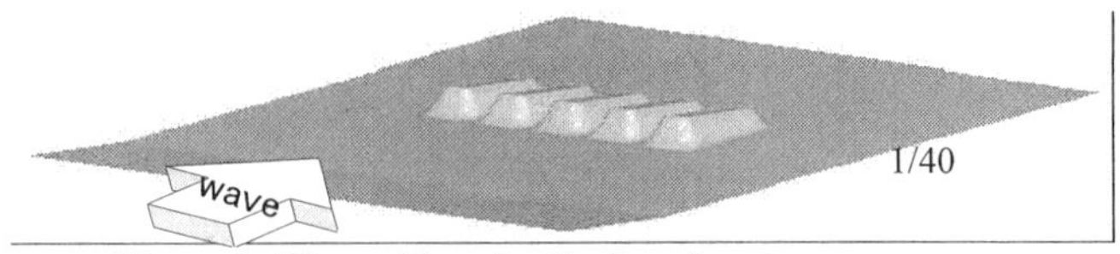

Figure 1: View of longitudinal reef system

2. Analysis of collective unconscious in coastal landscape----role of breaking waves

As a result of a local observation of the Japanese gardens (Koishikawa Korakuen and Hama Palace), which had been thought as a sight art, it is confirmed that Japanese gardens exhibit undeveloped, natural scenery of coastal area that contained the change in four seasons. One example is sand drawings slot to the sand surface as averaging technique of a Japanese garden with the broom. Some of the figures indicate the wave motion. Moreover, the Japanese poem was taken up as a language expression, and the meaning of 319 showplace song words, which condenses the scenery to one image, was analyzed. Here, "Waves" means the scenery including breaking. Total 47 paintings by which the coast scenery was drawn are sampled, and classified from the

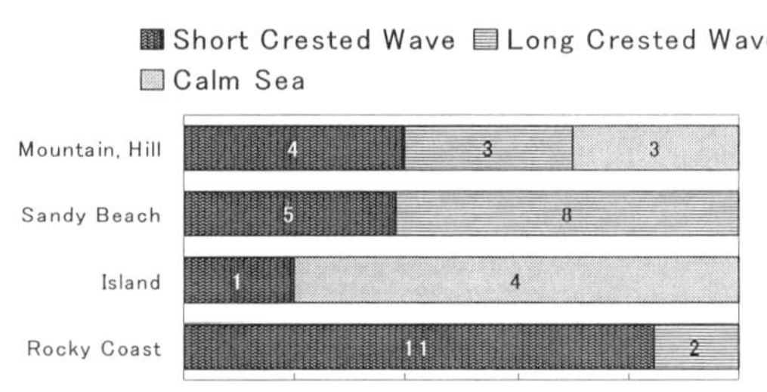

Fig. 2: Classification of geographical features

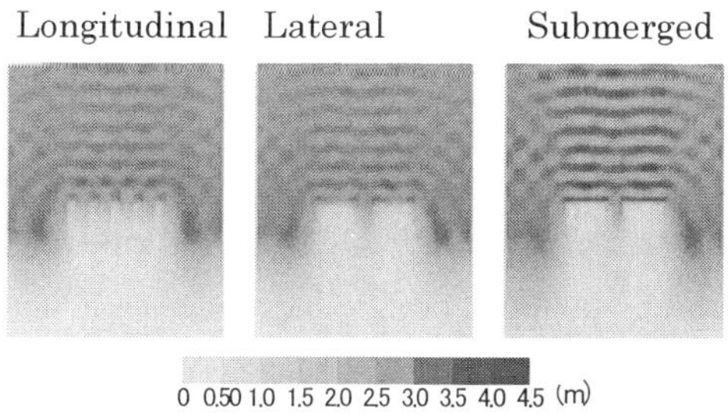

Figure 3: Wave field around three structures

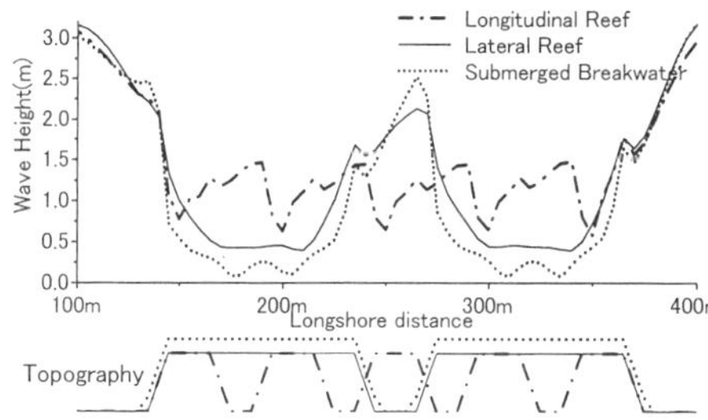

Fig. 4:Comparison of wave field (H_0: 3.2 m, T: 8s)

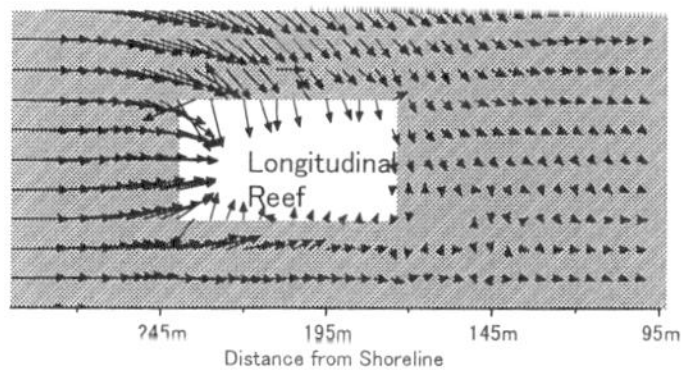

Figure 5: Wave refraction over longitudinal reef system (Length of arrow indicates wave height, H_0 : 3.2m)

condition of geographical features and waves (Figure 2). Not only a static element such as beaches and islands but also waves are included in the image of the sea, and it has been understood to be tied to the background image such as island or beach.

3. Simulation of wave view over three structures

Using an unsteady mild slope equation numerically simulates the wave behavior and image. Three kinds of structures are taken

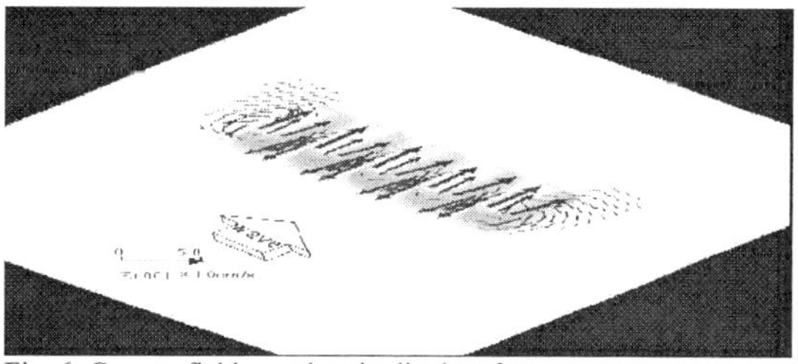

Fig. 6: Current field over longitudinal reef system

Figure 7: Image of wave breaking (H_0=3. 2m)

as examples and compared. The wave behavior and image are numerically simulated by using a unsteady mild slope equation. Three kinds of structures are taken as examples and compared. Fig. 3 shows the wave field around three structures. Fig. 4 is a comparison of cross-sectional distribution of wave height. Fig. 5 indicates wave refraction effect over longitudinal reef. Fig. 6 is current field over longitudinal reef system and Fig. 7 is image of wave breaking over longitudinal reef system. From the figures we can say that the longitudinal reef has advantages not only in wave protection but also for land-scope design from the view of breaking wave spectacle.

4. Conclusions

1) Main elements of the coast spectacle include dynamic wave motion as well as beaches and islands.
2) Three kinds of structures are compared by a numeric calculation. The longitudinal reef system is a structure, which remarkably shows the effect of refraction and wave breaking.
3) The longitudinal reef system looks like "Rough Sea" of the natural rock coast by the dynamic movement of water surface. It can control waves without giving the sense of incompatibility on the spectacle around.

References: Goda, Y.and H.Takagi (1998): Lateral versus longitudinal artificial reef system, Proc. of 26th. Coastal Engineering Conf., pp.2152-2165.

RISK ASSESSMENT ON STORM SURGE FLOODS

Fuminori KATO and Ken'ichi TORII
Coast Division, National Institute for Land and Infrastructure Management
Asahi 1, Tsukuba 305-0804, Japan
TEL: +81-298-64-3163, FAX: +81-298-64-1168, E-mail: katou-f92fr@nilim.go.jp

INTRODUCTION

Typhoons have caused catastrophic damage in Japan by flooding due to storm surges. In spite of implementation of shore protection works and warning notice systems, storm surge disasters have occurred still in recent years.

One effective measure against storm surge disasters is hazard maps on storm surge floods. Unfortunately, these maps are not available for all of Japan. What's more, the method of risk assessment for the map has not firmly been established.

To promote disaster mitigation through the nationwide distribution of hazard maps on storm surge floods, this study aims to establish a method for assessing the risk of storm surge floods.

FRAMEWORK OF RISK ASSESSMENT ON STORM SURGE FLOODS

There are two issues associated with these hazard maps.

One is the length of the return period of waves and tides that is assumed for the maps. Concurrent probability of high tide and high wave can not be evaluated precisely with combination of probability distributions of tide level and wave characteristics because wave characteristics depend on tide level. Besides observation of waves and tides have not been conducted long enough for extreme statistics analysis.

The second issue is whether coastal dikes are assumed in flood simulations to be broken by wave pressure or wave overtopping. In Japan, coastal dikes ensure safety against designed waves and tides, which are generally based on the highest records. It is possible that waves and tides exceeding a certain level may break coastal dike.

From the viewpoint of mitigating fatalities, risk assessment on storm surge floods can be conducted as shown in Figure 1.

(1) Probabilistic evaluation on tide level and wave overtopping rate

Extreme statistics analysis of waves and storm surges using a stochastic typhoon model was conducted by Yamaguchi et al. (1994). We developed a method of probabilistic evaluation on wave overtopping rates as well as tide levels based on the results of the Monte Carlo simulation with our stochastic typhoon model, taking astronomical tides into account.

(2) Judgment on coastal dike break

Although most coastal dikes in Japan are armored with concrete or asphalt, tide levels and waves higher than the designed level may break them and vastly increase the overflow rate there.

(3) Flood simulation

Inundation depth, velocity of flood flow, and flood arrival time in coastal zone can be estimated with a numerical model.

(4) Risk assessment

The results of the flood simulation can be used to assess the risk of storm surge flood in term of evacuation and risk to houses.

PROBABILISTIC EVALUATION ON TIDE LEVEL AND WAVE OVERTOPPING RATE

Probabilistic evaluation on tide level and wave overtopping rate by a stochastic typhoon model was conducted as shown in Figure 2.

(1) Making a stochastic typhoon model

Among 1,282 typhoons from 1951 to 1997, we selected those whose atmospheric pressures at the typhoon eye were 980 hPa or less. Then we analyzed characteristics of the selected typhoons such as the location of typhoon eye, atmospheric pressure at their eyes, and their speed and radius. Average values and standard deviations of the typhoon characteristics and their variation were calculated for two-degree grids in 0 to 50 degrees of north latitude and 100 to 180

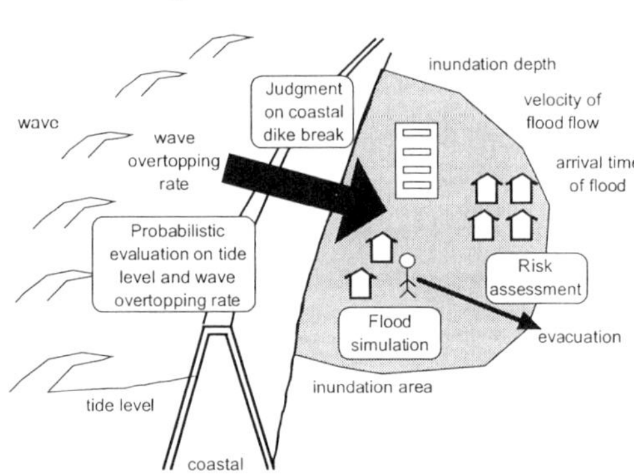

Figure 1 Schematic view of risk assessment on storm surge floods

degrees of east longitude.

It was supposed in the stochastic typhoon model that probability density function of typhoon characteristics and their variation at each mesh was standard normal distribution with the average values and the standard deviations calculated with the observed data.

Results of the typhoon simulation are almost equivalent to observed data. It was thus reasonable to conclude that typhoons can be simulated with the stochastic typhoon model.

(2) Calculating wave overtopping rates

Typhoons were simulated with the stochastic typhoon model in two thousand years. The number of typhoons each year was determined from the Poisson distribution. The distribution of atmospheric pressure and wind near each typhoon were estimated with Myers' Model. Storm surges and waves were estimated from the distribution of atmospheric pressure and wind. Random sampling of astronomical tides were used to calculate the wave overtopping rate due to each typhoon.

(3) Extreme statistics analysis

Extreme statistics analysis of wave overtopping rate was conducted by fitting an optimal distribution function to the calculated wave overtopping rate. Table 1 shows results of the extreme statistics analysis for coasts facing Ise Bay and Tosa Bay.

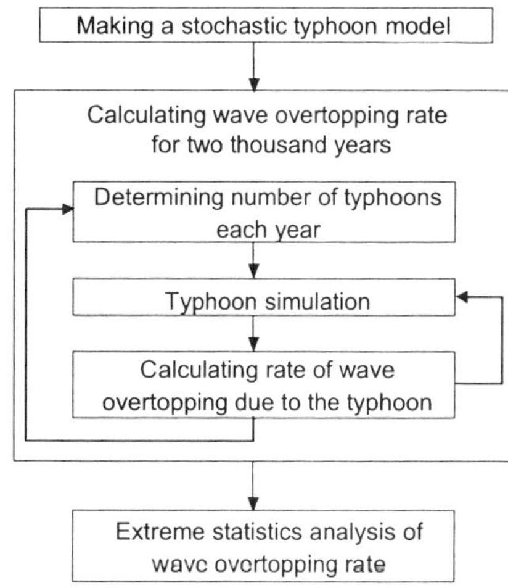

Figure 2 Flowchart of probabilistic evaluation on tide level and wave overtopping rate

Table 1 Results of probabilistic evaluation on wave overtopping rates

	return period (years)				
bay	50	100	200	500	1000
Ise	0.005	0.013	0.024	0.041	0.060
Tosa	0.166	0.221	0.270	0.361	0.427

(m^3/m/s)

PROBABILISTIC EVALUATION ON COASTAL DIKE BREAK

There are a few examples of break of coastal dike reinforced with concrete in Japan. Based on the precedents, it is reasonable to suppose that the beginning of the dike break in coasts facing inner bays is destruction of the concrete armor due to wave pressure and overflow.

The main forces acting on concrete armor are wave pressure at the seaward slope and crest of the coastal dike, and hydrodynamic pressure at the landward slope. The degree of destruction of the concrete armor can be judged with structural analysis.

The critical wave overtopping rate of the destruction will be shown at the conference.

RISK ASSESSMENT BASED ON FLOOD SIMULATION

Flood depth and flood arrival time are obtained with the flood simulation, taking coastal dike break into account. Based on acceptability, the return period of storm for hazard mapping should be determined. Time series of wave overtopping rate and tide level are set as boundary conditions in the flood simulation from the peak values of wave overtopping rate and tide level with the return period.

Hazard ranking criteria for storm surge floods has been proposed by Torii and Kato (2001). In their method, relative hazards to human are evaluated based on maximum flood depth and flood arrival time obtained from the flood simulation.

CONCLUSIONS

The method of the risk assessment established in this study can be used for making hazard maps on storm surge floods. Besides, our method of the probabilistic evaluation on tide level and wave overtopping rate with the stochastic typhoon model would be useful for design of shore protection works where typhoons are the dominant cause of high tides and waves.

REFERENCES

Torii, K. and F. Kato: Study on Storm Surge Disaster, Technical Memorandum of Public Works Research Institute, No.3803, 97p., 2001. (in Japanese)

Yamaguchi, M., Y. Hatada and Y. Nakamura: Probabilistic Typhoon Model and Its Application to Extreme Value Estimation of Storm Surges and Waves, Proceedings of Annual Conference of Japan Society for Natural Disaster Science, pp.135-138, 1994. (in Japanese)

TOWARDS BETTER COASTAL ZONE MANAGEMENT PLANNING – SRI LANKAN EXPERIENCES

Mangala Wickramanayake
Chief Engineer (Research & Designs), Coast Conservation Department, P.O. Box 556, Colombo, Sri Lanka
Phone: 94 1 449755; Fax: 94 1 445205
Email: mwickramanayake@hotmail.com

Indra Ranasinghe
Manager, (Coastal Resources Development), Coast Conservation, P.O. Box 556 Colombo, Sri Lanka

Abstract

As an Island Sri Lanka has always been closely related with coastal environment and maritime related activities. Its 1770-km long coastline is a critical lifeline that supports the country's socio economic and environmental development. Sri Lanka's coastal region contains 24 percent of the country's land area and 32 percent of the population within 67 coastal bounded administration units. Also it contains 65 percent of the urbanized land area and nearly 80 percent of the tourism related activities, and produces 70 percent of total industrial output. The coastal region produces 80 percent of the country's annual fish production from both coastal and brakish water areas. The main infrastructures high ways, rail link and Airports are located within the coastal zone and therefore the coastal areas are extremely important for the national economy.

The natural coastal ecosystems found within the coastal zone are diverse, and include lagoons, coral reefs, mangrove swamps, salt marshes, sea grass beds, barrier beaches and dune systems. Coastal resources play an important role in social and economic well-being of the people in terms of livelihood, economic output and food production. However these ecosystems are extremely susceptible to degradation of the coastal zone, while extremely valuable to the country's economy is also fragile and highly vulnerable to many dynamic process which occur both land and sea shaped by natural and man induced forces. As a result of this process and unsustainable extractive practices a number of critical issues have been appeared in the coastal environment. This includes coastal erosion, coastal pollution, loss and degradation of coastal habitats and coastal resources, loss and degradation of scenic resources within the coastal zone. The inability of the resource base to renew itself under heavy exploitation and growing population has led to create social and economic problems within the coastal zone during the past few decades.

As a policy response to the above situation, Sri Lanka government placed great emphasis on adopting more technical oriented strategies especially to curb coastal erosion problem. However, due to partial management intervention adopting through short term engineering solutions made no or little positive results with compared to the severity of the environmental problems prevailed within the coastal zone. With the realization of increasing capital expenditure on coast protection and low level of positive results of short term engineering solutions done on adhoc basis, Sri Lanka Government shifted its policy approach from coast protection to Integrated Coastal Resources Management by strengthening Institutional , legal and management capacities. Under this process the legal framework for coastal resource management was introduced with the enactment of Coast Conservation Act No. 57 of 1981. In addition, the "first generation" National Coastal Zone Management Plan was prepared and implemented in 1990 with a view to minimizing prevalent issues within the coastal zone. To achieve the desired objectives of the Coastal Zone Management Plan (CZM Plan), and considering the multi-dimension of the issues, management strategies were formulated in the fields of coastal construction (long term coastal stabilization schemes instead of short term coast protection structures) regulations, plan and policy development, coordination, monitoring and environmental education and awareness. In this process both hard and soft management strategies were introduced and implemented. In considering the interrelationship between coastal issues and the activities of the coastal communities, heavy emphasis was placed on public participation and awareness on decision making especially on hard engineering solutions such as outlet management, coast protection schemes, salinity barriers etc.

The "second generation" Coastal Zone Management Plan was produces in 1997 as a revision to the first generation plan of 1990 to address some coastal issues not covered in the first generation plan. Presently the Coast Conservation Department is in the process of preparing the "third generation" plan through the experiences and feed back gained through the implementation of Coastal Zone Management Plan of 1990 & 1997, along with the present development trends in the country. Some critical areas not covered in the previous CZM Plans such as sea level rise, coastal pollution from industrial discharges and urban wastes are also identified to be addressed. The new policies reflect the need of devolve responsibilities to local government and communities for sustainable Coastal Resources Management.

This paper discusses the achievements, constraints, problems faced and lessons learnt in implementation of the previous CZM Plans. It also discusses how the policies and strategies described therein have been tested and refined in the past decade since the initial plans were adopted in order to provide a better planning process towards the "third generation" Coastal Zone Management Plan.

References :

1. CZM Plan, 1990, Coast Conservation Department
2. CZM Plan, 1997, Coast Conservation Department
3. Master Plan for Coastal Erosion Management
4. Development Permit Compliance Survey 2000

PRESENT SITUATION OF COASTAL PROTECTION SYSTEM IN ISLAND COUNTRIES IN THE SOUTH PACIFIC

Paulo Vanualailai[1] and Nobuo Mimura[2]

[1]Department of Urban and Civil Engineering
Graduate School of Science and Engineering, Ibaraki University
[2]Center for Water Environment Studies, Ibaraki University
4-12-1, Nakanarusawa, Hitachi, Ibaraki, 316-8511, Japan
Tel: 0294-38-5169, Fax: 0294-38-5268
E-mail: [1] nd1515x@hcs.ipc.ibaraki.ac.jp, [2] mimura@hcs.ibaraki.ac.jp

(Send all correspondence to Nobuo Mimura)

1. Objectives

Small island countries in the South Pacific have been suffering from increased coastal inundation, flooding, and erosion. Fenney (1989) and Carter (1990) reported the occurrences of cyclones over the past century "to average about one cyclone per year with a severe cyclone once every four years" (SOPAC 1996). Loss of mangroves and increased bleaching of coral reefs (Wilkinson 1998) have been also reported. Moreover, IPCC (2001) predicted sea-level rises due to global warming by 2100 to be from 9cm to 88cm. These factors would induce increased threats on the coastal zone in these countries. Therefore, mitigation options to coastal disasters and sea-level rise are an issue of high priority for small island nations such as Fiji, Samoa, Kiribati, Cook Islands, Tuvalu, Solomon Islands, Federated States of Micronesia (FSM), Marshall Islands and Tonga, for almost 90% or more of the vital political, social, and economic zones are located in vulnerable coastal areas. These countries have paid many efforts by conventional or traditional means to protect their coastal areas against the natural disasters. However, it seems that these efforts are not effective and sufficient.

In this study, we carried out a technology assessment of coastal protection systems(CPSs) with the following objectives:

1) to review past practices of coastal protection systems and to identify major constraints for its effectiveness. It outlines how effective these structures are for the past and recent natural disasters such as storm surges, high waves, cyclones and tsunamis in the region;
2) to suggest relevant system or structures for such constraints using available local knowledge derived from the region combined with coastal engineering methodologies that are currently lacking in the design and construction of coastal structures;
3) to develop criteria for assessment of existing and desirable technologies for coastal protection appropriate for the South Pacific.

In this study, the examples are taken from Fiji, Samoa, Kiribati, Cook Islands, Tuvalu, Solomon Islands, Federated States of Micronesia (FSM), Marshall and Tonga.

2. Coastal Protection System in Fiji and the South Pacific

The two main types of CPSs are hard and soft measures. The hard measures includes seawalls, breakwaters, groins, revetments, dykes and causeways, while the soft structures are coastal vegetation, coral and mangrove protection management, integrated coastal zone management (ICZM) and policies such as setback and retreat strategies.

The appropriate measures for protecting coastal areas against natural disasters are very limited in the South Pacific. Through technology assessment in the present study, we identified the common problems prevailing throughout the region. These problems are; inappropriate material, placement, and design of CPSs, and their social, environmental and economic impact. The problems also included how to mitigate the vulnerability of the fragile island ecosystems to coastal disaster and sea level rise by using CPSs.

Furthermore, the complex nature and absence of political and legal framework makes ICZM and appropriate options difficult to implement. In the presentation of the conference, we will show these constraints in detail.

3. Needed Approaches to Problems

As solutions to the problems of CPSs in the South Pacific, this study identified three main areas, i.e. engineering, planning, and institutional,.

1) Engineering aspect

Three important areas were identified. They are: i) accumulation of scientific knowledge, ii) evaluation and assessment of structural design, and iii) appropriate guideline criteria. In the South Pacific, CPSs in situ are causing so many problems on the coastal environment, because of poor design and construction technologies. There is a lack of understanding on the basic scientific knowledge of coastal processes by the government and locals. This includes scientific and engineering understanding on wave motion, currents, littoral drift, the effect of sea level rise on coastal areas, the effect of structures, the importance of coral reefs and mangroves as natural defense system etc.

2) Planning aspect

Three main areas identified are i) land-use plan, ii) environmental protection plan and, iii) natural disasters mitigation plan. Since land-tenure systems are a very complex issue in the South Pacific region, it is very difficult to implement and facilitate proper coastal development and mitigation measures. Therefore, proper legal tools are needed at first to collectively present a clear goal and direction of all the responsible bodies involved. National plan for natural disasters mitigation should be incorporated into the traditional disaster mitigation plans of the locals. Setback strategies should include movement of settlement in the villages located in low-lying areas vulnerable to coastal flooding, and adoption of traditional houses that offers faster evacuation and recovery, and elevated high pole houses in low lying areas if land is limited. More and better telecommunications systems should be provided to very remote and isolated islands to consistently update and forecast weather patterns to the locals.

3) Institutional aspect

Many South Pacific Island Countries lacked institutional capacities to address vulnerability of coastal areas. It is evident that rules and policies to organize management and development of coastal zone are not well structured. For example in Fiji, setback policy regarding construction of infrastructure on vulnerable coastal areas is ineffective and even confusing. Due to this discrepancy, very little policing enforcing, and monitoring of coastal development, such as construction of tourism infrastructures, road and ports facilities, sand mining, waste disposal, are carried out without proper environmental impact assessment (EIA) from governmental bodies responsible for such developments.

4. Conclusion

Through this study, we reviewed the present situation of the coastal protection system in the South Pacific. Though many countries have paid many efforts to protect their coastal areas, the effects are limited and unsuccessful. This study identified three common constraints, and suggested approaches needed to overcome these constraints. Based on these consideration, we also developed a set of criteria for assessing existing and desirable technologies for coastal protection appropriate for the South Pacific, which will be indicated in our presentation.

The study also suggested that advanced technological methods should be imported from successful projects in similar coastal settings from developed countries. These artificial structures should be combined with the region's existing natural coastal defense system, such as mangroves and corals, to obtain optimum advantage of coastal protection.

SIMPLE MODEL FOR COST ALLOCATION BETWEEN CONSTRUCTION AND MAINTENANCE OF COASTAL FACILITIES

Hajime Tanji, Yoshinobu Araragi, Hirohide Kiri and Ryouichi Ohnishi
National Institute for Rural Engineering, Japan

1. Forewords Aim of the study is to show a simple model for cost allocation between construction and maintenance of coastal facilities, which is basically depend on LCC. The authors already suggested financial limitation and LCC 1). Here, the authors mainly concern about the following two problems.

1) Difference of percentage of Grants-in-Aid in construction and maintenance makes bias to reconstruction period that is most suitable by LCC.

2) To find the proper inspection cost by LCC calculation

2. Basic Concept of LCC LCC is the procedure to minimize the cost of design, construction and maintenance. Comparing B/C analysis and LCC method, LCC is under the assumption that benefit is always constant and the cost to endure the aimed benefit is calculated. Basic concept of LCC is shown as **Figure 1.** Under the simplest assumption, each term is described as follows 2),

$$e = a \times (1-j)^t$$

$$b = \sum_{n=1}^{t} b_n \cdot (1+i)^{-n}$$

$$d = d \cdot (1+i)^{-t}$$

$$f = a + b + d + e$$

Where, a: initial cost, b: maintenance cost, e: rest of initial cost, f: total cost, d disposal cost, t: year after construction, i: inflation ratio, j: write-down ratio. In this case, maintenance cost b is assumed as constant per year and an inflation ratio is considered.

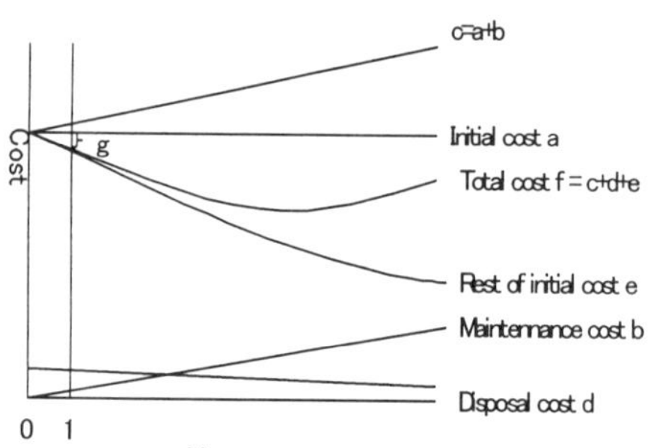

Figure 1 Basic concept of LCC

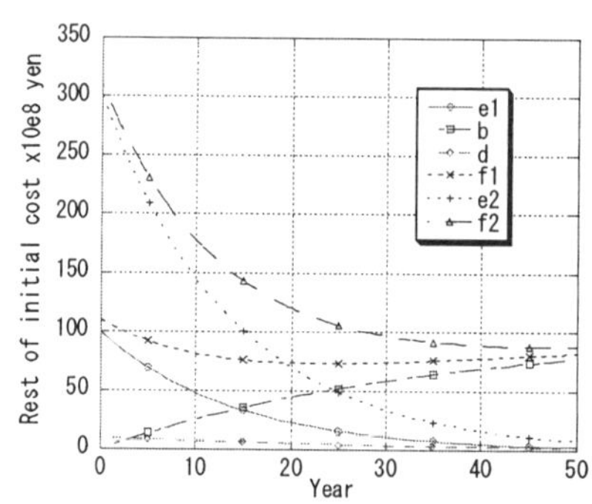

Figure 2 LCC of Grant-in Aid

3. Models, Results and Discussion

3.1 Grants-in-Aid problem Coastal protection facilities can get Grant-in-Aid of 67% at the construction, 33% at repairing construction and 0% at maintenance by the central government. In case of recovery of disaster damage, almost 100% of Grant-in-Aid can be given. Therefore there are very small incentives for the local government to renew coastal facilities at the best period based on LCC.

In a simple calculation of LCC, Grant-in-Aid for construction can be expressed as the change of initial cost. If the construction cost of coastal facilities is 10 billion yen and grant-in-aid rate is 67%, the net cost for construction for the local government is 0.33 billion yen. The effect of Grant-in-Aid to LCC is calculated just as shown in **Figure 2**. In this case, the difference of Grant-in-Aid is expressed as the difference of an initial cost. The actual initial cost is 10 billion yen (e1 in **Figure 2**). But considering the Grant-in-Aid of 67%, the initial cost will influence as 30 billion yen (e2 in **Figure 2**) to the suitable period of reconstruction based on LCC. Here, inflation ration of 3%, write-down ration of 7%, maintenance cost of 0.3 billion yen per year and disposal cost of 1 billion yen were assumed. In case of the initial cost of 10 billion yen without Grant-in-Aid, the profitable reconstruction period was 23 years. But in case of the initial cost of 30 billion yen with Grant-in-Aid of 67%, the profitable reconstruction period was 49 years. Therefore the grant made the reconstruction period 2.1 times long.

3.2 Inspection cost problem In the basic LCC model, maintenance cost is treated as a constant value per a year. But, for considering suitable cost for inspection, maintenance cost should be modeled in detail. The authors make an assumption that maintenance cost consist of the following parts.

(1) Inspection cost (2) Operation cost (fuel, changing of easy ageing parts) (3) Repairing Cost (renew a broken portion of facilities)

There three parts represents the following characteristics. Operation cost is independent to inspection cost and repairing cost. Repairing cost depend on the elapsed time between when the something is out of order and the finding of that fact. Generally thinking, the repairing cost will become larger if the elapse time becomes larger. This relation is shown as **Figure 4**. The relation between inspection cost and this elapse time is shown as Figure 3. If inspection cost becomes higher, which means the interval of inspection becomes shorter. Therefore inspection cost and the elapse time between the errors and those findings have minus correlation. Main factor is the ratio between repairing cost to inspection cost. If locally, this value is approximately –3, the LCC simulation results showed the profitable reconstruction period of 34 years under the condition of 40% increase of inspection cost.

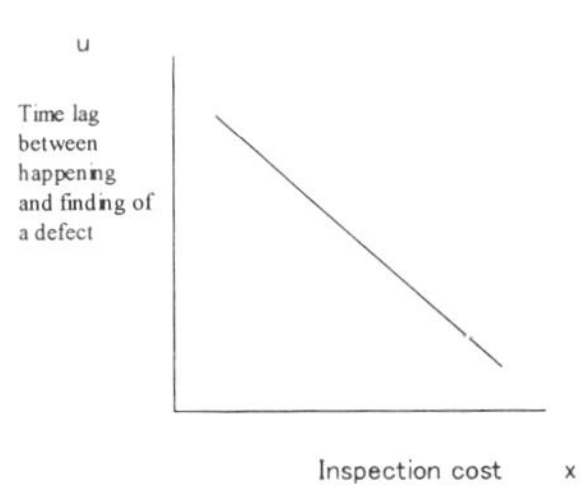

Figure 3 Inspection cost and error finding

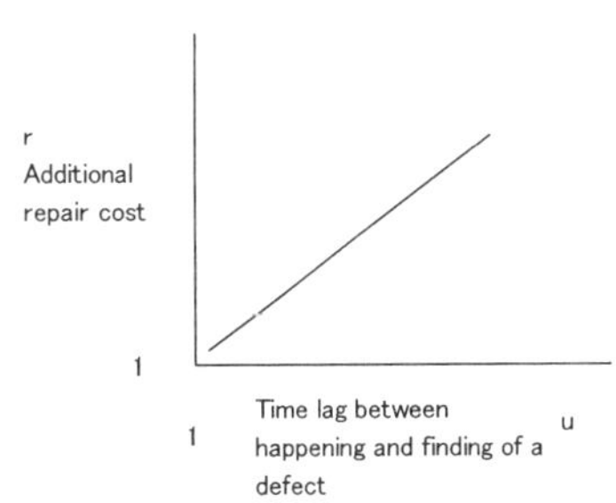

Figure 4 Error finding and repair cost

4. Conclusion The following points made clear here.

1) If there is the difference of Grant-in-Aid, which makes a bias to reconstruction time. In the case study, this difference was approximately 2.1 times long when Grant-in-Aid of 67 % was supported.

2) A simple model of inspection cost and maintenance cost was constructed based on LCC a model. This model can calculate sensitivity analysis of inspection cost and repairing cost.

5. References 1) Tanji, Hajime and Yoshinobu Araragi (2002): A Proposal on Renewal Time and Order of Old Pumping Stations, Journal of the Japanese Society of Irrigation, Drainage and Reclamation Engineering, 70(12), pp.1-5, in Japanese 2)MAFF (2001): Guideline of Renewal of Agricultural Facilities, p.284, in Japanese

Multi-agent Simulation (MAS) Techniques for Coastal Land Use Change Modeling

Maria Cecilia D. Rubio and Mario R. delos Reyes
School of Urban and Regional Planning
University of the Philippines, Diliman, Quezon City, 1101 Philippines
E-mail: mcdrubio@wv.mei.titech.ac.jp

ABSTRACT

Introduction and Objectives

The coastal area contains diverse and productive habitats important to human settlements, development and local subsidence. Planning the coastal land use plays a vital role in coastal zone management. Human-based parameters such as social values, land prices, available natural resources, etc., must be given importance thru the design of pre-determined steps to bring a situation that is mutually beneficial to all concerned entities. Land use change analysis provides a picture on how people react or adapt to environment depending on their needs and values. If these things are fully captured as basis of the land use plans, the people will feel important, thus, will uplift their sense of commitment in the planning and management of coastal areas

This study seeks in general, to explore the utility of multi-agent system (MAS) as an alternative to traditional analytical tools such as map sieving and GIS and as dynamic geographic modeling tool to coastal land use change analysis in the coastal area of Ormoc City, Philippines. The main focus of this study is to apply MAS techniques to model land use change in an attempt to understand how land-based human activities brought change to the current state of land. It is impossible for a real-world environment such as Ormoc City to be subjected to a set of controls and do experiments typical of a laboratory research. MAS enables the researcher to do experiments with it and thus provide understanding of the interactions of dynamic system components such as land uses. The expected result for this study is a simulation model to show the dynamics, particularly in land use changes, in coastal zone, which could help improve its management. The simulation model may be used to give definitive clues on the implications of proposed policies made in coastal zone management prior to their implementation.

Study Area

The coverage of the selected study area, Ormoc City, falls inside a square bounded by the following geographic coordinates (11°5'22.30"N, 124°32'16.56"E) and (11°0'34.18"N, 124°37'41.19"E). It has a total area of 10,000 hectares which is about 24 percent of the total land area of the City. It contains the coastal portion and the urban core of the city.

Methods and Materials

First, a set of Landsat multi-temporal images was analyzed in terms of spectral responses from different coastal land cover types. Actors responsible for changes detected were identified together with their associated transition rules in land cover or use change. The transition rules were organized to come up with a simulation model that produces the spatial distribution of land uses through time given a set of agent and knowledge of their apparent decision. The images eventually classified according to change were used to validate the results of the simulation model. (See figure 1.)

Results

About 50 percent of the total study area or 5,052 hectares has not undergone any change according to the overlaying of the 1989 and 2000 classified images. Only 35 percent or 3,509 hectares are being taken into account in the discussion of the change process, while the remaining percentage are considered either as unconventional changes or as unaccounted changes due to presence of clouds and shadows. Unconventional changes refer to the built-up areas in 1989 data sets that are classified as either water, bareland, forest, agricultural or fishponds in 2000 data set. Also, it is also seemingly rare that water will change to built-up, agricultural and fishpond. This can be proven by low probability values in table 1 . In comparing the transition matrices from different scenarios versus the transition matrix from the images in table 2, probabilities are high along the diagonals or the no change process. In some portion of the table, there are some inconsistencies in degree of probabilities. This can be attributed to the fact that there is the presence of undocumented disturbances in the images, which are not considered in the model such as presence of clouds and shadows in the recent image. Looking at the pattern of high and probabilities, in both instances despite the disturbances, they occur in almost the cell position. The simulated and the actual data show a defined correlation thus validating the model. With improved satellite data classification, it might provide better correlation result.

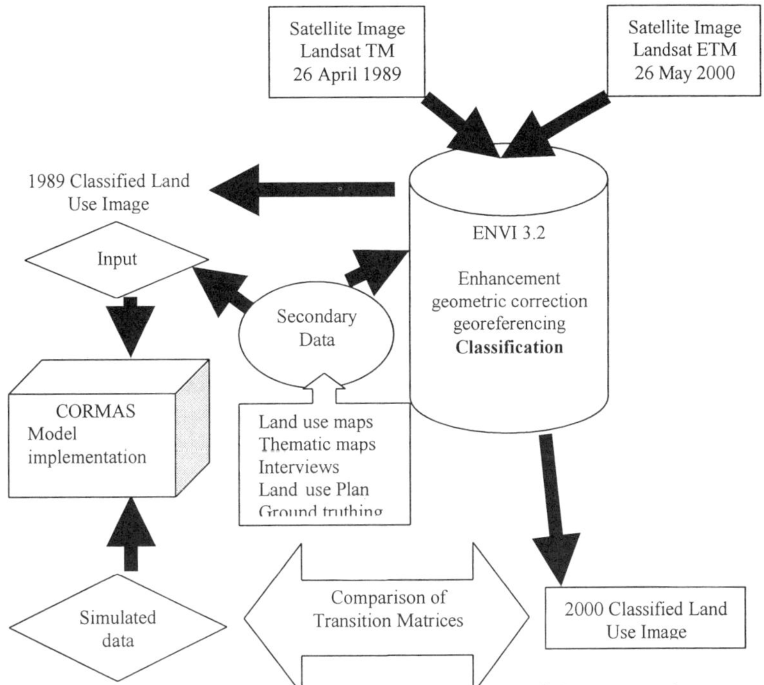

Figure 1. Conceptual Framework

Table 1

		April 1989					
		Water	Built-up	Bareland	Forest	Agricultural	Fishpond
May 2000	Water	**90**	2	0	0	0	0
	Built-up	1	**57**	9	5	**12**	7
	Bareland	0	10	**36**	6	**16**	5
	Forest	1	3	1	**41**	4	8
	Agricultural	1	9	**43**	**37**	**55**	**26**
	Fishpond	7	10	8	5	8	**45**

Table 2

	NoChange	Initial State					
	Scenario	Water	Built-up	Bareland	Forest	Agricultural	Fishpond
Simulation Result	Water	**100**	0	0	0	0	0
	Built-up	0	**100**	2	1	2	3
	Bareland	0	0	**53**	14	36	2
	Forest	0	0	21	**70**	24	13
	Agricultural	0	0	24	10	**38**	11
	Fishpond	0	0	0	5	0	**72**

COAST PROTECTION IN SRI LANKA: SHIFT IN PARADIGMS FROM HARD SOLUTIONS TO SOFT SOLUTIONS

Nimal Wijayaratna and Channa Fernando

Abstract:

The western and southern coastal belts of Sri Lanka are both socially as well as economically important stretches due to high population density and high concentration of fisheries and tourism infrastructures. This coastline has become more vulnerable to sea erosion, partly due to seasonal variation in the wave climate and partly due to the human interference in the natural sediment movement process. Human interference is in the ways of sand mining in the rivers (thus reducing the sand supply to the coast), destruction of nearshore coral reefs for economic activities, utilizing coastal zone for unwarranted development activities and ad-hoc construction of hard structures to arrest the coastal erosion.

Until recently, the protective measures adopted were mainly evolved around the construction of hard structures such as revetments and groynes. Further, erosion problem was seen as isolated to a local area and the solutions were focused at a limited stretch where the erosion was severe. The impacts of such structures on the adjacent coastlines were less addressed. As a result, in most of the situations, the problem was shifted to elsewhere. Moreover, such hard structures were unpleasant and denied the free access to the beach for economic and recreational activities. Hence the need for shift in paradigms in analysing and providing solutions for coastal erosion problem has been stressed.

Present day policy makers and designers are more inclined towards soft solutions such as beach nourishment supplemented with strict enforcement of laws to limit the unwarranted human interference on the coastal zone. Moreover a holistic approach is being adopted in analysing an erosion problem by taking a longer stretch of the coastline for consideration through accommodating the concept of 'coastal cell' and thus providing protective measures for the coastal cell, instead of only for the localized eroding area, while taking all existing and future developments into consideration. In providing the soft solutions such as beach nourishment, loss of sand during the 'design life' will be artificially supplied, with hard structures only to retain the supplied sand.

Even though the aforesaid approach is seen as prudent and providing sustainable solutions to erosion problems, it has given rise to the crop up of some socio-political and technical issues that need to be addressed by the policy makers and designers. In socio-political point of view, beach nourishment and associated periodic replenishment is looked as a temporary solution to overcome an erosion problem, and merely recognized it as putting the lost sand during erosion, back on the beach from time to time to be carried away during the subsequent eroding cycle. Further, absence of any hard structures (or minimal number of structures) in front of human settlements and other coastal development projects, is seems to be psychologically not acceptable as capable of providing protection during severe wave attack. Criticism is levelled against the designers for the longer time taken in analysing the erosion problem (which is necessary to address the problem in a more holistic manner), and for

extending the 'solution area' beyond the 'problem area'. Need for allocating funds for periodic replenishments is also found to be a major political issue.
In technical point of view, lack of institutional capacity to maintain necessary equipments (dredgers etc.) for undertaking periodic replenishments, limited knowledge on the availability of offshore sand deposits and their physical and chemical characteristics are found to be the main obstacles that impede the implementation of the new 'soft solution' concept. The uncertainty associated in the prediction of replenishment frequency is identified as both socio-politically and technically sensitive issue.

The paper will describe in detail some of the main coastal protection projects recently being designed / concluded having the main focus on the concept of 'soft solutions', while giving due consideration to the associated socio-political as well as technical issues faced during planning / implementation.

Authors:

1) Nimal Wijayaratna
Senior Lecturer
Department of Civil and Environmental Engineering,
Faculty of Engineering,
University of Ruhuna,
Hapugala, Galle,
Sri Lanka

Phone: +94-9-45764
Fax: +94-9-45762
Email: nwijayaratna@yahoo.com
nimal@cee.ruh.ac.lk

2) Channa Fernando
Senior Research Engineer
Lanka Hydraulic Institute,
John Rodrigo Mawatha,
Kaubedda, Moratuwa,
Sri Lanka
Phone: +94-1-650409
Fax: +94-1-650470
Email: channa@lhi.lk

Correspondances:

Please forward all correspondences to the first author, Nimal Wijayaratna.

PLANNING OF FISHERY HARBOUR FACILITIES IN SRI LANKA: AN EXAMPLE OF A RATIONAL APPROACH ADOPTED IN A DEVELOPING COUNTRY

Channa Fernando

Abstract:

Being an island with a total of 1770 km long coastline, the aquatic environment which endowed with rich aquatic habitats plays an important role in Sri Lanka's economy. The exploitable fishery resources of the country could be broadly categorized into fresh water, brackish water and marine water, of which marine waters provide more than 80% of the total fishery resources.

Fishing crafts used by Sri Lankan fishermen range from indigenous log crafts to modern multi-day boats. The concentration of fishing activities in Sri Lanka has expanded from the traditional fishing grounds in lagoons, estuaries and inshore waters towards the oceanic waters offshore. The general trend during the past decades shows that larger crafts continuously tried to extend their radius of operation while smaller crafts tried to fill up the void brought about by this and likewise operated farther from shore.

It has been evident that expansion in the larger multi-day boats is impeded by the lack of safe anchorage facilities and other related infrastructure facilities. This has led to the over-exploitation of marine resources in the nearshore coastal waters by the undesired expansion of the smaller boats while the much valued offshore resources kept untouched. Review of the existing harbour facilities indicated that the shortfall in the harbour facilities is not mainly due to the lack of investment in the area, but due to the ad-hoc planning and short sighted investment decisions made. This has led to the under utilization of some of the harbours in spite of the vast investments done, while some highly deserving areas are not provided with any facility. Further more, siting of harbours at technically undesirable locations has given rise to the severe coastal morphological and environmental problems. Hence the importance of a more rational approach is stressed in planning such fishery harbour facilities which would reap the optimum benefits from the investment made.

Under the Fisheries Harbour Component of the ongoing Coastal Resources Management Project (CRMP) funded by Asian Development Bank (ADB), Sri Lanka has embarked on developing a more rational policy which encompasses technical, economical, social and political considerations to support the decisions making process.

As the first step in the policy development, an inventory of the geographical distribution of different types of fishing crafts engaged in the fishing, and boat accommodating capacities of existing harbours along the coastline of Sri Lanka was taken. This has helped to identify the geographical locations of over-crowded and under-utilized harbour facilities and highly deserving areas which currently have no facility at all. As the second step, potential future increase in fishing activity in different coastal stretches (geographically) was assessed. This was done at two levels; by comparing total number of fishermen currently engaged in the industry and number of potential members in their families who may enter the industry in the future. While the statistics on total number of current day fishermen gave an indication on the present trend in increase of fishing crafts

(if proper facilities are provided), the number of potential members gave an indication of the trend of future demand for such facilities. As the third step, geographical distributions of current day fish production at each stretch was assessed. This gave good qualitative understanding of the potential economic/ financial benefits that can be reaped from siting a harbour at a particular geographical location, as presence of a harbour ensures the safe navigation, hygienic fish handling facilities (better quality fish), and development of competitive regional markets.
By taking these three factors in integration, a list of potential harbour sites along the coastline was developed. Since this potential list was prepared mainly based on the socio-political and partially on economic factors, a more detailed evaluation of these sites with respect to technical, environmental and financial point of view was necessitated. Hence a comprehensive evaluation matrix was developed to assess these sites more objectively and prioritize for investment. The evaluation matrix composed of technical (physical shape of coastline, nature of see bottom, wave impacts, morphological impacts, presence of estuaries, navigability, availability of construction material etc.,), socio-political / economic factors (already available infrastructure facilities and demand for new facilities etc.,) and environmental factors (effect on the environmentally sensitive areas, archeological sites and biodiversity etc.,). Each of these factors were assigned with a total mark depending on its importance in the functionality of the harbour. Then each site in the potential list of harbour locations were assessed with respect to each of these factors and assigned a mark. The summation of these marks for each individual site was done and ranking of the sites was based on such arrived summation.

The paper will describe step by step of the such rationally developed criteria and how the evaluation and ranking of sites were done when some of the evaluation factors in the matrix interact/ influence each other. The benefits and future use of the evaluation matrix will also be addressed.

Author:

Channa Fernando
Senior Research Engineer
Lanka Hydraulic Institute,
John Rodrigo Mawatha,
Katubedda, Moratuwa,
Sri Lanka.
Phone: +94-1-650409
Fax: +94-1-650470
Email: channa@lhi.lk

An Integrated Approach of Remote Sensing and GIS to Poverty Alleviation and Coastal Development in Cox's Bazar, Bangladesh

Md. Shahadat Hossain and Yusuf Sharif Ahmed Khan
Institute of Marine Sciences, University of Chittagong, Chittagong 4331, Bangladesh
Tel. 88-031-710347, Mobile: 88-0171-720950, Fax: 88-031-726310
Email: hossainms@yahoo.com

Abstract

Coastal natural resources in the Cox's Bazar coast of Bangladesh have been used for multi purposes and strongly influenced socio-economic development. Remote sensing imageries provide cost-effective support in compiling the latest information about the environment and natural resources, while GIS brings new functions for multidisciplinary technology and decision support. The integrated analysis can serve as vital inputs for better planning, policy formulation, and decision-making in national development. Landsat TM image and topographic maps of some coastal islands namely Kariadia, Matarbari, Moheshkhali and Sonadia in Cox's Bazar coast were analyzed; and hill, plain land, tidal land, mangroves and waterways were identified and mapped. The opinions of the local community are considered based on the results of a participatory rapid appraisal carried out among the coastal stakeholders. The major sources of livelihood for coastal communities are agriculture, aquaculture, fishing, fish drying, sea salt production and trading. Proper resource utilization and integration among stakeholders activity can increase environment-friendly production efficiency, which have a positive role in successful poverty alleviation program and socio-economic development of coastal rural communities. Finally, this study demonstrates the utility of extracted coastal information to coastal zone managers.

References

Badarch, M. 1997. Mongolia's plan for sustainable development, poverty alleviation and improving the status of women. Space applications for sustainable development. United Nations Publication, ST/ESCAP/1817, pp. 43-50, New York.

BSCIC, 2001, Local climatological effects on solar salt production in Cox's Bazar coastal zone of Bangladesh. Bangladesh Small and Cottage Industries Corporation, salt project, Cox's Bazar.

ESCAP, 1989. Coastal Environmental Management Plan for Bangladesh. Final Report, ESCAP, Bangkok, Thailand.

Hu, Q. 1997. Uses of space technology applications for sustainable rural development in China. Space applications for sustainable development. United Nations Publication, ST/ESCAP/1817, pp. 15- 26, New York.

Jianbo, L. 1997. Remote sensing applications for agricultural resource management in China. Space applications for sustainable development. United Nations Publication, ST/ESCAP/1817, pp. 160-169, New York.

Pido, M. D., Pomeroy, R. S., Carlos, M. B., & Garces, L. R. (1996). A handbook for rapid appraisal of fisheries management systems (version 1). Manila, Philippines: ICLARM.

Rao, R.S., 1996. Integrated mission for sustainable development: a case study of Anantpur District. Technical volume, National Workshop on Application of remote sensing and GIS techniques to integrated rural development, Hyderabad, India, June 1996.

Sharma, T. 1997. An integrated approach to sustainable development of a watershed in India, using remote sensing and GIS. Space applications for sustainable development. United Nations Publication, ST/ESCAP/1817, pp. 35-42, New York.

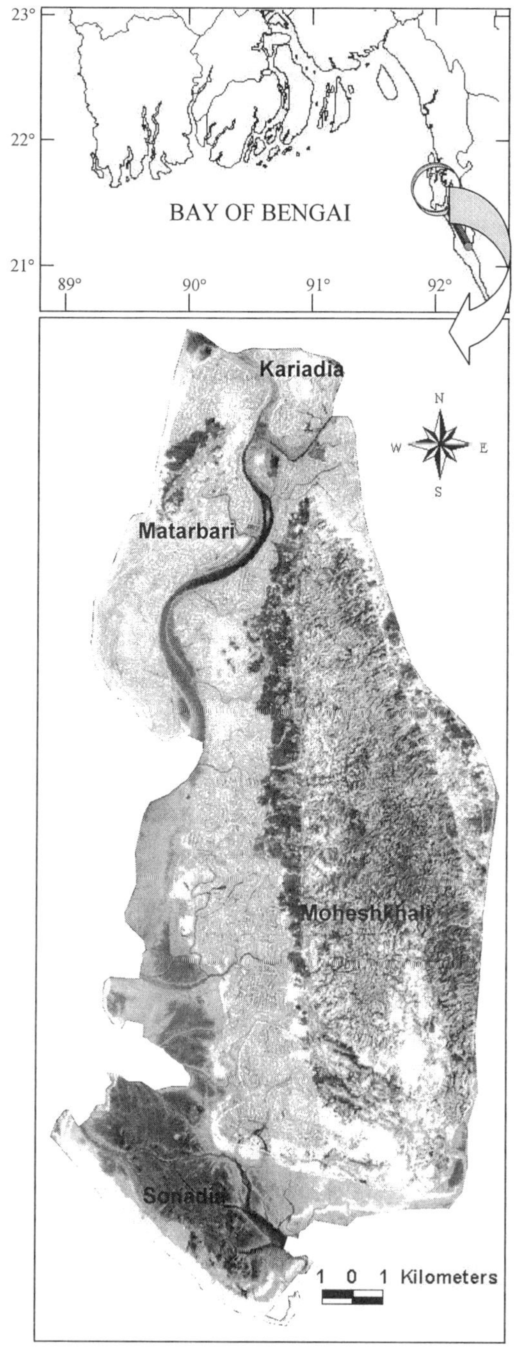

Figure 1. Geographical location of Bangladesh coast and Landsat TM image showing the study area of Kariadia, Matarbari, Moheshkhali and Sonadia islands.

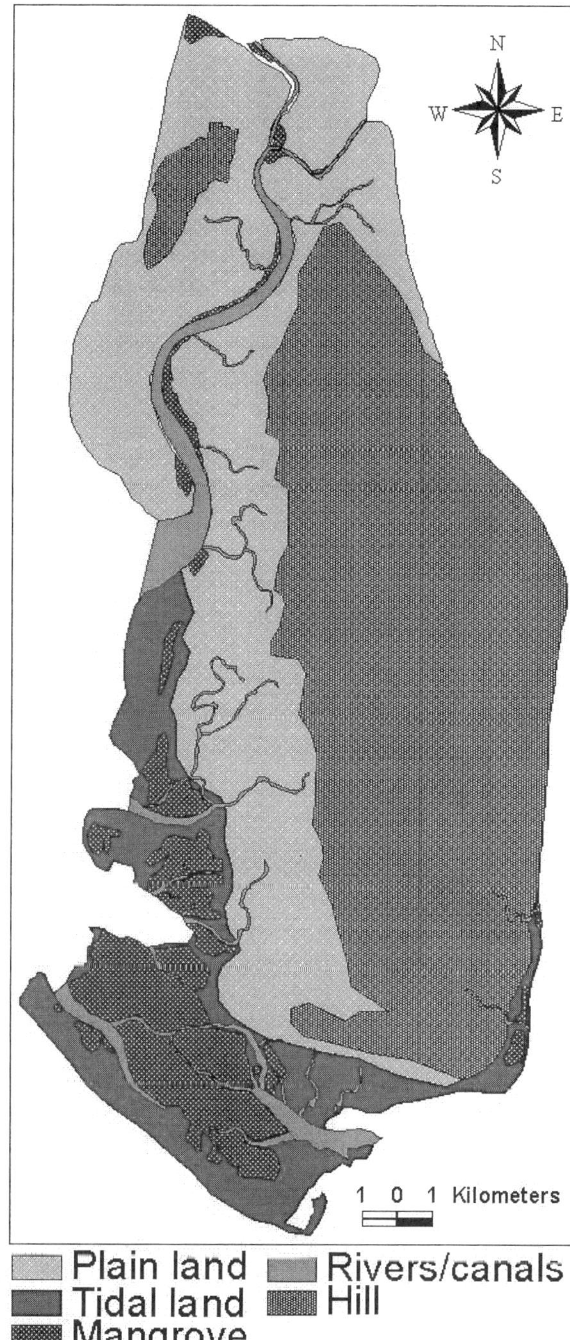

Figure 2. Coastal natural resources of the Kariadia, Matarbari, Moheshkhali and Sonadia islands at Cox's Bazar, Bangladesh.

WAVE AND MORPHOLOGY OBSERVATION DURING A STORM EVENT WITH A NAUTICAL X-BAND RADAR

Satoshi Takewaka and Hitoshi Nishimura
Institute of Engineering Mechanics and Systems, University of Tsukuba, Tsukuba, Ibaraki, Japan 305-8573
Email: takewaka@kz.tsukuba.ac.jp, Tel: 0298-53-5361, Fax: 0298-53-5207

Introduction

A X-band nautical radar system has been employed to observe sea surface patterns during a storm event. Wave gauges and current meters are employed in a conventional field survey, however, the deployment of the instruments is limited, especially in the surf zone. In this context, a radar system is introduced here, which can remotely and broadly infer the sea surface state even during a severe sea conditions.

The radar system was operated at the research pier HORS and radar echo images were collected during a passage of a storm in the vicinity of the pier. Analyses on radar echo images were done and the results demonstrate the potential of the system in observation of sea states and estimation of morphological features under severe sea condition.

X-band radar system

The radar system provides images of sea surface over an area of several kilometers. Figure 1 is an example of a single image with coordinate system used in this study. White pixels in the image correspond to regions where reflection or echo of the emitted radar beam were intense. Vertical streak in the middle of the figure is the pier, which is 400 m in length, and slanted lines extending in longshore direction are positions where wave crests indicating oblique wave incidence to the shore. Images were sampled with an interval of 2 second, which roughly corresponds to the time resolution of the radar system.

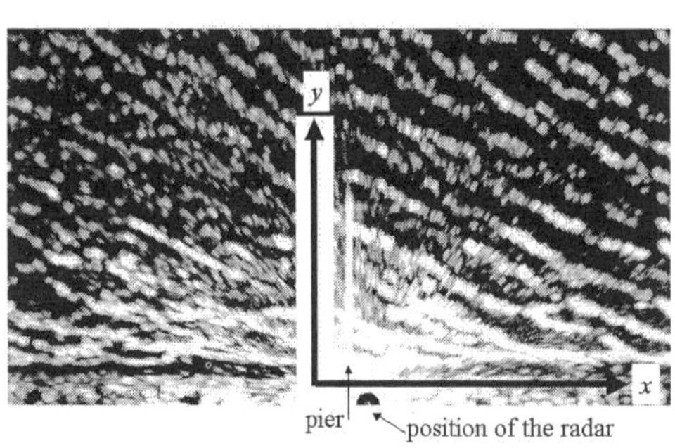

Figure 1 Radar echo image. Horizontal extent = 1800 *m*. Length of the pier = 400 *m*.

Storm event

Super Typhoon CHATAAN (T6, 2002) traveled through western edge of the Pacific Ocean on 10th and 11th of June, 2002. Figure 2 shows changes of mean water level (MWL), significant wave height ($H_{1/3}$) and period ($T_{1/3}$) measured at the pier for 20 hours during the passage of the typhoon. Bottom profile along the pier changed slightly due to this storm event as shown in Fig. 3. The off shore bar (y ~ 350 m) reduced slightly its height and migrated shorewards.

Averaged images

Radar images of every hour were averaged to show the change of locations of the shoreline and distributions of mean echo intensities. Figure 4 are averaged images for t = 13 h and t = 20 h,

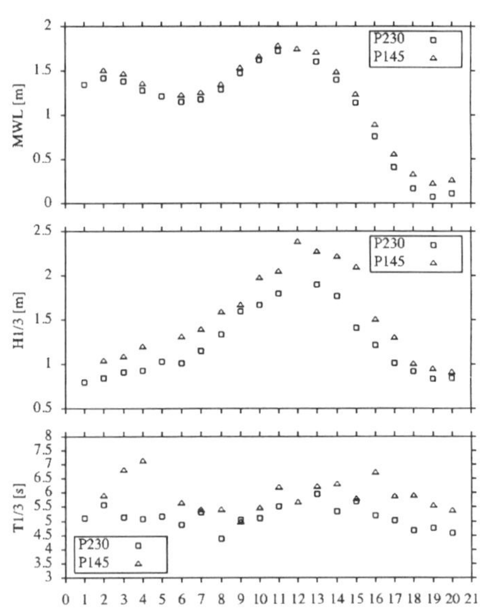

Figure 2 Changes of mean water level (MLW), significant wave height ($H_{1/3}$) and wave period ($T_{1/3}$) during a storm measured at the pier (P230: y = 230 m and P145: y = 145 m).

respectively. Locations of the shoreline varied due to tidal level change and brightness of the pictures differs remarkably, where a brighter image corresponds to a more energetic wave field. The variation of shoreline position at $x = 50\ m$ is shown in Fig. 5. The diagram captures a good correspondence between MWL change measured at the pier and cross shore shift of the shoreline. Longshore distributions of foreshore slope along the shore were estimated from the change of shoreline positions and tidal levels during the storm event.

Estimation of the bottom profile

Local water depths were estimated providing phase speed and wave period to linear dispersion relationship, where both items are estimated from echo images. Figure 6 shows a result of bottom profile estimation along the pier. Discrepancy between the measured and estimated bottom profiles near the trough of the bar is apparent. Change of phase speed in this region does not reflect directly the change of bottom profile, so further refinements accounting on wave and current dynamics are necessary to improve the accuracy of the estimation.

Conclusions

Nautical radar measurement on surf zone waves during a passage of a storm was conducted. Shoreline positions and foreshore slopes were estimated reasonably well from the analyses of radar echo images. Bottom profiles were inferred from the analyses of wave propagation speed, but further refinements are necessary to improve the accuracy of the estimation.

Acknowledgments

The authors are grateful to the staffs of research pier HORS who provided wave and morphological data, and enormous support in radar measurements. Part of this is supported by a grant of JSPS.

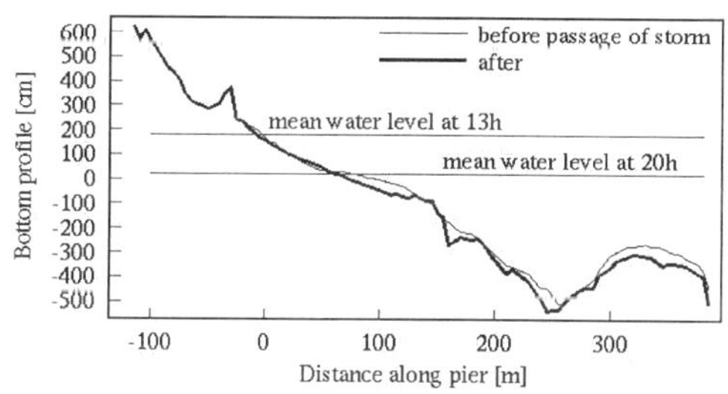

Figure 3 Change of bottom profile along the pier

(1) $t = 13\ h$

(2) $t = 20\ h$

Figure 4 Averaged images at $t = 13\ h$ and $t = 20\ h$.

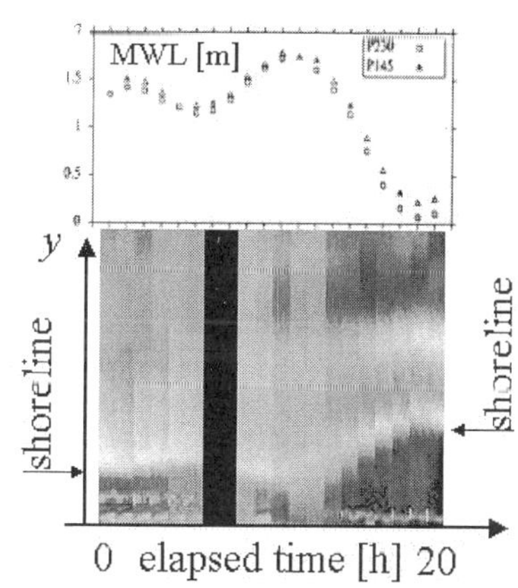

Figure 5 Change of shoreline positions. Upper panel: MWL, lower panel: time stack of averaged image at $x = 50\ m$. Time stack image is a montage of cross shore slices of averaged images as shown in Fig. 4.

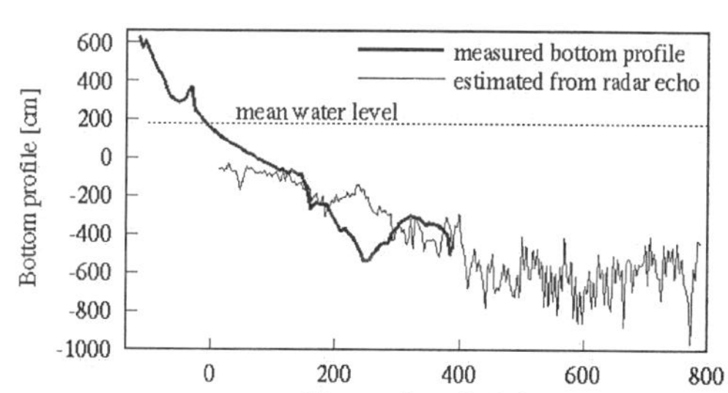

Figure 6 Result of bottom profile estimation along the pier

Development of Profiler System for Suspended Sediment Concentration

Ryoji Sugawara*, Hitoshi Tanaka and Hiroto Yamaji

*Corresponding author
Department of Civil Engineering, Tohoku University
06 Aoba, Sendai 980-8579, Japan.
E-mail: ryoji@kasen1.civil.tohoku.ac.jp
Phone & Fax (81)-22-217-7541

INTRODUCTION

There have been several methods to measure sediment transport rate as suspended load, as well as bed load (e.g., Horikawa, 1988; Seymour, 1989). As for the former mode of sediment movement, the methods proposed up to now can be classified into two groups: (1) light transmission meters and (2) sand samplers. Both type of the measuring system has provided a number of observed data that has been utilized for understanding mechanism of sediment movement and resulting beach topography change. However, all of these existing systems are not suitable for simultaneous data acquisition in vertical direction with sufficiently high spatial resolution, because of the size of a sensor. In the present study, a profiler system is developed for measuring system, which enables us to determine spatial and temporal variation of sediment concentration under waves and/or unidirectional flow.

CONCENTRATION MEASURING SYSTEM

Photo 1 shows a concentration measuring system developed in the present investigation. This system has originally been used for detecting sand surface elevation. In this study, the system is improved to be able to detect sediment concentration in suspension, not only sediment layer on the bottom. The transducer consists of a pair of light source and a phototransistor, and they are mounted with interval of 1cm in vertical direction along a thin rod as seen in Photo.1.

Due to low intensity of a light from the source, output from the phototransistor is easily affected by natural light. For this reason, five types of attachment shown in Fig.1 are tested. However, it should be noted that too long attachment might prevent sand particles from reaching the measuring section between the light source and sensor.

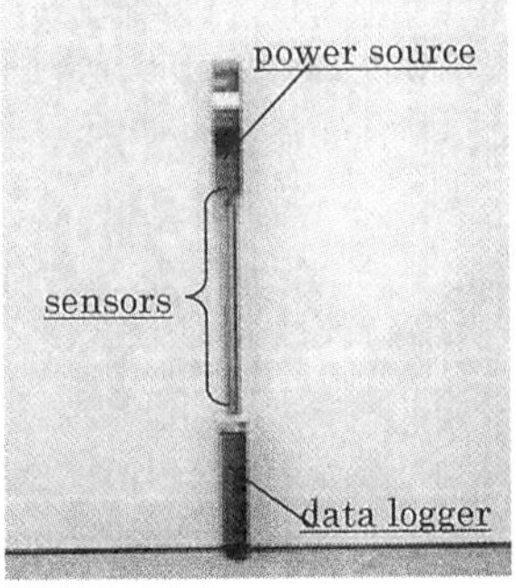

Photo 1 Concentration measuring sensor.

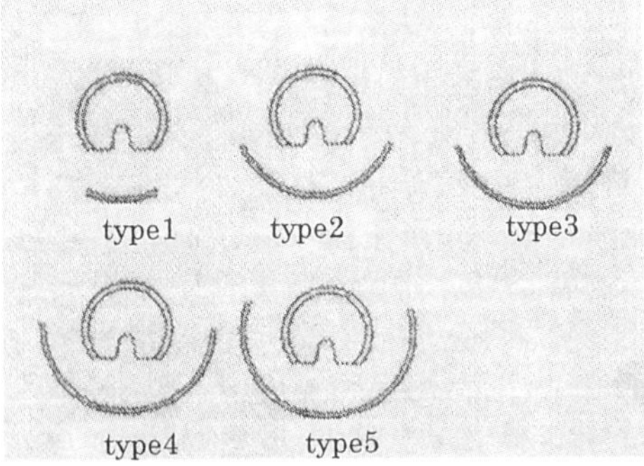

Figure1 Attachment for measuring sensor.

RESULT OF FIELD OBSERVATIONS

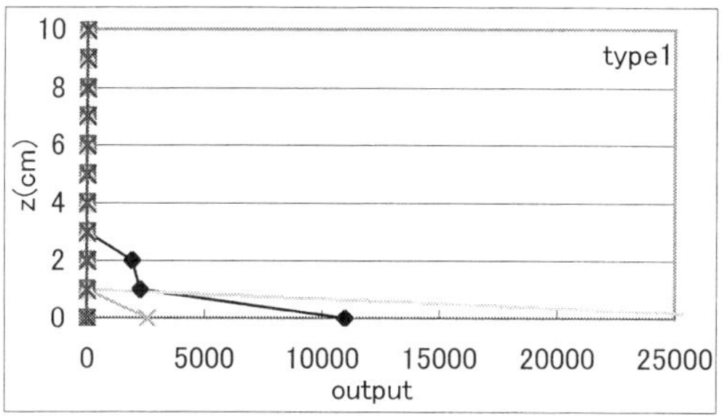

(a) Type 1

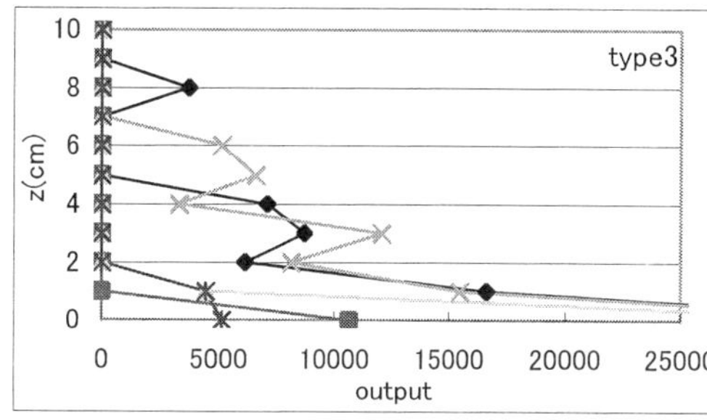

(b) Type 3

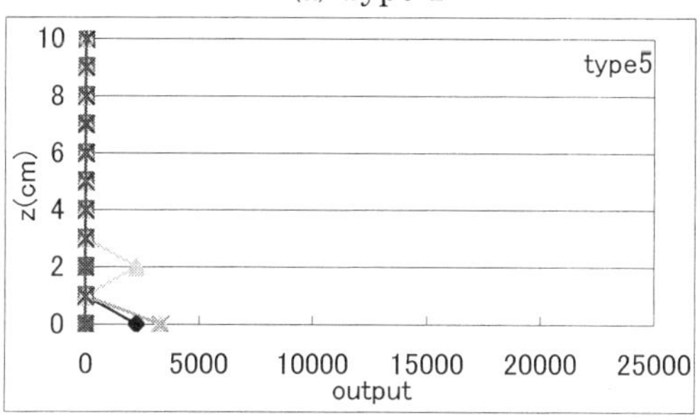

(c) Type 5

Figure 2 Vertical profile of measured concentration

Field test has been carried out at the mouth of Nanakita River, Japan under the combined effect of waves and current. The profile data was obtained every one second and was stored in the data logger unit.

Figure 2 shows an example of measured output signal with the interval one second. Using Type 2 attachment, output from the phototransistor is affected by natural light, although slight variation can be observed in vertical direction. Type 5 also gives similar profile, but the lower intensity of output signal is not due to natural light, but due to too much covering of the sensors by the attachment, which caused less amount of sediment intrusion into the measuring section. In contrast to these two types of attachment, Type 3 gives more realistic variation temporally as well as spatially.

REFRENCES

Horikawa, K. (ed.): *Nearshore Dynamics and Coastal Processes,* University of Tokyo Press, 522pp, 1988.

Seymour, R.J. (ed.): *Nearshore Sediment Transport,* Plenum Press, 418pp, 1989.

EXPERIMENTAL STUDY OF AIR BUBBLE DISTRIBUTIONS INDUCED BY WIND-WAVE BREAKING

Nobuhito Mori

Department of Fluid Mechanics,
Central Research Institute of Electric Power Industry (CRIEPI),
1646 Abiko, Abiko, Chiba 270-1194, JAPAN.
email:mori@criepi.denken.or.jp

1 Introduction

An accurate estimation of gas transfer velocity at the air-sea interface is very important to understand environmental mechanisms of the ocean and coastal area. A wind dependent gas transfer model proposed by Liss and Merlivat model has widely used the last decade. The gas transfer velocity of the Liss and Merlivat model increases rapidly when the wind speed exceeds 13m/s. This rapid increase of gas transfer velocity is explained by several reasons such as enhancements of wave breaking induced turbulence, breaking wave induced air bubbles, and sea spray. However, quantitative roles and detail mechanisms of the enhancement are not well known due to the lack of detail measurements. To understand the gas transfer enhancement in high wind speed region, the bubble mediate gas transfer is one of important phenomenon.

In order to measure bubble behavior induced by wind wave breaking, the phase-Doppler anemometry (PDA) measurement was conducted in wind-wave flume The diameter and velocity components of the air bubbles can were measured using PDA.

2 PDA measurement and Experimental Setup

An experiment was conducted in an acrylic-made small scale wind-wave tank at Abiko Res. Lab., CRIEPI. The wind generator and wave paddle are installed at the end of tank. To enhance breaking probability, mechanically generated waves were generated with wind-wave at wind speed of 9.2m/s (mean velocity). The surface displacements and velocities were measured by wave gage and two velocity component LDV system. The both of LDA/PDA measurements were carried out 11-20 points vertically at 6m downstream of the wind inlet.

PDA measurement can be classified into two types, one is a forward scattering and another is backward scattering measurement[3]. The forward scattering measurement was difficult for the wind-wave tunnel due to the geometrical limitation. Therefore, PDA measurements were carried out with backward scattering method. The 600mm and 500mm focal distant lenses were used for transmitter and receiver, respectively. The measurement period was two hours at each locations.

3 Results and Discussion

column using the PDA and a CCD camera was conducted pre-examine of the measurements in the wind wave flume for the calibration. Instantaneous images of the bubbles about 500mm above the aerator was measured. The differences between PDA and image processing measured diameter were dependent on the bubble diameter. The differences between them were small for small size bubble but were increased with bubble diameter become larger. The one of the reason to explain this discrepancy is that a large size bubble has not spherical shape(r >3mm). Therefore, the error of measurements are increased both PDA and the image processing for large size bubble.

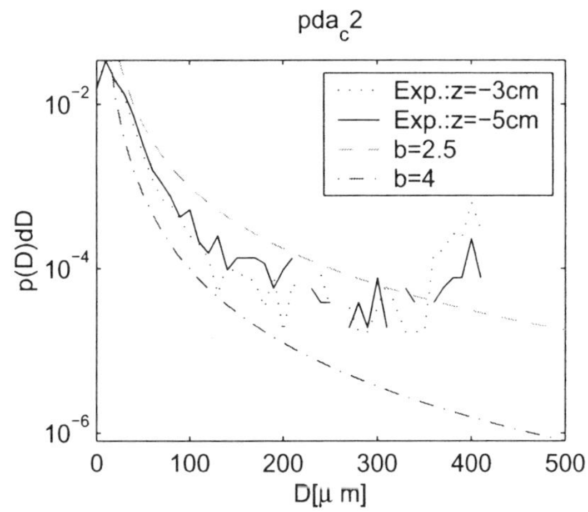

Figure 1: Comparison of the PDF of bubble diameter with log lows

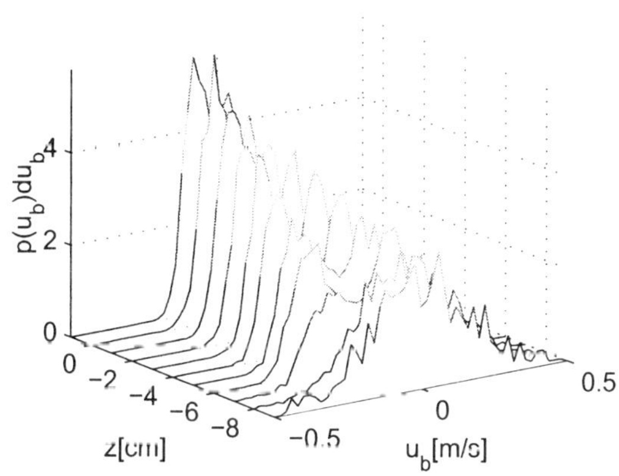

Figure 2: Vertical distribution of vertical bubble velocity

Figure 1 shows comparison of the PDF of bubble diameter with log lows. Monahan[2] acoustically investigated bubble diameter distribution with $D^{-2} \sim D^{-4}$. The experimental data shows D^{-3} approximately and is qualitatively agree with his results.

Figure 2 shows vertical distribution of vertical bubble rising velocity below MWL. The existence of mechanically generated waves, the bubble rising velocity has two peaks near the free surface. However, it becomes the Gaussian like profile deeper than 8cm.

The detail of results and theoretical considerations of bubble mediate gas transfer will be discussed at the conference.

References

[1] David Lee Black and Mardson Q. McQuay. Laser-based particle measurements of spherical and nonspherical particles. *International Journal of Multiphase Flow*, 27:1333–1362, 2001.

[2] E.C. Monahan. Occurrence and evolution of acoustically relevant subsurface bubble plumes and their associated, remotely monitorable surface whitecaps. In B.R. Kerman, editor, *Natural Physical Sources of Underwater Sound.* Kluwer Academic Pub., Dordrecht, 1990.

[3] Shao Lee Soo. *Instrumentation for fluid particle flow.* William Andrew Publishing, NY, 1999.

BUBBLE SIZING AND COUNTING IN A SURF ZONE BY AN ACOUSTIC TECHNIQUE

Yasunori Watanabe[1], Yuji Yamauchi[2] and Hiroshi Saeki[3]
[1]Hokkaido University, Hydraulic Engineering Lab., N13 W8, Sapporo 060-8628, Japan
Tel &Fax: +81-11-706-6185; Email: yasunori@eng.hokudai.ac.jp
[2]ditto for Address and Tel. Email: y-yuji@eng.hokudai.ac.jp
[3]ditto for Address. Tel & Fax: +81-11-706-6183; Email: h-saeki@eng.hokudai.ac.jp

In a surf zone, numbers of bubbles are entrained through a splash-up cycle of plunging jets, possibly impacting on physical and biological nearshore environment – sediment transport, vegetation and water quality. Especially, size and number distributions of the bubbles are predominant factors to determine the drags for the surf zone flow and also gas transfer rate in this region. The distributions may depend on the local fluid motion under breaking waves and surfactant contained in seawater. In this paper, size spectra of the entrained bubbles in the surf zone and the effect of surfactant to the spectrum are investigated on the basis of an acoustic experiment. An interconnecting relation between the jet motion and the bubble size is also examined. The so-called micro-bubbles, suspended in water for long time, were trapped in situ experiments, and the void fractions under seawater breaking waves were also estimated.

In situ experiments were conducted at Ishikari beach, Hokkaido, Japan to estimate a cross-shore distribution of the void fraction for seawater containing the micro-bubbles (c.f. figure 1). The void fraction increases in a transition region up to a bore region, while the bore region contains almost constant volume of gas. Maximum increase of the void fraction can be seen in a swash zone. The distribution of the void fraction under freshwater waves in a laboratory experiment tends to be qualitatively analogous to the in situ experiment while the freshwater has much smaller void fraction than the seawater. It is also found that the void fraction in the freshwater containing surfactant increases up to 50%, compared with the freshwater surf zone.

A resonant frequency of pulsating single bubble was derived by Strasberg (1956), which exhibits a relation between the inherent frequency of sound emitted from the bubble and the bubble size. Based on the resonant relation, the bubble sizes under breaking waves can be estimated by a frequency spectrum of the bubble sounds measured by hydrophones which set in the laboratory surf zone. The process of bubble generation is found to depend on local motion of plunging jets in the transition region, which can be classified into the following three types in terms of the frequencies of the bubble sounds (See figure 2). (i) Rebound type; when the overturning jet rebounds at the plunging point, the bubble sounds with higher frequencies are emitted from very small bubbles in the vicinity of an interface between the jet and undisturbed water surface. (ii) Penetration type; the following water penetrates the undisturbed water surface, and then the entrained bubbles with various sizes emit multi-frequency sounds over a broad range. (iii) Bubble clouds type; the entrained bubbles are trapped by organized large-scale vortices, forming the so-called bubble clouds. The clouds emit low frequency sounds associated with a collective bubble oscillation rather than the single bubble pulsation.

Figure 3 shows the number density for the bubble sizes in the transition and bore regions. While large numbers of bubbles for certain discrete sizes are produced in the transition region, there are various sized bubbles with relatively low number density in the bore region. This result suggests that a certain large-scale fluid motion on the surface produces inherent sized bubbles at the inception of wave breaking, and the turbulence developing through wave breaking process induces the various sized bubbles in the bore region.

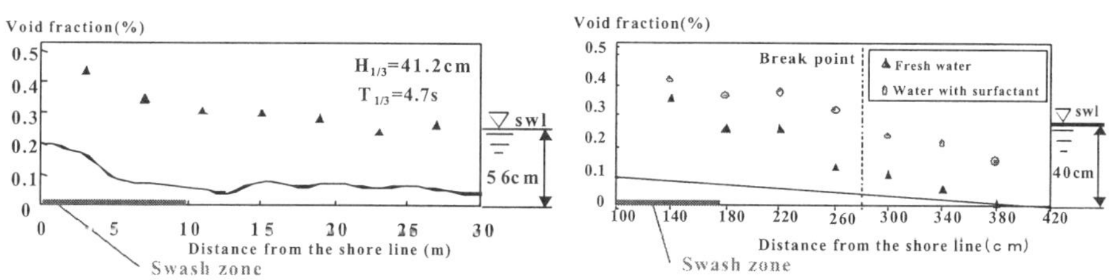

Figure 1. cross-shore distribution of the void fraction (left: field experiment, right: laboratory experiment : Wave period: 2 sec, Breaking wave height: 12.6cm).

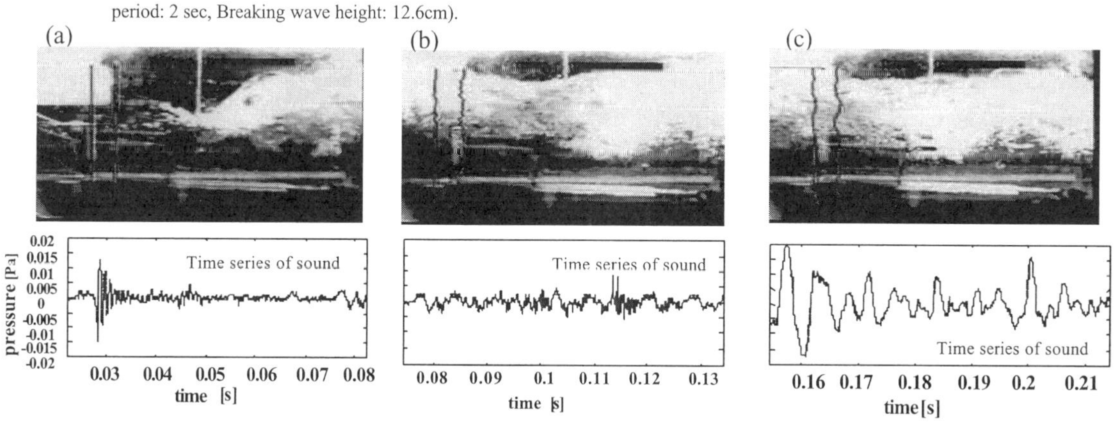

Figure 2 Typical types of the bubble sounds in the transition region (left: Rebound type, center: penetration type, right: Bubble clouds type)

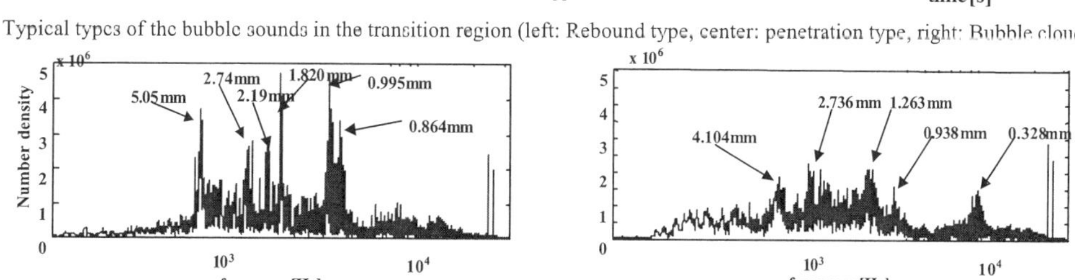

Figure 3 Estimated number density of the entrained bubbles (left: transition region, right: bore region, Wave period: 2 sec, Breaking wave height: 12.6cm, The arrows indicate the radius of bubble).

Remote Sensing of Chlorophyll Concentration in a Bay from Landsat TM Data

Shin-ichi AOKI
Associate Professor, Dept. of Architecture and Civil Eng., Toyohashi University of Technology
Tempaku-cho, Toyohashi, Aichi 441-8580 JAPAN
Phone: +81-532-44-6850, FAX: +81-532-44-6831
e-mail: aoki@jughead.tutrp.tut.ac.jp

1. INTRODUCTION

Major inner bays in Japan are in chronic eutrophication and suffering from so-called red tide and blue tide or anoxia in summer, which often cause serious damages to coastal fisheries and environments. In order to improve the water quality and reduce the damage, various kinds of countermeasure have been carried out, which include reducing inflow load of nitrogen and phosphorus, removing organic bottom mud and covering the bottom mud with sand etc. Continuous monitoring of water quality is very important to grasp the effects of these countermeasures and to take proper procedures. Mikawa Bay (area = $604km^2$, mean water depth = 9.2m) is a shallow and closed inner bay connected to Ise Bay, where the water quality problems have increasingly been serious due to eutrophication. The prefectural government has continued water quality measurement every month at about 20 locations in the bay. More frequent measurements at more locations are however necessary to understand the dynamic behavior of water quality in the bay.

In the ocean, satellite images have been used for analyzing distributions of water surface temperature, chlorophyll concentration (Chl-a) and turbidity etc. The remote sensing of water quality in inner bays are also expected because satellite images regularly supply simultaneous information of wide area. Water quality of inner bays is usually evaluated through various indices such as COD, SS and Chl-a etc. Among these, Chl-a is the most important index for eutrophic waters. Estimate of Chl-a in the ocean from satellite images has been tackled by many researchers. In particular, various types of algorithm has been developed and verified for the CZCS sensor on NIMBUS7. Most of them relate Chl-a to a ratio between radiance values of two different bands (Gordon et al., 1983). These algorithms are however applicable only to relatively clear water in the ocean where Chl-a is around several mg/m^3. On the other hand, eutrophic water in inner bays sometimes shows very high concentration more than 100 mg/m^3 with high SS and dissolved organic compound, where these algorithms cannot be applied.

In this paper, an algorithm that enables us to estimate Chl-a from Landsat TM data is proposed based on field data obtained in Mikawa Bay for several years. Atmospheric effects that deteriorate the quality of information of Chl-a are considered in the algorithm.

2. FIELD MEASUREMENTS

Field measurements of water qualities in Mikawa Bay were carried out on nine different days with clear sky from 1995 to 2000 when the Landsat satellite took the images of the bay at the same time. At about 30 locations in the bay, surface water was sampled and Chl-a was analyzed by a spectro-photometer in the laboratory. The vertical distribution of chlorophyll concentration up to -6m was detected by an optical chlorophyll meter. Turbidity, temperature and transparency (secchi disk depth) were also measured.

3. COMPARISON WITH TM DATA

The field data of Chl-a were first compared with Landsat TM data at corresponding locations. After investigating correlations between the field data and radiances of the bands 1 to 6, it was found that the difference in radiance between bands 1 and 3 gives the best correlation with Chl-a. However, the magnitude of the difference value seemed to be influenced by the aerosol concentration in the atmosphere. The atmospheric disturbance was eliminated in the algorithm by knowing that the aerosol concentration can be related to the averaged radiance values of band 1 over the bay. Finally an empirical formula was proposed for estimating the Chl-a distribution in the bay only fromTM data.

Figure 1 shows the comparison between measured and estimated Chl-a for all the data for 6 years. Figure 2 shows the distribution of estimated Chl-a over Ise and Mikawa Bays. High concentration is seen near the coast and at the head of the bay.

REFERENCES

Gordon, H. R. et al. (1983): Phytoplankton pigment concentrations in the Middle Atlantic Bight: comparison of ship determinations and CZCS estimates, Appl. Optics, 22(1), pp.20-36.

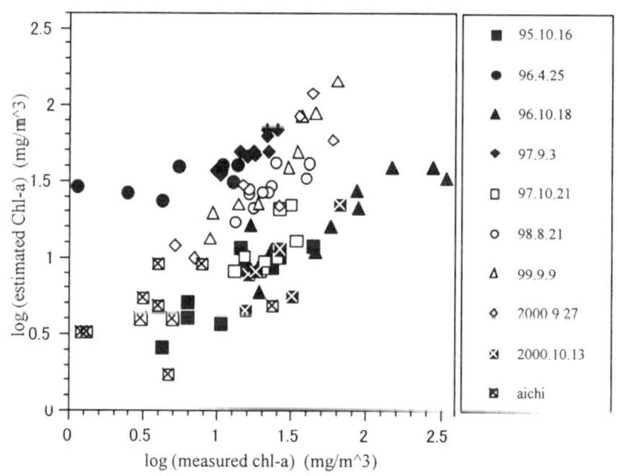

Figure 1 Correlation between measured and estimated Chl-a for all the data

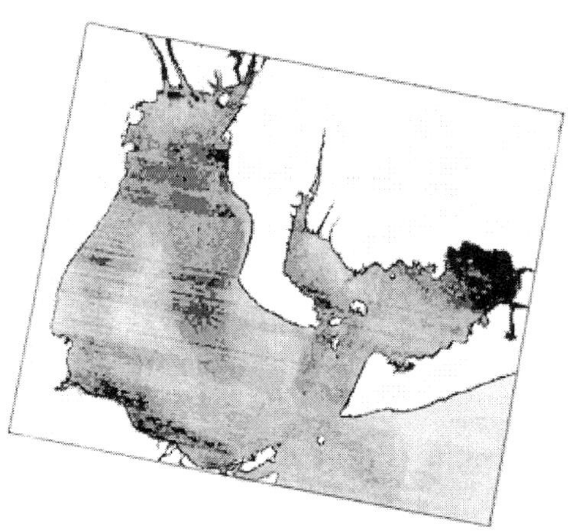

Figure 2 Estimated Distribution of Chl-a in Mikawa and Ise Bays (Sept. 9, 1999)

DERIVING RELATIONSHIPS BETWEEN REEF SEDIMENTATION AND INLAND EROSION CHARACTERISTICS BASED ON FIELD OBSERVATION DATA, HYDROLOGIC MODELING AND REMOTE SENSING ANALYSIS

Enrico C. Paringit and Kazuo Nadaoka
Department of Civil Engineering
Tokyo Institute of Technology
Meguro-ku, Tokyo, JAPAN
e-mail: ecp@wv.mei.titech.ac.jp, nadaoka@mei.titech.ac.jp

ABSTRACT

I. Introduction and Objectives

Coral reefs frequently exist in complex and dynamic environments. Although anthropogenic effects such as introduction of terrigenous sediments are often assumed to be primary stressors of coral reef communities, a quantitative link between these potential stressors and the decline of reefs is seldom established. This study aims to derive meaningful relationships between data gathered from inland field surveys, hydrologic modeling and results of remote sensing data analysis in monitoring the sedimentation in a coral reef environment. In demonstrating a direct coupling between land and ocean processes, remote sensing acted as the integrating method of analysis.

II. Study Area, Methods and Materials

The study area covers Shiraho Reef, Ishigaki Island, Southern Ryukus (24° 21' N, 124°15' E) typifies a subtropical fringing reef. Shiraho Reef is host to over 359 known coral species but perennially suffers from substantial erosion of local red silt, called *akatuchi* discharged through the 11.5 sq km Todoroki watershed which drains eastward to the Shiraho Reef area due to intense cultivation activities inside the catchment.

The Todoroki watershed was continuously gauged for depth and turbidity in three points along the Todoroki River and precipitation (4 locations) for a period of 1.5 years (Jun 2000-Oct 2001). For the Todoroki watershed, a spatially-distributed, time continuous sediment discharge model was developed using a mini-database of geomorphic and physical characteristics where numerical solutions of overland flow simulations and sediment continuity equations were applied. In the process of computing for runoff and erosion rates, remotely-sensed data (aerial photographs and Landsat image) was used to parameterize land cover and vegetation abundance by using a constrained inverse spectral mixture model wherein the fractions of different land types are determined for each image pixel based on its detected spectral response (called "unmixing").

For the reef area, the extent and distribution of sediment accumulation on benthic cover types (coral, seagrass and algae), substrates (sand and rubble) and the deposited sediment were also determined using the principle of spectral unmixing. The method was applied to four Landsat images, covering the study area. coral-rich Shiraho Reef located in southwest Ryukus to assess the multitemporal nature of sediment deposition in the area. Results are validated by comparing satellite-derived sediment fraction maps with field data obtained from a reddish soil pollution assessment protocol.

III. Results

Based on turbidity meter and aided by the sediment discharge modeling, total sediment load was estimated at 1084 tons while rainfall measured amounted to 3312 mm for the field observation period (See Figure 1) for the 70 rainfall events identified. Special attention is drawn to the May 31 rainfall event lasting for 25 hrs (vs. 12 hrs ave.). Downpour was heavier than usual (197 mm vs. 39.41 mm ave.). Peak rainfall rate was higher (36 mm/hr vs 28 mm/hr ave.). However, it is as intense as usual average 7.88 mm/hr vs 7.03 mm/hr ave. peak 36.34 mm/hr vs 26.28 mm/hr ave. Highest recorded depth from sensor (3.11 m) was deepest in 1.5 years of monitoring. This resulted in an equivalent discharge of 101.76 cu.m/s. or a total load of 159.6 tons with highest concentration reaching 1.1 g/L x cu.m/s.

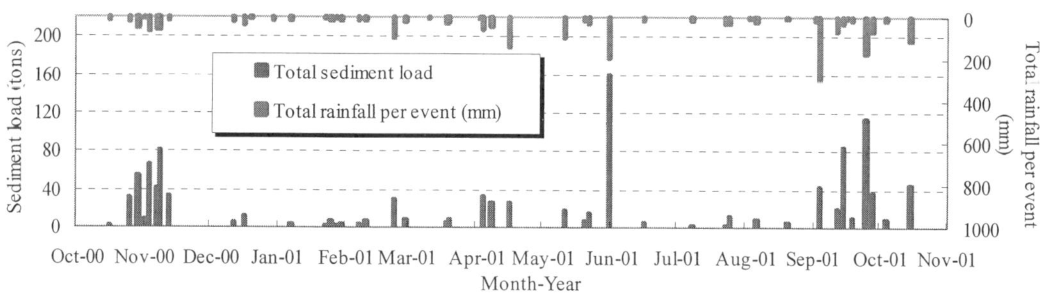

Figure 1. Year-round rainfall and sediment discharge data

The broadening histogram distribution for June 2001 implies that sediment coverage has widened after May 31 event. Mean coverage values were also highest for June 2001 results, which likewise indicates heavier/thicker accumulation at that time. Decreasing maximum coverage coupled with equivalent increase in mean suggest that high coverage only affect limited areas specifically that fronting the river mouth. Coverage increase was reflected in Jan 2001 results from heavy sediment load in Nov 2001 (227 ton monthly total) that occurred after satellite image acquisition. Distinctively, June 2001 reflected an increase in sediment accumulation south of Todoroki River which can be explained by the wind deflected southward direction (as opposed to the predominantly northerly) the during the storm.

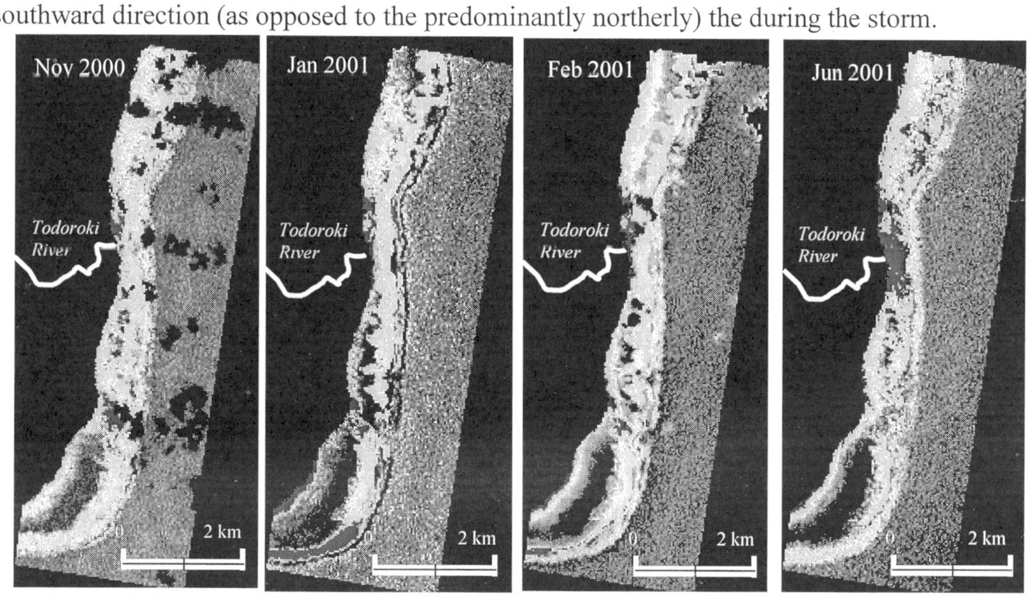

Figure 2. Results of the spectral unmixing technique as applied to the coral reef area.